Resource Management and Environmental Uncertainty

Volume

11

in the Wiley Series in

Advances in Environmental Science and Technology

Robert L. Metcalf, Series Editor

Resource Management and Environmental Uncertainty

Lessons from Coastal Upwelling Fisheries

Edited by

Michael H. Glantz

National Center for Atmospheric Research
Boulder, Colorado

J. Dana Thompson

Naval Ocean Research and Development Activity
National Space Technology Laboratory
Bay St. Louis, Mississippi

A WILEY-INTERSCIENCE PUBLICATION
JOHN WILEY & SONS, New York • Chichester • Brisbane • Toronto

Library of Congress Cataloging in Publication Data

Main entry under title:
Resource management and environmental uncertainty.

   (Advances in environmental science and technology
ISSN 0065-2563)
   "A Wiley-Interscience publication."
   Includes index.
   1. Fishery management. 2. Fishery resources —
Forecasting. 3. Fisheries — Hydrologic factors.
4. El Niño Current. 5. Anchovy fisheries —Peru.
6. Peru—Industries. I. Glantz, Michael H.
II. Thompson, J. Dana. III. Title: Coastal upwelling
fisheries. IV. Series.

TD180.A38   [SH328]   628s [333.95′6]   80-16645
ISBN 0-471-05984-6

Printed in the United States of America

10 9 8 7 6 5 4 3 2 1

To my daughter Mica Glantz and her friend
Janet Underwood

and

to Scott and Helen Nearing, sources of inspiration

. . .

To my mother, Ernestine Thompson

Contributors

BUSALACCHI, ANTONIO, Department of Oceanography, Florida State University, Tallahassee

CALIENES Z., RUTH, Instituto del Mar del Perú, Lima

CAVIEDES, CÉSAR, Department of Geography, University of Florida, Gainesville

CRAM, DAVID, Institut für Seefischerei, Hamburg, Federal Republic of Germany

ENFIELD, DAVID, School of Oceanography, Oregon State University, Corvallis

GLANTZ, MICHAEL H., Environmental and Societal Impacts Group, National Center for Atmospheric Research, Boulder, Colorado

GUILLÉN, OSCAR, Instituto del Mar del Perú, Lima

HAMMERGREN, LINN, Department of Political Science, Vanderbilt University, Nashville, Tennessee

INSTITUTO DEL MAR DEL PERÚ, Lima

KINDLE, JOHN C., Department of Oceanography, Florida State University, Tallahassee

MURPHY, ROBERT C. (deceased), American Museum of Natural History, New York

O'BRIEN, JAMES J., Department of Meteorology, Florida State University, Tallahassee

PAULIK, GERALD J. (deceased), College of Fisheries, University of Washington

RADOVICH, JOHN, California Department of Fish and Game, Sacramento

RAMAGE, COLIN S., Department of Meteorology, University of Hawaii at Manoa

THOMPSON, J. DANA, Naval Ocean Research and Development Activity, National Space Technology Laboratory, Bay St. Louis, Mississippi

TOMCZAK, MATTHIAS, JR., Department of Oceanography, University of Liverpool, U.K.

VONDRUSKA, JOHN, Fisheries Development Division, National Marine Fisheries Service, NOAA (Washington, D.C.)

Introduction to the Series

Advances in Environmental Science and Technology is a series of multi-authored books devoted to the study of the quality of the environment and to the technology of its conservation. Environmental sciences relate, therefore, to the chemical, physical, and biological changes in the environment through contamination or modification; to the physical nature and biological behavior of air, water, soil, food, and waste as they are affected by man's agricultural, industrial, and social activities; and to the application of science and technology to the control and improvement of environmental quality.

The deterioration of environmental quality, which began when man first assembled into villages and utilized fire, has existed as a serious problem since the industrial revolution. In the second half of the twentieth century, under the ever-increasing impacts of exponentially growing population and of industrializing society, environmental contamination of air, water, soil, and food has become a threat to the continued existence of many plant and animal communities of the ecosystem and may ultimately threaten the very survival of the human race.

It seems clear that if we are to preserve for future generations some semblance of the existing biological order and if we hope to improve on the deteriorating standards of urban public health, environmental sciences and technology must quickly come to play a dominant role in designing our social and industrial structures for tomorrow. Scientifically rigorous criteria of environmental quality must be developed and, based in part on these, realistic standards must be established, so that our technological progress can be tailored to meet such standards. Civilization will continue to require increasing amounts of fuel, transportation, industrial chemicals, fertilizers, pesticides, and countless other products, as well as to produce waste products of all descriptions. What is urgently needed is a total systems approach to modern civilization through which the pooled talents of scientists and engineers, in cooperation with social scientists and the medical profession, can be focused on the development of order and equilibrium among the presently disparate segments of the human environment. Most of the skills and tools that are needed already exist. Surely a technology that has created manifold environmental problems is also capable of solving them. It is our hope that the series in Environmental Science and Technology will not only serve to make this challenge more explicit to the established professional but will also help to stimulate the student toward the career opportunities in this vital area.

The chapters in this series of Advances are written by experts in their respective disciplines, who also are involved with the broad scope of environmental science. As editors, we asked the authors to give their "points of view" on key questions; we were not concerned simply with literature surveys. They have responded in a gratifying manner with thoughtful and challenging statements on critical environmental problems.

ROBERT L. METCALF, Editor

Foreword

The subject of this volume is a classic example of the complexity of interactions that characterize marine affairs. In addition to the unique blend of physical and biological sciences found in oceanography, the book analyzes economic, political, and social factors involved in the interaction between science and society.

Consider the plot in its bare essentials. The surface circulation of the ocean is largely wind driven. Along the eastern boundaries of the major ocean basins, there is an important vertical component—upwelling. Because of this upwelling, such regions are highly productive and are the location of major fisheries. The west coast of South America, especially that of Peru, is the most dramatic example of a coastal upwelling region; until 1972, when the fishery collapsed, the Peruvian anchovy (*Engraulis ringens*) supported the world's largest fishery.

The stock of anchoveta had been under heavy fishing pressure for several years, and the collapse might reasonably be attributed to over-fishing. Similar disasters have happened in other fisheries, especially those for sardines, anchovies, and herring. However, in the case of the 1972 fishery collapse, an aperiodic oceanic perturbation known as El Niño occurred at about the same time and has been blamed by many for the disaster. Such disturbances had occurred before, for example, in 1957–1958 and 1965, and had not been associated with serious trouble in the anchoveta fishery; they had, however, adversely affected important competitors for the anchoveta, the several species of guano-producing seabirds.

Several subplots now arise. El Niño itself is an interesting physical phenomenon that has been studied for at least 50 years. It appears to be associated with major changes in the atmospheric circulation (the primordial cause is unknown) and affects at least the entire tropical Pacific.

The mechanism whereby El Niño arises and eventually decays is beginning to be understood, and El Niño may soon be amenable to prediction.

But surely physical perturbations alone do not cause enormous stocks of fish to disappear! Niños have presumably been occurring for millennia, and only in 1972 did the Peruvian anchoveta fishery collapse (it was, of course, trivial until the early 1960s). In the past, only when the stock was being "over-heavily" fished did the environmental disturbance have such a disastrous effect (a similar linkage in the case of the Pacific sardine may be instructive). Just how the unusual environment affected the stock, presumably at the recruitment level, is poorly understood, and successful prediction of another El Niño would not mean that its biological consequences could be foreseen.

There is also the dimension of social consequences, not of El Niño *per se*, but of the perturbations in the anchoveta fishery that may or may not derive from El Niño. The main product of the Peruvian anchoveta fishery, fish meal, competes in the world market with other protein supplements for animal feed and is a source of foreign exchange rather than of food for Peru. Capital is invested (usually overinvested) in fishing boats and processing plants; fishermen must find other work when they cannot fish; governments establish bureaucracies that thrive on dealing with such affairs.

How then should one manage a fishery to accommodate not only environmental uncertainties but also their biological and social consequences? This is especially difficult where the interactions and feedbacks are interwoven throughout a multidimensional fabric whose intricate structure is only dimly understood. Although responsible attempts were made to cope with the Peruvian problem, neither the data nor the understanding were commensurate with its magnitude. The articles in this book set the stage for learning to deal more knowledgeably with such dilemmas, in Peru and elsewhere.

WARREN S. WOOSTER

Institute for Marine Studies
University of Washington
Seattle, Washington
November 1980

Acknowledgments

We would like to thank Maria Krenz for her editorial and administrative assistance without which this book might not have been possible. We would also like to express our appreciation to Ursula Rosner, Verlene Leeburg, Jan Stewart, Karen Lynch, Suzanne Parker, Brenda Thompson, and Betty Wilson for their help in the preparation of the manuscript. We wish to thank others at the National Center for Atmospheric Research (such as Elmer Armstrong of the publications office, printing and reproduction, and the graphics group) for their consistent support.

MICHAEL H. GLANTZ
J. DANA THOMPSON

Boulder, Colorado
Bay St. Louis, Mississippi
November 1980

Contents

Part II   Scientific Aspects of El Niño

Part III   Societal Implications of El Niño

Part IV The Future

Resource Management and Environmental Uncertainty

PART I
BACKGROUND

CHAPTER 1

Introduction

MICHAEL H. GLANTZ

This book brings together research results of closely related but not often interacting disciplines on a topic of mutual interest—fisheries management under conditions of environmental uncertainty. The Peruvian anchoveta fishery case study has been emphasized. Whereas only a decade ago this was considered the largest fishery in the world (by tonnage, not value), today it appears to be in a precarious condition, with the annual number of anchoveta landings approximating only 10% of that taken during the fishing industry's heyday.

This set of extensively referenced papers is concerned with topics directly and indirectly related to the geophysical phenomenon known as El Niño, and its effect on the Peruvian anchoveta fishery. El Niño could be defined as an invasion of warm nutrient-poor water from time to time into the eastern equatorial Pacific, an area normally manifesting upwelling processes that bring nutrient-rich deep cold water to the sunlit zone. This invasion adversely affects the relatively high primary productivity which, in turn, affects higher trophic levels culminating (in the Peruvian case) with the fishermen, guano birds, and other predators.

In reality, there are scores of definitions of El Niño, and each author defines this phenomenon differently. The following characteristics, singly or in combination, have been used at one time or another to describe the El Niño event[1]:

- Natural event
- Lack of upwelling
- Warm coastal waters
- Not a periodic phenomenon
- Large or global in scale
- Large-scale fluctuation

3

- An eastern boundary phenomenon
- Anomalously high sea surface temperatures
- Rapid appearance of anomalously high sea level
- Accumulation of water in the western equatorial Pacific
- Relaxation of the southern trades
- Pacific manifestation of a global phenomenon
- Atmospheric-ocean coupling
- Commonly appears near Christmas time
- Reduction in the Peruvian anchoveta
- Redistribution of tropical and subtropical fauna
- A large-scale event of which El Niño is a coastal manifestation
- An invasion of abnormally warm water that persists for a year or more
- Any substantial oceanic temperature increase (in the eastern equatorial Pacific)
- A large-scale weakening of the interior zonal wind stress field in the equatorial band
- Induced by anomalous behavior of the large atmospheric circulation cells over the South and North Pacific Ocean
- Just one of various side effects of disturbances of the global tropical circulation
- Warm countercurrent flowing south from the coast to at least 100 km offshore
- An invasion of anomalously warm, nutrient-depleted water into the coastal regions of Ecuador and Peru

On the problem of definition, Prohaska wrote that

Whatever the phenomena associated with the El Niño abnormality, threshold values of the meteorological and or oceanological (sic) parameters and their extension in space and time have never been objectively defined.[2]

Also clear from a collection of definitions from various sources, however, is that there is a sometimes explicit, sometimes implicit, societal component to many of the definitions of El Niño. For example, some have suggested that any invasion of warm water that does not affect the established economy of the region (in the Peruvian case, the anchoveta fishery) is a non-El-Niño event. Thus, what might be considered an El Niño event by Ecuadorian observers might not be considered one by the Peruvians. What might be considered an El Niño by Peruvian fishermen might not be considered as such by Chilean fishermen. What might be considered an El Niño by scientists might not be considered one by fishermen. The reader must be aware that as the definition of El Niño may vary slightly from one article to the next in the literature, it may also vary from one chapter to the next here.

El Niño events have occurred from time to time (quasiperiodically) in past centuries.[3-5] The most recent ones have had major socioeconomic, environmental, and political consequences for Peru. This was especially the case in 1972-1973 (see, for example, Caviedes,[6] Miller and Laurs,[7] and Wooster and Guillen[8]). According to one researcher, there have been at least eight *major* El Niño events in Peruvian waters since the turn of this century, events he defined as having encompassed two summers and one winter (in the Southern Hemisphere).[5] The most recent ones occurred in 1957-1958, 1965-1966, and 1972-1973 (see, for example, Bjerknes,[9] Paulik,[10] and Miller and Laurs,[7] respectively). Also, Wooster and Guillen have distinguished major El Niño events from minor ones.[8] Paulik noted that each El Niño seemingly had a different effect on the fisheries and, thus, on the Peruvian economy.

> There is considerable evidence that an El Niño is not an all or none phenomenon, but rather a condition whose degree of severity differs from year to year. The biological damage caused by the El Niño is dependent on its duration, magnitude and the speed with which it invades the southern waters.[10]

The most devastating El Niño event, which took place in 1972-1973, attracted the most attention, not only from scientific researchers interested in El Niño as a geophysical phenomenon to be investigated and understood, but also from those concerned with the exploitation of the ocean's living marine resources as a potential food source for expanding world populations (and, in some cases, increasing affluence).

1972: THE YEAR OF CLIMATE ANOMALIES

To put the 1972-1973 El Niño event in proper perspective, it must be kept in mind that El Niño was not the only (or even the most important from a global standpoint) climate-related anomaly to occur in 1972 but was one of several that adversely affected the global food production system as well as the optimistic perceptions about the system's ability to feed its members. The Soviet Union, for example, registered one of its worst shortfalls in grain production as a result of drought,[11] and resorted to major grain imports from the United States.[12] This in turn led to a scarcity of wheat on the world market.

Droughts also occurred that year throughout the Sahelian zone of West Africa, in India, Central America, the People's Republic of China, and in parts of Australia, Kenya and Ethiopia.[13] In addition to these events, there was a general decline in fish landings in other regions (as well as off the coast of Peru). This raised questions about the then-prevailing view that the oceans could be a future source of food that could supplement agricultural production to feed increasing populations world-wide.[14]

The anomalies in 1972 also heightened speculation about the future trend of global climate and its implications for global food production in the future. Although some suggested that we may have been entering a new cooler climate regime, others agreed that those anomalies were simply fluctuations within the existing climate regime. With respect to climate change and the oceans, a recent U.S. National Academy of Sciences report suggested that "improved knowledge of climate change in the oceans and the effect of this change on living marine resources is essential."[15]

A new controversial theory, emerging now among some members of the scientific community, is that increases of carbon dioxide in the atmosphere resulting primarily from the burning of fossil fuels and wood can lead to a warming of the atmosphere, possibly shifting precipitation and temperature regimes on a worldwide scale (see, for example, World Meteorological Organization[16]).

It is not difficult to see that the anomalous weather events of 1972 precipitated a greater awareness, among scientific researchers as well as government decisionmakers, of the direct and indirect implications of climatic fluctuations for society and the environment.

CHAPTER OVERVIEW

In the early stages of the preparation of this book, some of the then-proposed chapters were challenged, for one reason or another, as not having "a place" in a book on Peru and El Niño: Why include South West African fisheries? Why discuss the Pacific sardine? Why include that ideological perspective? In spite of such concerns, the original topics (and authors) were retained with only two of the original science chapters withdrawn by their authors. Thus, it would be useful to readers to understand why the existing chapters were selected.

The book is divided into four parts: Background (including contemporary case studies of other fisheries), Scientific Aspects of El Niño, Societal Aspects, and Considerations for the Future.

In Part I (Background) Dana Thompson, an oceanographer, presents an overview of upwelling processes in the five major (highly productive) upwelling regions of the world. He discusses some of the physical processes in the atmosphere and in the ocean associated with upwelling and with the relatively high biological productivity in these regions. Making special reference to El Niño and the Peruvian fisheries, he outlines recent research results that suggest El Niño is not simply a local phenomenon, locally forced and locally manifested.

The chapters by Murphy and Paulik provide historical background to the 1972–1973 El Niño event. The Murphy chapter was written in 1954 and until now was published only in Spanish. The English manuscript was held by the American Museum of Natural History (New York). Murphy had researched the Peruvian littoral for decades (see, for example, Murphy[3,17]). His report on "Guano and the Anchoveta Fishery" was sponsored by the Guano Administration (then responsible for overseeing the guano industry). He discusses the significance of the anchoveta as the key food resource to the guano bird population as well as to various kinds of fish and sea mammals. He then discusses linkages between the exploitation of the anchoveta on a commercial basis and the ability of the guano birds to survive the competition with the fishermen for their (the birds') source of food. Murphy's report, although written more than 25 years ago, exposes questions about the anchoveta fishery many of which are still being asked today.

Paulik's chapter gives insight into various aspects of the anchoveta fishery and, more generally, into the fishing industry with references throughout to fisheries outside Peruvian waters. He presents a scenario (in 1970) that approximated what transpired a few years later in 1972–1973. Paulik raises an interesting rhetorical question with respect to living marine resources under conditions of environmental and biological uncertainties:

> . . . thinking about unthinkable catastrophes is one way of preventing them. Destruction of the world's greatest single species fishery is unthinkable. Could it really occur?

One should be indebted to Paulik for his documentation and analysis of the "Anchovies, birds, and fishermen in the Peru Current."

The chapters by Radovich and Cram are concerned with the management of fisheries in coastal upwelling regions off the coasts of California and South West Africa/Namibia, respectively. Radovich's chapter deals with the well-known, but still controversial, collapse of the Pacific sardine fishery (a source of inspiration for John Steinbeck's *Cannery Row*). He discusses in historical perspective the national versus state divergence of views on how to manage that state's fisheries, with emphasis on their opposing views about the cause of the decline of the Pacific sardine fishery off the coast of California.

Radovich suggests that although considerable attention has been focused on the complementary role of sardines and anchovies as a single biomass while competing with each other as part of that biomass, those relationships still require demonstration. He concludes with specific suggestions about what a model for the sardine fishery must include.

Cram presents the historical development and the scientific, administrative, political, and economic aspects of the South West African/Namibian pilchard fishery. In this case study he concludes that the public view of fisheries management is that of a purely scientific exercise and that conservation policy is derived in an orderly way on a scientific basis. Although arguments are centered on scientific matters, other equally important political, administrative, legal, and economic elements, normally hidden from public view, dominate fisheries management decisions and those elements outweigh scientific arguments, which lead to the decline of the stock and the fishery.

Section II (Scientific Aspects of El Niño) presents scientific considerations related to upwelling in general and the phenomenon known as El Niño in particular. O'Brien, Busalacchi, and Kindle present in their chapter recent theoretical research in physical oceanography that links in a physical way remote forcing in the central Pacific to the invasion of warm water along the western coast of South America. The dynamic mechanism is the propagation of internal Kelvin waves that propagate eastward from the central Pacific. They review recent research activities and explain how anomalous winds in the central Pacific (on which predictions of El Niño are based) can influence physical and biological processes in the eastern equatorial Pacific.

Enfield, recently the Liaison Officer for the Coastal Upwelling Ecosystems Analysis (CUEA) JOINT-II experiment in Peru, discusses observations that verify many of the theoretical predictions discussed by O'Brien, Busalacchi, and Kindle. Guillen and Calienes' chapter provides a link between the physical oceanographic anomalies and the distribution of nutrients and biological productivity, emphasizing the influences of El Niño on primary productivity in the eastern equatorial Pacific.

Some of the many potential societal aspects of El Niño and its effect on the anchoveta and other fisheries in Peru and Chile are presented in

Section III. Fish meal is a cost-effective, high-protein poultry, livestock, and farmed fish feed ingredient made by cooking, pressing, drying, and grinding fish (or shellfish). Vondruska, an economist, assesses fish meal production for the United States and the world. Postwar price trends of fish meal are discussed and analyzed with emphasis on price peaks and depressions. Vondruska concludes that whereas

> . . . the effects of El Niño events can be observed in market variables (such as fish meal price, production, consumption, trade and stocks), El Niño is only one of the factors that must be considered.

The IMARPE Report, written at the end of 1969 by fisheries economists, notes that pelagic species, such as the anchoveta, seem particularly susceptible to reduction of fish stocks as a result in part of excessive, uncontrolled fishing. The Report, alluding to the 1969–1970 anchoveta catches, notes that some of the excess was allowed in order to meet economic difficulties at that time. The report presents possible methods for the reduction of fleet and plant capacities specifically and a plan for the economic rationalization of the entire industry.

Hammargren's chapter presents political, administrative, and ideological aspects affecting operations of the fishing sector and the government with respect to Peruvian anchoveta fishery management between 1960 and 1979. She focuses on administrative and policymaking processes within the fishing sector, under different types of government. She assesses constraints and pressures acting on policymakers in resource-related crises, like that of the 1972–1973 El Niño, concluding that the recent "expropriation was clearly not the only solution [to the El Niño-related anchoveta fishery crisis], but it was most compatible with the government's political and ideological preferences."

Caviedes, a geographer, compares some aspects of the development of the Chilean fisheries to the Peruvian situation. He comments on scientific and management aspects of the fishery under four recent (and different) political regimes in Chile, noting that

> . . . the experiences gained from El Niño 1972–73, even though that event coincided with the occurrence of significant political events both in Chile and Peru, did not serve to change substantially the economic approaches toward an important resource for both nations and did very little to alter long-established attitudes of environmental disregard.

Tomczak's appraisal of the "El Niño problem" represents a viewpoint that can be found in various quarters throughout the Third World. His analysis demonstrates how the same set of events can be interpreted in varying ways depending on one's ideological perspective. Even though Tomczak's perspective (which he makes explicit) may not be shared by all the contributors, his thought-provoking discussion and assessment are important in any attempt to understand the political, economic, and scientific problems confronting a Third World government (or any government for that matter) seeking to exploit its living marine resources for the purposes of social and economic development (see also Quijano[18]).

The final section presents some future considerations of forecasting El Niño events and their implications for the rational management of the Peruvian anchoveta fishery. The reader is asked, however, to keep in mind that each chapter considers to some extent the future implications for the anchoveta fishery from the disciplinary and ideological perspective of its author, and that the views in these chapters are also not necessarily shared by other contributors.

Ramage, a tropical meteorologist, critically assesses the potential for forecasting the phenomenon known as El Niño. He concludes that it has an unpromising prospect, in spite of some of the more recent seemingly encouraging preliminary scientific results, arguing that essential physical linkages between forecast indicators and El Niño conditions are not well understood. Since El Niño is an integral part of the ocean-atmosphere system on very large scales, successful predictions of El Niño may only be possible in the unlikely event that worldwide seasonal forecasts are successful.

The chapter by Glantz presents a preliminary assessment of the societal value of an El Niño forecast. The study suggests that, although in theory many good uses can be attributed to the development of a reliable El Niño forecast, in reality that forecast will be only one input into the decisionmaking process. The chapter also points out that the social, economic, and political dislocations caused by the collapse of a fishery (contributed to by overcapitalization and excessive exploitation, as well as by El Niño) may prove more costly than whatever temporary benefits might be derived from exploitation of the fishery based primarily on short-term economic interests.

REFERENCES

1   M. H. Glantz, "El Niño forecast value study: Background paper," National Center for Atmospheric Research, Environmental and Societal Impacts Group, Boulder, 1977.

2   F. J. Prohaska, "New evidence on the climatic controls along the Peruvian coast," in D. H. K. Amiran and A. W. Wilson, Eds., *Coastal Deserts,* University of Arizona Press, Tucson, 1973.

3   R. C. Murphy, "Oceanic and climatic phenomena along the west coast of South America during 1925," *Geographical Review,* **16,** 26–54 (1926).

4   E. R. Gunther, "Variations in behavior of the Peru coastal current—with an historical introduction," *Geographical Journal,* **88,** 37–65 (1936).

5   C. Ramage, "Preliminary discussion of the meteorology of the 1972–73 El Niño," *Bulletin of the American Meteorological Society,* **56,** 234–242 (1975).

6   C. Caviedes, "El Niño: Its climatic, ecological, human and economic implications," *Geographical Review,* **65,** 493–509 (1975).

7   F. R. Miller and R. M. Laurs, "The El Niño of 1972–1973 in the Eastern Tropical Pacific Ocean," *Inter-American Tropical Tuna Commission Bulletin,* **16**(5), 403–416 (1975).

8   W. S. Wooster and O. Guillen, "Characteristics of El Niño in 1972," *Journal of Marine Research,* **32**(3), 387–404 (1974).

9   J. Bjerknes, "A possible response of the atmospheric Hadley calculations to equatorial anomalies of ocean temperatures," *Tellus,* **XVIII**(4), 820–829 (1966).

10   G. J. Paulik, "Anchovies, birds, and fishermen in the Peru current," published in 1971, reprinted in this volume.

11   A. L. Katz, *The Unusual Summer of 1972.* L. A. Hutchinson, Trans., Gidrometeorizdat, Leningrad, 1973.

12   J. Trager, *The Great Grain Robbery,* Ballantine, New York, 1975.

13   U.N. World Food Conference, Summary of Principle Documentation, (CSD/74/34), 5-16 November, U.N. Food and Agriculture Organization, Rome, 1974.

14   L. Brown, *By Bread Alone,* Praeger, New York, 1974.

15   National Academy of Sciences, *Food and Climate,* Washington, D. C., 1976.

16   World Meteorological Organization, *Report of the Scientific Workshop on Atmospheric Carbon Dioxide,* WMO Publication 474, Geneva, 1977.

17   R. C. Murphy, "The oceanography of the Peruvian littoral with reference to the abundance and distribution of marine life," *Geographical Review,* **16,** 26–54 (1923).

18   A. Quijano, *Nationalism and Capitalism in Peru: A Study of Neo-Imperialism,* Monthly Review Press, New York, 1972.

CHAPTER 2

Climate, Upwelling, and Biological Productivity

Some Primary Relationships

J. DANA THOMPSON

Abstract

A general overview of the physical mechanisms linking climate, upwelling, biological productivity, and fisheries is presented in anticipation of the detailed studies of El Niño to follow. The dynamic relationships between the large-scale atmospheric circulation, vigorous vertical currents in the upper ocean, and above-average primary productivity are outlined. The major coastal upwelling ecosystems of the world are examined, and their influence on the coastal climate is assessed. The combined effects of natural variability and overfishing on an upwelling ecosystem are described. In particular, the case of El Niño and the Peruvian anchoveta is outlined. Recent observational and theoretical evidence supporting the hypothesis that forced oceanographic anomalies propagate into the eastern Pacific during El Niño is briefly reviewed.

INTRODUCTION

The purpose of this chapter is to review the primary relationships between climate, upwelling, and biological productivity and to provide an appropriate background for the detailed studies of El Niño to follow.

Because the interpretation of El Niño as an anomalous event implies knowledge of the mean state, we are obliged to attempt to define clearly that state—a difficult, if not impossible, task. Nevertheless, we can, based on present knowledge, outline the physical mechanisms responsible for high biological productivity in coastal upwelling ecosystems. We can link above-average productivity to the local physical oceanography of the ecosystem and in turn relate the local circulation to the large-scale behavior of the atmosphere and ocean. There are reasonable hypotheses that link disruptions in the "normal" upwelling ecosystems with changes in the large-scale circulation patterns of the atmosphere and ocean.

In the first part of this chapter, the physical oceanographic basis for the high biological productivity existing in upwelling ecosystems is established, and the locations of those ecosystems are linked to the climate of the overlying atmosphere. It is shown how the atmosphere itself may be influenced by the ocean in these regions. The relationship between upwelling intensity, biological productivity, and fishery yields is then investigated, emphasizing the natural variability of upwelling eco-systems. Finally, the upwelling ecosystem of Peru and Chile, that is dramatically influenced by El Niño, is surveyed.

MECHANISMS FOR UPWELLING

Phytoplankton form the basis for the food chains in the sea. Given sufficient light, the availability of nutrients is generally the most important factor in the growth of these plants. They are said to be "nutrient-limited." In the upper 100 m or so of the ocean (in the sunlit or euphotic zone), nutrients are assimilated by the phytoplankton, which grow, die, and sink out of the zone, carrying their nutrients with them. If those nutrients are not replaced, large phytoplankton populations cannot be maintained. In turn, extensive marine communities dependent on the phytoplankton cannot exist. While run-off carrying sediment from the land into the coastal waters may provide some nutrients for phytoplank-ton growth, it is largely the upward vertical currents (upwelling) and the turbulent mixing that return decomposed organic material and nutrients into the euphotic zone. It is not surprising, therefore, that a map of intense upwelling areas in the world's oceans also adequately serves as a map of the areas of high organic productivity.[1]

Estimates of the average photosynthetic (primary) production can be quantitatively measured in terms of the amount of inorganic carbon converted to organic carbon by photosynthesis ("fixed") per unit of ocean area per unit of time. Those estimates yield interesting information about the heterogeneity of ocean productivity. For example, Ryther calculated that in 90% of the ocean (areas where upward vertical currents are absent or are but a few meters per year), the average primary production is only about 50 g/m²·yr, ranging from 25 to 75 g/m²·yr.[2] In about 10% of the world's ocean areas the rate of primary production is considered to be moderate, at about 100 g/m² · yr. These areas include shallow coastal zones (waters less than 180 m deep) where strong upwelling currents are absent but where nutrients are provided by water run-off from land or by vertical mixing down to the shallow ocean bottom. Also included are open ocean areas where dynamic conditions produce vigorous upwelling, as in oceanic fronts and equatorial divergences.

Only a small fraction of the world's ocean area, about 0.1% of the total, exhibits high primary production, fixing as much as 1 g/m²·day of carbon for brief periods.[3] These areas of coastal upwelling exist along the west coasts of the United States, Chile, and Peru, northwest and southwest Africa, Somalia, and the Arabian coast (Figure 1). Despite their small areal extent, they contribute a significant fraction (in recent years almost half) of the world's total commercial fish catch.

Since the correlation between primary production and fish catch is high in coastal upwelling areas, a physical relationship might seem readily quantifiable. However, such is not the case. First, the number of "links" or trophic levels in the food chain between phytoplankton and harvestable fish must be identified, and the efficiency of organic conversion from level to level must be established. The specific relationship between vigorous upwelling and phytoplankton growth, the rate at which nutrients

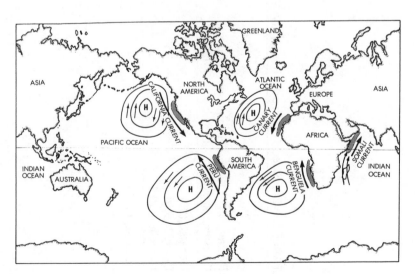

Figure 1 Idealized schematic diagram of the major coastal upwelling regions of the world and the summertime sea-level atmospheric pressure systems (anticyclones) that influence them.

TABLE 1. Estimated Fish Production of the World's Oceans

Regime (% ocean area)	Trophic Levels	Conversion Efficiency (%)	Fish Production (fresh wt. metric tons/yr)	Fish Production Per Unit Area Per Year (relative to the open ocean)
Oceanic (90)	5	10	16×10^5	1
Coastal (9.9)	3	15	12×10^7	660
Upwelling (0.1)	1.5	20	12×10^7	66,000

From Ryther.[2]

move up the food chain, and the distribution of the mobile fish predators must also be determined.

Although there are large scientific uncertainties, Ryther, Cushing, and Gulland, among others, have attempted to relate fish production to primary production using educated guesses concerning the number of trophic levels and conversion efficiencies involved in each of the three ocean areas mentioned earlier.[2,4,5] As indicated in Table 1, using Ryther's 1969 estimates, the variations in the number of trophic levels and conversion efficiencies from one ocean regime to another are considerable. The table highlights two important characteristics of coastal upwelling ecosystems. First, fewer trophic levels are involved compared to coastal or open ocean regimes. Harvestable fish are not far up the food chain from the primary producers. Second, the highest conversion efficiencies are realized in coastal upwelling areas. Thus we calculate that fish production per unit area per unit time in coastal upwelling zones is roughly 66,000 times that of the open ocean. Although this is an uncertain figure, it does suggest that the narrow coastal upwelling ecosystems are highly productive.

CLIMATE AND UPWELLING REGIMES

The physics related to oceanic upwelling processes is clearly important to this discussion, because the biological evidence overwhelmingly supports a link between ocean productivity and intensity of upper ocean vertical motions, the sources of which will be explained in some detail.

The surface wind systems of the world are largely responsible for driving the circulation of the upper ocean (here, "upper ocean" refers to the upper 500 m of the water column). Upwelling and marine productivity are significantly influenced by these wind patterns. We cannot present in this chapter a comprehensive review of the wind-driven ocean circulation or its influence on marine productivity; however, fundamental physical mechanisms important to this discussion are outlined briefly, using a highly idealized physical model.

Centers of high atmospheric pressure at sea level are generally situated at subtropical latitudes over the world's oceans. In the absence of the earth's rotation, one might expect surface winds near these centers to blow outward from high to low pressure, perpendicular to lines of constant pressure (or isobars). Because of the earth's rotation, however, these winds tend to turn to the right in the Northern Hemisphere and to the left in the Southern Hemisphere, aligning themselves roughly parallel to the isobars. For example, the large North Pacific high off the west coast of North America near 30°N and 150°W in summer is characterized by surface winds extending thousands of kilometers from the high center, blowing clockwise around it. Due to frictional drag at the sea surface, the winds are directed at an angle slightly outward from the isobars (see Figure 1).

Surface currents in the ocean are moved by these winds, and, in the absence of the earth's rotation, would move in the same direction as the winds. This rotation is of considerable influence, even to ocean currents, and the surface waters tend to be driven to the right of the wind direction. In fact, a careful mathematical analysis by Ekman showed that the net effect on a rotating earth of winds driving ocean surface waters was to produce a net transport of water (later termed "Ekman Drift") directed 90° to the right (left) of the wind in the Northern (Southern) Hemisphere.[6] Under a surface high, the current has a component directed inward toward the center, and the waters begin to pile up, tending to raise the sea level under the high.

Clearly, surface water cannot converge indefinitely under the high's center. Instead, it must move downward, then outward at some lower level, then upward again far from the high center to provide mass balance. Downwelling is thus established in the upper ocean under a surface high. Generally, this downwelling is persistent, encompasses thousands of square kilometers of ocean area, and is quite weak, only a meter or so in a month. Nevertheless, it inhibits the vertical transport of nutrients into the upper ocean and helps explain the low, desert-like primary productivity of the open ocean. Some vertical exchanges of water and nutrients still occur, particularly during the fall season when the surface waters become cooler than subsurface waters and grow convectively unstable, producing turbulent vertical exchange in the upper ocean. Consequently, these open ocean regions are not totally devoid of plant life.

Occasionally a cyclonic low pressure system, such as a hurricane, will produce a strong diverging surface current over the open ocean. Surface water will spiral outward from the low center and a vigorous upwelling of several meters per day will result. Since these systems usually move rapidly and are short-lived, they do not substantially increase the nutrient content of the upper ocean.

One region of the open ocean where persistent, vigorous upwellings are present is within a few degrees of the equator. The equatorial perimeter of the circulation around the oceanic highs marks a zone of fairly steady easterly trade winds. Precisely on the equator, the direct influence of the earth's rotation is absent and winds blowing westward tend to drive westward surface currents. Just to the north of the equator, the earth's rotation is felt by the currents and they tend to turn to the right of the wind. Just to the south, they turn to the left. The net result is a divergence of surface water away from the equator, with replacement water coming from a few hundred meters below the sea surface. Nutrients and cooler subsurface waters are lifted up into the euphotic zone and lead to increased marine productivity. This productivity is generally higher in the eastern equatorial ocean than in the western ocean, since the thermocline, which effectively separates the warm upper ocean from the cool deep water, deepens toward the west. As noted earlier, equatorial divergence zones may have primary production rates as high as some coastal areas. These areas have received increased study, both observationally and mathematically, in recent years.[7-9]

The eastern flanks of the subtropical oceanic highs are generally situated near the west coasts of continents, as shown in Figure 1. This is not simply coincidental, since land-sea contrasts of temperature and surface topography influence the wind systems. An extreme example is the low-level atmospheric winds off the coast of western Peru, which tend to be channeled along that coast by the high Andes mountain range that straddles the coast. Even in the absence of topographical features, there is generally some time during the year at mid latitudes when the predominant surface winds blow equatorward along the western coastline of continents. The Ekman theory, as extended by Thorade[10] and later Sverdrup,[11] predicts an offshore turning of the coastal surface currents to the right (left) of the wind in the Northern (Southern) Hemisphere, and a net offshore transport of surface water. To replace the diverging surface water, there must be compensating subsurface currents and upward vertical motion, or coastal upwelling. Thus, coastal upwelling may be defined as an ascending motion of some minimum duration and extent by which water from subsurface layers is brought to the surface near the coast and removed by horizontal flow.[1]

Recent theoretical and observational studies suggest that not all coastal upwelling processes are locally wind driven. Remotely forced wave motions in the ocean have also been shown to influence coastal upwelling circulation and are presently an important area of research, particularly for the problem of El Niño (see Chapters 7 and 8).

Despite evidence that some coastal upwelling activity is not driven locally by winds, the Ekman-Sverdrup theory has led to a useful index for coastal upwelling—the Ekman transport away from the coastal boundary. Its magnitude is proportional to the surface wind stress divided by the sine of the latitude. This index has been used by Wooster and Reid,[12] Bakun,[13] and others to identify regions of potentially strong coastal upwelling. In the Wooster and Reid study, seasonal estimates of

Ekman transport of 5° intervals of latitude along west coasts of continents were calculated. Smith has summarized their findings as follows[1] (maximum values of the index suggest maximum coastal upwelling rates):

1 The maximum values are usually observed in the spring or summer, with the exception of the distinct winter maximum for the Peru Current region equatorward of 30°S.

2 The maximum values of the index migrate from south to north, from spring to summer, in the California, Benguela, and Canary Current regions.

3 The maximum values for the index are in the Southern Hemisphere. The smallest values are off the west coast of North America.

4 Negative values are observed at high latitudes (poleward of 50°).

One region not examined in the Wooster and Reid study was the western Indian Ocean, along the Arabian and Somali coasts. During the Southwest Monsoon, winds blowing parallel to the coast produce extremely vigorous upwelling. Also, although their index predicts upwelling to occur off the western coast of Australia, observations have not shown it to be significant.

It is now necessary to review some basic facts concerning the primary coastal upwelling areas of the world. As previously discussed (see Figure 1), there are five primary coastal upwelling regions: off the west coast of South America (primarily Peru and northern Chile); northwest Africa; southwest Africa; the northeast coast of Africa from Somalia northward into the Arabian Sea; and the western coast of North America from Baja California to the Canadian border. Each of them has identifiable coastal current systems within several hundred kilometers of its coast. There is usually one season in which coastal upwelling predominates, during which sea surface temperatures within 100 km of the coast may be several degrees centigrade colder than off-shore waters. This tends to suppress atmospheric convection and helps to stabilize the marine boundary layer. As a result, the coastal climates along upwelling zones tend to be cooler, more arid, and (somewhat paradoxically) more humid than inland areas at the same latitude. Each area is notable for its fishery, Peru's having been one of the most productive. Despite their many general similarities, each area is in fact unique on closer scrutiny.

Peru

The most well known coastal upwelling region in the world lies in an area roughly 300 × 33 miles off the Peruvian coast.[11] During the seven-month fishing season of 1969–1970, 11 mmt of a single fish species, *Engraulis ringens* (the Peruvian anchoveta), were harvested in that region. For comparison, the total U. S. catch for all species of fish and shellfish in 1969 was about 2.5 mmt. Since the first major oceanographic survey of the region by the *William Scoresby* in 1931, a host of researchers have studied this oceanographic system.[14–17] The system is composed of four distinct currents, two flowing to the north, two flowing to the south. The Peru Coastal Current, which flows next to shore, and the Peru Oceanic Current (or Humboldt Current), which is situated farther out to sea, flow northward. Between them flows the near-surface Peru Countercurrent. Beneath the three flows the Peru Undercurrent. The anchoveta tend to congregate in massive schools in the northern part of the coastal current. The oceanic current stretches southward from Peru several thousand miles and extends to depths of 700 m. This current was first thought to carry enough cold sub-Antarctic water into the coastal region to explain the low sea surface temperatures observed there. Unfortunately, that theory could not explain why the coldest coastal waters were often found at low latitudes. Only after the emergence of the Ekman theory was there a general recognition of the importance of upwelling to the thermal structure of the coastal waters. Gunther, for example, found that, although the appearance of upwelling was somewhat irregular in time and space, it had a very definite relation to the wind; this finding is in general agreement with the Ekman theory.[14] Wyrtki, using hydrographic data, deduced that upwelling along the coast is restricted to depths of less than 100 m but that ascending motions at greater depths and further offshore are probably important to the upwelling process.[18] Over large areas, Wyrtki determined vertical motions to be of the order of 10^{-5}–10^{-4} cm/sec at 100 m. He also concluded that areas of coastal upwelling extend toward the equator and gradually merge with regions of equatorial upwelling. The Peru upwelling ecosystem and El Niño will be considered again in a later section.

North America

The North Pacific high has a major influence on coastal upwelling off the west coast of North America from Baja California to British Columbia. The equatorward winds are strongest off Baja in April and May, off southern California in May and June, off northern California in June and July, and off Oregon and Washington in July and August.[1] The primary current system is the equatorward-flowing California Current.[19] During fall and early winter, a poleward flowing current near the coast, called the Davidson Current, is also observed. Associated with active coastal upwelling is a highly time-dependent equatorward surface jet within 50 km off the coast. This coastal jet has also been observed off northwest and southwest Africa, off Peru, and off the Somalia coast.

Perhaps the most systematically studied coastal upwelling region in the world lies off the Oregon coast. Scientists at Oregon State University (and more recently other groups participating in the Coastal Upwelling Experiments CUE-I, 1972, and CUE-II, 1973) have carefully studied, for more than a decade, the biology, physical oceanography, and marine meteorology of a 2500 km² area off Newport, Oregon. Mathematical models of the ocean circulation,[20,21] the low-level atmospheric circulation,[22] and the primary biological productivity,[23] have recently been designed to help organize and explain the large amount of data obtained by the field projects. One important result of the studies was the discovery of occasional intense wind-driven upwelling "events" lasting a week or 10 days. From this research, pioneering attempts to aid Coho salmon fishermen by predicting areas of potentially good fishing using wind and current information have met with moderate success.[24]

Northwest Africa

The Canary Current moves equatorward from about 36°N to about 18°N, and then joins the Equatorial Current moving southeastward along the Guinea coast toward the equator. There it merges with the northward-flowing Benguela Current. The richest upwelling area tends to center off ex-Spanish Sahara and Mauritania. Apparently, upwelling is most intense in spring and summer and migrates northward from winter to summer.[12] In the summer the upwelling may extend northward to the Straits of Gibraltar. Less intense and less regular upwellings occur in the tropical region along the Gulf of Guinea.

A nearshore equatorward jet, a poleward undercurrent, and high correlation between currents and surface wind stress appear as persistent features of the reported observations.[25] Despite the similarities between the northwest Africa and Peruvian upwelling areas, the annual fish production off Peru is 10 times that off northwest Africa. This difference may be attributed to the single-step food chain and extremely high primary productivity found off Peru, compared to the relatively complex food web and moderate primary productivity found off northwest Africa.[26]

Southwest Africa

A region of coastal upwelling extends off the west coast of southern Africa from about 15°S to 34°S. To the west, in the eastern branch of the

South Atlantic anticyclonic gyre, flows the Benguela Current.[27] Smith identifies the coastal (Benguela) current as analogous to the Peru Coastal Current and the offshore flow as analogous to the Peru Oceanic Current.[1] Defant found the region of coldest water to be in the coastal current between 23°S and 31°S.[28] From the charts of Bohnecke,[29] there appears to be most intense upwelling in summer and fall with an equatorward migration of maximum upwelling in summer and fall with an equatorward migration of maximum upwelling from the Southern Hemisphere summer to winter.

The upwelling off Namibia (South West Africa) appears to have the highest primary production in this region and the largest fish stocks. Until recently, the fisheries in this upwelling region were confined to the coastal states of Angola, Namibia, and South Africa, under the jurisdiction of the International Commission for the Southeast Atlantic Fisheries (ICSEAF). During the past decade, distant-water operations from Japan, Spain, the Union of Soviet Socialist Republics, and several other foreign nations have become common. Some fisheries experts believe part of this regional fishery have already been fully exploited, particularly the southern hake stocks.[30] Other species have yet to be fully exploited (also, see Chapter 6).

Somalia and Arabia

The upwelling off Somalia and Arabia is unique in that it occurs on the *east* coast of a continent. This is due to the fact that the most vigorous winds blow parallel to the coast from *south* to *north* in late spring and summer. Consistent with the Ekman theory, this situation also produces coastal upwelling. Thus, during the Southwest Monsoon, very intense coastal upwelling occurs along Somalia and Arabia, roughly from the equator to 25°N. Although the sea surface temperatures (SSTs) in the open oceans of the Northern Hemisphere are warming during this season, SSTs may drop as much as 16°C in the upwelling areas off Muscat and Oman and the Somali coast. It is believed that these low SSTs influence the atmospheric circulation and may be linked to rainfall distributions over India.[31] The fishery potential for this area is largely underutilized, although this situation may now be changing.[32] The region has not been intensively studied in the manner of northwest Africa or Peru, although we learned much from the International Indian Ocean Expedition (1960–1965). Recently INDEX (the Indian Ocean Experiment), a part of the Global Atmospheric Research Program, has begun to reexamine the ocean circulation of this upwelling region.

OCEAN-ATMOSPHERE INTERACTIONS

The mean large-scale atmospheric circulation that forces coastal up-welling also tends to produce the terrestrial coastal deserts that exist in Peru and Chile and northwestern and southwestern Africa (see, for example, Amiran and Wilson[33]). The cold surface waters produced by the upwelling itself may locally influence the climate of the adjacent land. Additional evidence for this influence can be found off the west coast of the United States and perhaps, on a larger scale, off the Somali coast and over the Western Indian Ocean during the Southwest Monsoon. Not surprisingly, anomalous patterns in the large-scale atmospheric circula-tion lead to anomalies in the large-scale ocean circulation. These anom-alies may be dramatically manifest in the response of the upwelling ecosystem.

Concerning local climatic effects, the most intense aridity on earth probably occurs along the coasts of Peru and Chile. On this point Trewartha has noted that

> . . . Because of the general sparseness of weather stations in arid regions, it is almost impossible to indicate with any degree of certainty where the absolutely driest part of the earth is located. Without fear of contradiction, however, it can be stated that the Chilean-Peruvian desert is a strong contender for this honor of being the earth's most arid region over such a large range of latitude. From Piura in northern Peru at about 5°S or 6°S to Coquimbo in northern Chile at about 30°S the highest average annual rainfall (1944–1955) at any coastal station is under 50 mm and a number of stations in northern Chile have recorded no rainfall over periods of one or more decades. In the Chilean desert even the drizzle is largely absent, so that here an intensity of aridity prevails that, so far as is known, is not equaled in any other desert of the earth.[34]

Certainly the primary physical mechanism responsible for this coastal desert is the strong subsidence associated with the eastern flank of the South Pacific anticyclone. However, this mechanism alone does not appear sufficient to reduce rainfall to the minimal levels observed. This is evidence that the most intense aridity is concentrated in a narrow strip along the coast and that rainfall increases both seaward and inland. Where points of land project westward into the ocean, rainfall is lowest. Cold upwelled waters (as much as 8°C colder than water 100 km seaward), occurring within 50 km of the coast of Peru and Chile, stabilize the lower marine boundary layer, intensify and lower the temperature inversion that exists there, and enhance the development of fog, low stratus, and sometimes drizzle along the coast. The cold water and the contrast of temperature and frictional drag between land and sea may also contribute to the strength of the subsiding air over this coastal desert. Unquestionably, active coastal upwelling is an important factor in the creation of the Peru-Chile coastal desert.

Along the coast of southwest Africa lies another desert adjacent to a coastal upwelling zone. The Namib has a climate characterized by intense aridity, negative temperature anomalies, small annual and diurnal ranges of temperature, and high frequency of fog, low stratus, and drizzle. The Namib has an extensive latitudinal range of about 15° along a narrow coastal strip extending only a few tens of kilometers inland. Several coastal stations average less than 25 mm of annual rainfall,[34] with the most intense aridity concentrated around 22–28°S. As along the Peru-Chile coast, rainfall amounts increased sharply inland.

The coastal upwelling region off northwest Africa also adjoins the western edge of the Sahara desert. The cool Canary Current and the intense coastal upwelling influence the coastal climate southward at least to 15°N. At Nouadhibou (formerly Port Etienne), Mauritania, at 21°N, for example, rainfall is about 28 mm/yr.

Along the west coast of the United States, the major climatic influence is the North Pacific subtropical high. Since the North American coastline is situated at considerably higher latitudes than the other coastal upwelling regions we have examined, the influence of this high is most important during summer when it extends farthest poleward. Along the coast, south of 50°N, cool surface temperatures and a precipitation minimum are commonly observed in summer. During winter, when the Pacific high migrates southward, major storms provide the coastal region with significant rainfall and coastal upwelling is not observed. As in the other major coastal upwelling regions, the primary mechanisms suppressing atmospheric convection are large-scale subsidence from the oceanic high and the stabilizing influence of the cold upwelled waters. Further south, under the strengthening influence of the subtropical high, precipitation amounts decrease. From northern California southward to Baja California, the coastal climate ranges from cool summer marine to mediterranean to steppe to subtropical desert.

The coldest coastal waters during summer occur near 40°N. For example, during August 1973, sea surface temperatures as low as 7°C were observed within 20 km of the Oregon coast.[35] Occasionally the cool, moist marine air along the coast is advected inland to the warm, dry interior valleys by intense sea breeze circulations.[36] Perhaps the most well-known climatic influence of these upwelled cold waters is in San Francisco (northern California). That city has one of the coldest mean summertime temperatures in the continental United States. The

July mean temperature is 59°F and the daily summertime maximum is 65°F. Precipitation is also at a minimum during this active upwelling season. The cool upwelled waters and the high incidence of stratus clouds that reduce insolation are responsible for this remarkable climatic anomaly.

The most dramatic anomalous behavior in the climate of an upwelling ecosystem occurs off the Peru-Chile coast, during El Niño. In the mean state the southeast trades drive Peruvian coastal upwelling, during the latter half of the year. During December, there is an annual warming of the coastal waters originally termed El Niño. However, from time to time (for example during 1941, 1957, 1972, and 1976), this warming continues unabated for much longer than usual. El Niño has now come to be associated with this anomalous longer-term warming. During anomalous years the warming spreads southward during the southern summer and fall until the whole coast of Peru and northern Chile is affected, which can mean major variations in the climate and the ecosystem. For example, El Niño events often produce torrential rains in deserts where there had been no rain in years and can also cause widespread dislocation of the anchoveta fishery.

Originally it was believed that El Niño occurred when the local winds over the coastal upwelling zone became anomalously weak. However, there now is considerable evidence suggesting that basin-wide atmospheric and oceanographic anomalies are at least as important as the local winds in creating El Niño. Wyrtki, for example, has shown that equatorial trades were unusually strong in 1970 and 1971, but were unusually weak in 1972.[17,37] Furthermore, the eastward flowing North Equatorial Countercurrent intensified and the South Equatorial Current weakened during the 1972 El Niño. It appears that warm water accumulated in the eastern Pacific, deepening the usually shallow thermocline and covering the cooler upwelled waters. Enfield (Chapter 8) has shown that the local coastal winds along the Peru coast during the same period were normal or only slightly below normal.

It is possible that El Niño events may be produced by several different dynamic processes. Certainly the weakening of the local winds along the coast can result in lessened upwelling. However, the "wave-dynamical" hypothesis—in which large-scale changes in the wind field far out in the Pacific perturb the internal thermal structure of the ocean, exciting Kelvin waves that propagate eastward impinging on the coastal zones of Ecuador, Peru, and Chile—appears to be strongly supported by theoretical and observational evidence. For example, McCreary,[9] Hurlburt et al.,[8] Kindle,[38] and O'Brien (see Chapter 7) have examined mathematical ocean models of El Niño, which yield results consistent with some of the observations. Enfield (see Chapter 8) presents considerable oceanographic evidence to support the "wave-dynamical" hypothesis for the 1972 El Niño. Philander, however, has suggested that if the time scales for changes in the wind field are long compared to the time scale for the Kelvin waves to be excited, then the waves will not be explicitly evident and the adjustment of the upper ocean to the atmosphere will appear to be instantaneous.[39] Furthermore, he suggests that it may not be possible to predict El Niño on the basis of sudden changes in the surface winds, unless it is known beforehand that the new wind conditions will persist for a considerable time.

NATURAL VARIABILITY, OVERFISHING, AND EL NIÑO

Natural fluctuations in the population and distribution of plant and animal species in the sea are common. Some fluctuations are clearly correlated with observable changes in the physical environment. Other natural fluctuations, such as blooms of the "red tide" or the disappearance of large populations of the Pacific sardine are not so obviously correlated to changes in the physical environment. Both the natural variability of the ecosystem and overfishing have been cited as reasons for the disastrous decline in the Peruvian anchoveta stocks during (and since) the 1972 El Niño.[40] Can the relative importance of these two factors be assessed? Perhaps not. As Dickie has suggested,

. . . The primary difficulty faced by the biological oceanographer in analyzing changes in fisheries is that the effects that arise from both economic and environmental causes look very nearly the same as those that would result if excessive fishing were damaging the stock productivity.[41]

Consider, for example, the case of the Pacific sardine (*Sardinops sagax*) (also see Chapter 5). The fact that fishery scientists are still debating the relative impacts on the decline of this fishery of natural variability as opposed to overfishing, may lend support to the view that there is a lack of knowledge concerning population dynamics in marine species. From 1916 to 1936 the annual catch of Pacific sardines rose from 28,000 to nearly 800,000 tons.[42] During the following 10 years, the annual catch fluctuated between 500,000 and 700,000 tons. In 1945, it dropped below 200,000 tons, rose slightly for several years, then collapsed in 1952 to about 20,000 tons, where it remains today.

Murphy suggests that there is little likelihood that the sardine population would have declined in the absence of fishing pressure. The biological niche vacated by the Pacific sardine was filled by the northern anchovy and the sardine population has remained low.[43] Dickie argues that sediment records indicate that the sardine had been replaced by the anchovy in earlier times and that overfishing may have only served to increase the probability that this replacement would occur again.[44]

The case of the Peruvian anchoveta fishery is even more dramatic. Until World War II the guano industry of Peru greatly overshadowed the fishery. However, by 1963 Peru, with a population of 13 million, had become the leading fishing nation of the world, harvesting 15% of the world's total catch. As the fishery grew, the bird population declined. In 1957 about 28 million guano birds were estimated, but by 1970 they had reduced to about 2 million. The population has not significantly increased in this decade.[45]

As the fish harvest soared in the early 1960s, the United Nations Food and Agriculture Organization (FAO) and the Peruvian government (through the Instituto del Mar del Peru, IMARPE), began working together to study, monitor, and regulate the fishery. Increased fishing effort and catch capacity were controlled by regulations such as closed seasons and catch quotas. Since quotas were rapidly taken by competing fish companies, the days at sea each year were reduced. However, these days were continually more concentrated in periods when new recruits joined the fish stocks, hence fishing pressure on fish in their first year increased.[46]

In 1970, under FAO and IMARPE sponsorship, a panel of fisheries economists (see Chapter 13) recommended that the maximum sustainable yield for the Peru Current anchoveta stock was about 9.5 mmt. However, by 1970, the harvesting and processing capabilities of the Peruvian fishery were far above the capacity required to land efficiently 9.5 mmt. Although 9.5 mmt was harvested for the 1969–1970 season by April 28, 1970, an additional 10 days of fishing yielded an additional 1 mmt. Before the season was closed, yet another 0.5 mmt of anchoveta was landed. The total catch for the 1970 calendar year was about 12.5 mmt.

Although a mild El Niño had forced a brief pause in catch growth during 1965–1966, the fishing industry was hardly prepared for the events of 1972 (Figure 2). As early as March 1971, surface waters in the equatorial Pacific east of 110°W became significantly warmer than in the previous year.[16] On the northern Peruvian coast the warming first became evident in February 1972, when low salinity surface water was found as far south as 10°S. The catch statistics taken in the latter half of 1971 indicated anchoveta concentrations close to shore (within 36 km). In March 1972, when the fishing season reopened, these concentrations

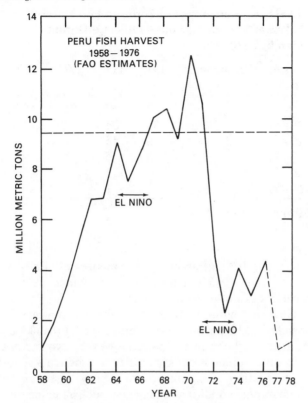

Figure 2 Peru fish harvest based on FAO and U.S. Department of Agriculture data. The dashed line represents estimates from other sources. Dashed line is maximum sustainable yield estimate (see Paulik, Chapter 3).

remained near shore, but decreased latitudinally with captures occurring only south of 10°S. During March, with El Niño already evident, a record catch rate (over 170,000 tons/day) was realized within a few kilometers of the coast. As shown in Table 2 (from Valdivia[46]) the percentage of the total catch taken within 16 km of the coast during March doubled in 1972, compared to the percentages for 1970 and 1971. By April the schools of fish had retreated south of 12°S, then to 14°S, and by June fishing operations became uneconomical and were halted.

Between June 1972 and mid-1976, the Peruvian fleet was permitted to fish in only 18 of the 48 months, with the total anchoveta catches at 2 mmt for 1973, 3.6 mmt for 1974, and 3.1 mmt for 1975. Yet even after a 30% reduction in fleet size in 1974, the potential fleet catch was still

TABLE 2. Percentage of Peruvian Fish Harvest
Taken at Various Distances Offshore During
March, 1970–1974

Distance from Coast (miles)	1970	1971	1972	1973
0–10	42	47	91	88
10–20	33	33	7	11
20–30	16	13	1	1
30–40	5	5	1	—
40–50	4	1	—	—
50–60	—	1	—	—

From Valdivia.[46]

estimated at about 10 mmt annually. A "moderate" El Niño similar in intensity to that of 1965 occurred in 1976. Since 1976, the annual anchoveta catch has not exceeded 2 mmt.

It seems clear that natural variability of the ocean-atmosphere system was responsible for the decline in the Peruvian anchoveta harvest during 1972–1973, as it has been in the past. However, two factors appear to have acted in catastrophic harmony in 1972 to produce a severe, perhaps permanent, dislocation of the anchoveta fishery. First, during El Niño, natural oceanographic conditions changed in such a manner as to increase the apparent concentration of anchoveta very near the coast. Second, at the time of El Niño, the fishing fleet had grown far too large for controlled exploitation of anchoveta under prevailing conditions of high concentrations. Although one does not know why the fishery has failed to make a rapid recovery, one does know that during El Niño the fish driven close to shore were landed in record numbers just prior to the decline and collapse of the fishery. One might argue, however, that there is some reason to believe that man and nature acting simultaneously may have decimated a productive fishery, the recovery of which is still in question.

SUMMARY

As a background to the more detailed studies of El Niño and its environmental and societal impacts, this chapter provides a general overview of the relationships between climate, vertical currents, productivity, and fisheries in coastal upwelling ecosystems. We have described the dynamic mechanisms through which low-level atmospheric wind systems drive vertical motions in the upper ocean, and have demonstrated how those wind systems influence the climate of the adjacent land and how the upwelling itself can reduce sea surface temperatures and modify local climate.

Since upwelling carries nutrient-rich waters into the sunlit upper ocean, abundant phytoplankton populations may exist in upwelling ecosystems. These populations form the first link in a food chain that supports harvestable fish. Therefore, although coastal upwelling areas represent less than 0.1% of the world's ocean areas, they provide a substantial fraction of the world's total fish catch.

We have discussed the natural variability of the ocean-atmosphere system, emphasizing the combined effects of natural variability and overfishing in reducing potential fish yields. The case of El Niño and the Peruvian anchoveta fishery was offered as a particularly dramatic example of this interaction. Finally, we have outlined recent research results that suggest that El Niño is not simply a local phenomenon, locally forced and locally manifested, but it is a basin-wide (perhaps global) phenomenon in which remotely forced oceanographic anomalies are of crucial importance to the Peruvian upwelling ecosystem.

ACKNOWLEDGMENTS

I greatly appreciate the assistance and encouragement of Dr. Michael Glantz, George Wooten, Dr. Joseph Wroblewski, Dr. Steve Piacsek, and my wife Brenda during various phases of the preparation of this manuscript.

REFERENCES

1   R. L. Smith, "Upwelling," *Oceanography and Marine Biology Annual Review*, **6**, 11 (1968).

2   J. H. Ryther, "Photosynthesis and fish production in the sea," *Science*, **166**, 72 (1969).

3   J. H. Ryther, E. M. Hulburt, C. J. Lorenzen, and A. Corwin, *The Production and Utilization of Organic Matter in the Peru Coastal Current*, Texas A & M University Press, College Station, 1969.

4   D. H. Cushing, "Upwelling and fish production," FAO Fishery Technical Paper 84, Rome, 1969.

5   J. A. Gulland, "Population dynamics of the Peruvian anchoveta," FAO Fishery Technical Paper 72, Rome, 1968.

6   V. W. Ekman, "On the influence of the earth's rotation on ocean currents," *Arkiv foer Matematik Astronomi och Fysik*, **12**, 1 (1905).

7   K. Wyrtki, "Oceanography of the eastern equatorial Pacific Ocean," *Oceanography and Marine Biology Annual Review*, **4**, 33 (1966).

8   H. E. Hurlburt, J. C. Kindle, and J. J. O'Brien, "A numerical simulation of the onset of El Niño," *Journal of Physical Oceanography*, **6**, 621 (1976).

9   J. P. McCreary, "Eastern tropical ocean response to changing wind systems: With applications to El Niño," *Journal Physical Oceanography*, **6**, 632–645 (1977).

10   H. Thorade, *Annual Hydrographic Bulletin*, **37**, 17 and 63 (1909).

11   H. U. Sverdrup, "On the process of upwelling," *Journal of Marine Research*, **1**, 155 (1938).

12   W. S. Wooster and J. L. Reid, "Eastern Boundary Currents," in M. N. Hill, Ed., *The Sea*, Vol. 2, Wiley, New York, 1963, pp. 253–280.

13   A. Bakun, "Coastal upwelling indices, west coast of North America, 1946–1971," Technical Report NMFS SSRF 671, NOAA, Seattle, 1973.

14   E. R. Gunther, "A report on oceanographical investigations in the Peru coastal current", *Discovery Reports*, **13**, 107 (1936).

15   W. S. Wooster and M. Gilmartin, "The Peru-Chile undercurrent," *Journal of Marine Research*, **19**, 97 (1961).

16   W. S. Wooster and O. Guillen, "Characteristics of El Niño in 1972," *Journal of Marine Research*, **32**, 357 (1974).

17   K. Wyrtki, "Equatorial currents in the Pacific 1950 to 1970 and their relations to the trade winds," *Journal of Physical Oceanography*, **4**, 372 (1974).

18   K. Wyrtki, "The horizontal and vertical field of motion in the Peru current," *Bulletin of the Scripps Institution of Oceanography*, **8**, 313 (1963).

19   J. L. Reid, G. I. Roden, and J. G. Wyllie, "Studies of the California current system," Progress Report of the California Cooperative Oceanic Fisheries Investigations, July 1, 1956 to January 1, 1958, pp. 27–56, 1958.

20   J. D. Thompson, "The coastal upwelling cycle on a β-plane: Hydrodynamics and thermodynamics," Ph.D. Thesis, Florida State University, 1974.

21   J. B. Peffley and J. J. O'Brien, "A three-dimensional simulation of coastal upwelling off Oregon," *Journal of Physical Oceanography*, **6**, 164 (1976).

22   R. M. Clancy, H. E. Hurlburt, J. D. Thompson, and J. D. Lee, "A model of mesoscale air-sea interaction in a sea-breeze, coastal upwelling regime," *Monthly Weather Review*, **107**, 1476 (1979).

23   J. S. Wroblewski, "A model of the spatial structure and productivity of phytoplankton populations during variable upwelling off the coast of Oregon," Ph.D. Dissertation, Florida State University, 1976.

24   D. J. Wright, B. M. Woodworth, and J. J. O'Brien, "A system for monitoring the location of harvestable coho salmon stocks," *Marine Fishery Review*, **38**, 1 (1976).

25   E. Mittelstaedt, D. Pillsbury, and R. L. Smith, "Flow patterns in the Northwest African upwelling area," *Deutsche Hydrographische Zeitschrift*, **28**, 145 (1975).

26   S. A. Huntsman and R. T. Barber, "Primary production in the upwelling region off Northwest Africa—A Comparison with Peru," *CUEA Newsletter*, **5**, 2 (1976).

27 T. J. Hart and R. I. Currie, "The Benguela Current," *Discovery Reports*, **31**, 123 (1960).

28 A. Defant, "Das Kaltwasserauftriefsgebiet vor der Kuste Sudwest-Afrikas," *Landerkundliche Studien (Festschrift W. Krebs)*, pp. 52–66 (1936).

29 G. Bohnecke, "Temperatur, Salzgehalt und Dichte an der Oberfläches des Atlantischem Ozeans," *Atlas "Meteor" Report* 1936.

30 J. A. Crutchfield, and R. Lawson, *West African Marine Fisheries: Alternative for Management*, Resources for the Future, Washington, 1974.

31 J. Shukla, "Effect of Arabian sea-surface temperature anomaly on the Indian summer monsoon: A numerical experiment with the GFDL model," *Journal of the Atmospheric Sciences*, **32**, 503 (1975).

32 Somali Democratic Republic, *Feasibility Study on the Fishery Resettlement Programme*, State Planning Commission, Mogadiscio, July 1976.

33 D. H. K. Amiran and A. W. Wilson, Eds., *Coastal Deserts: Their Natural and Human Environments*, University of Arizona Press, Tucson, 1973.

34 G. T. Trewartha, *The Earth's Problem Climates*, University of Wisconsin Press, Madison, 1961.

35 C. G. Holladay and J. J. O'Brien, "Mesoscale variability of sea surface temperatures," *Journal of Physical Oceanography*, **6**, 761 (1975).

36 A. Johnson and J. J. O'Brien, "A study of an Oregon sea breeze event," *Journal of Applied Meteorology*, **12**, 1267 (1973).

37 K. Wyrtki, "Sea level during the 1972 El Niño," *Journal of Physical Oceanography*, **7**, 779 (1977).

38 J. C. Kindle, "Equatorial Pacific ocean variability—seasonal and El Niño time scales," Ph.D. Dissertation, Florida State University, 1979.

39 S. G. H. Philander, "Variability of the tropical oceans," *Dynamics of Atmospheres and Oceans*, **3**, 202 (1974).

40 J. D. Thompson, "Ocean Deserts and Ocean Oases," in M. H. Glantz, Ed., *Desertification: Environmental Degradation in and around Arid Lands*, Westview, Boulder, CO., 1977.

41 L. M. Dickie, "Problems in prediction," *Oceanus*, **18**, 30 (1975).

42 G. I. Murphy, "Population biology of the Pacific sardine (Sardinops caerulae)," *Proceedings of the California Academy of Sciences*, **34**, 1 (1966).

43 T. Joyner, "Resource Exploitation–Living," in D. W. Hood, Ed., *The Impingement of Man on the Oceans*, Wiley, New York, 1971.

44 L. M. Dickie, "Interaction between fishery management and environmental protection," *Journal of the Fishery Research Board of Canada*, **30**, 2496 (1973).

45 G. L. Kesteven, "Recovery of the anchovy and El Niño," *CUEA Newsletter*, **5**, 17 (1976).

46 J. Valdivia, "Biological aspects of the 1972–73 El Niño—Part 2: The anchovy population," presented at the IDOE Workshop on the Phenomenon Known as "El Niño," Guayaquil, Ecuador, December 4–12, 1974.

CHAPTER 3

Anchovies, Birds, and Fishermen in the Peru Current*

GERALD J. PAULIK

Editor's Abstract

This chapter, *published in 1971,* traces the commercial exploitation of the anchoveta fishery and the development of the fish meal industry in Peru. References are made to other fisheries throughout the world. He describes the physical setting, including the geophysical, biological, and fish population dynamic aspects of the anchoveta fishery. Guano bird stocks are also discussed as part of the physical setting for the fishery. Paulik's discussion of the industrial and economic setting is concerned with the problems associated with the overcapitalization of the fishing fleet and of the fish meal industry and its impact on the standing stock of anchoveta. The susceptibility to overexploitation of the anchoveta fishery is compared with pelagic fisheries elsewhere. The capstone of the chapter is a scenario he presents in the concluding section in response to his own question: "Destruction of the world's greatest single species fishery is unthinkable. Could it really occur?"

INTRODUCTION

The human mind has difficulty in grasping the true enormity of extremely large numbers. Peru's anchoveta *(Engraulis ringens)* catch of 11 million metric tons (mmt) is such a number. This catch was taken in the 1969–1970 fishing season of a little over 7 months. More than 10 trillion anchoveta are needed to make 11 mmt.

---

*Reprinted from *Environment: Resources, Pollution and Society,* William W. Murdoch, Ed., Sinauer Associates, Stamford, Conn., 1971: with permission of Mrs. Gerald Paulik.

One way of comprehending the meaning of large numbers is to compare them to more familiar quantities or to numbers for which we have some intuitive feeling. The total catch by U. S. fishermen of all species of fish and shellfish was approximately 2.5 mmt for 1969; thus Peru's catch is about 4.5 times larger than that of the United States.

For all practical purposes Peru's catch is composed of only one species. Comparing the Peruvian catch to the U.S. catch of a single species, the yellowfin (*Thunnus albacares*), which is the most common canned tuna sold in the United States, show the quantity of anchoveta to be a hundredfold greater than the quantity of tuna. The total weight of the annual U. S. catch of five species of Pacific salmon—chinook, chum, coho, pink, and sockeye—is about 250 thousand metric tons or 1/44 of the Peruvian catch.

Fortunately for the United States and unfortunately for Peru, sheer biomass does not measure economic worth. Salmon and anchoveta may be equally nutritious, but the ex-vessel of wholesale price of anchoveta is about $11/ton and that of salmon about $770/ton, making salmon 70 times as valuable per unit weight.

One-tenth is a commonly accepted ecological efficiency factor for stepping up one trophic level in a linear food chain. If it were possible to feed the anchoveta to a predator species as valuable as the salmon and to salvage 1/10 of their annual productivity, the income to Peruvian fishermen could be seven times greater.

Another way of viewing the 11 mmt taken in the 1969–1970 season is to compare it to the sustainable natural productivity of the anchoveta stock in the Peru Current. The secret of successful and continued harvesting of a renewable natural resource is to use the surplus produced by the resource without reducing the size of the stock enough to damage its productive capability.

During January of 1970 the Food and Agriculture Organization (FAO) of the United Nations, the United Nations Development Fund Program (UNDP), and the Instituto del Mar del Peru (IMARPE) sponsored a meeting of a group of distinguished international experts on population dynamics to determine the maximum sustainable yield (MSY) for the anchoveta stocks in the Peru Current. This panel concluded the best estimate for MSY to be about 9.5 mmt.[1] It is ironic that the ink had hardly dried on the panel's recommendation before it was substantially exceeded by the catch of 11 mmt in the 1969–1970 season.

The true capacity of the Peruvian fleet is staggering. On April 28, 1970 the total catch reached 9.5 mmt. The fishery continued for 10 more days, taking 100,000 metric tons per day. This fantastic catching power could have taken the whole U.S. yellowfin tuna catch in one day, or the entire

U.S. catch of all Pacific salmon in 2.5 days. Obviously, this type of destructive power must be handled most carefully, and precise management and regulation are essential.

As if to illustrate that extremely large numbers may still be too small, part of the Peruvian industry expressed dissatisfaction with the catch of 10.5 mmt taken by May 13. Smaller and less efficient producers petitioned the government for a special season to take advantage of availability of the fish and the high price of meal. They were granted an additional 300,000 tons, demonstrating most vividly the political power of a large number of marginal operators. The remaining 200,000 tons in the total catch of 11 mmt were taken near the southern boundary of Peru, where the anchoveta stocks are fished by both Peru and Chile and neither can be assured that its fishermen and factories would be able to realize the fruits of unilateral conservation efforts.

We have compared 11 mmt to other numbers and we have seen that it can be either too much or too little. In actuality it is a statistical fiction. In the process of unloading the purse seiners, or *bolicheras* as they are known in Peru, water pumped into the hold is mixed with fish and the entire slurry transported to the factory through a pipeline that may extend a quarter of a mile or so from the offshore unloading platform. Considerable losses of fish are sustained in this process, especially when the young fish or *peladilla* form the bulk of the catch in Janaury, February, and March. Further losses are sustained in the pursing operation and occasionally because of intentional dumping of excess catches. Under-reporting of the actual quantities landed is another source of bias during the *peladilla* season, when the yield of meal per unit of fish is low. Although some observers believe these losses may be as high as 40%, a conservative estimate of the actual quantity of fish removed from the sea during the 1969–1970 season is between 13 and 14 mmt.

Mathematicians sometimes think about a large number of things by assigning each thing to an entity in a known set. Enough anchoveta were taken from the sea to provide each of the 13 million or so citizens of Peru with 1 ton of high-quality protein food. This capability rests in a fishery that generates about $11/person · yr in terms of current ex-vessel prices. In terms of fish meal exports (assuming $165/ton for meal), foreign capital of about $33/person · yr is generated for the country. However, under current regulatory policies as the fleet and meal factories accumulate excess capacity, the length of the *veda* or closed season must increase, and as idle equipment deteriorates, scarce capital is oxidized into rust. Although much of the factory equipment and nearly all of the *bolicheras* are manufactured in Peru, purchase of electronic gear and engines is a serious drain on foreign capital reserves.

The remarkable development of the fish reduction industry has completely bypassed the food fish industry, which continues to operate at an extremely low level of production largely for local consumption and struggles along with obsolete equipment and methods. The Peruvian government recently declared development of a major food fish industry to be a national goal and has inaugurated an ambitious program to accomplish this objective. However, the greatest potential for a food fish industry may not lie in the direction of starting new fisheries for those species traditionally used for food, but rather in transforming the immense and already proven production of anchoveta from the Peru Current into a high-quality product suitable for human consumption. Substantial social and economic benefits would result from using Peru's surplus labor in such an enterprise.

From the catch of 11 mmt in 1969–1970, Peru will produce approximately 2.1 mmt of fish meal. This quantity of meal fed in the poultry, swine, and cattle industries of the industrialized nations of Western Europe and North America represents a significant transfer of nutrients from the Southern to the Northern Hemisphere.

HISTORICAL DEVELOPMENTS

The Peruvian fish industry began life as a war baby. During World War II the United States relied upon Peru for canned and salted fish products as well as fish liver and oil. This food fish industry, with bonito, skipjack tuna, swordfish, and shrimp as its prime products, fell victim to the United States' transfer from a war to peace economy and the post-war replacement of some fishery products by synthetics. However, the fertility of the Peru Current could not be denied, and in 1950 a fish meal factory was set up surreptitiously in a remote Peruvian bay. Secrecy was essential for this prototype operation to conceal it from the guano industry, which then represented a powerful vested interest firmly opposed to any use of anchoveta that might reduce the amount of fish available to the guano birds. Before the guano industry could react to this threat to their source of new material, the fish meal boom was out of control. The industry grew at an astonishing rate; from 1956 to 1963 Peru leaped from obscurity to the largest fish meal producer in the world. Fish meal production statistics from 1951 to 1968 are shown in Table 1.

Fish meal became the number one producer of foreign exchange for Peru. Fishery products and copper together account for 54.1% of the value of Peruvian export products; fish meal leads copper by 0.1%. This fantastic industrial expansion brought employment to approximately 25,000 people; a fleet of 1,700 *bolicheras* and 150 fish meal and oil

TABLE 1. Peruvian Fish Meal Production

Year	Thousand Metric Tons
1951	7.2
1952	9.2
1953	12.1
1954	16.5
1955	20.0
1956	30.9
1957	64.5
1958	126.9
1959	332.4
1960	558.3
1961	863.8
1962	1,120.8
1963	1,159.2
1964	1,552.2
1965	1,282.0
1966	1,470.5
1967	1,816.0
1968	1,922.0

reduction factories suddenly materialized; the only ingredient missing was John Steinbeck to record the lusty excitement as fishing towns sprouted from nothing along the barren coast of Peru.

Figure 1 shows yearly catches, in live weight, of the world's leading fishing nations for the decade from 1958 to 1968, the latest year for which total world fish production statistics are available. Clupeoids made up about a third (20.5 mmt) of the total 1968 catch. Besides the Peruvian anchoveta, this figure includes sardinella, pilchard, herring, and sardine. Codlike species accounted for about 9.5 mmt; mackerels, 2.3 mmt; flatfish, 1.2 mmt; and various tuna species, 1.4 mmt. While total world landings rose from 33 mmt in 1958 to 64 mmt in 1968, Peru's total catch increased from 0.96 mmt in 1958 to 10.5 in 1968.

Today guano birds and large predatory fish must compete with human fishermen for their share of the anchoveta stocks. The guano industry, however, existed long before man began to manufacture fish meal. Over a thousand years ago, the Incas used guano to fertilize the barren coastal soil of Peru. Intense mining of the guano deposits occurred during the latter half of the nineteenth century, when 150-foot high mountains of

Gerald J. Paulik

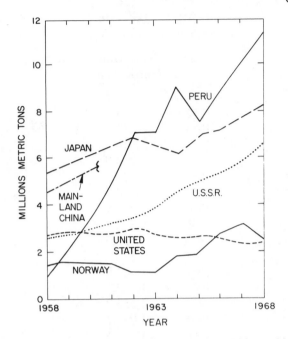

Figure 1   Annual catches in live weight by leading fishing nations of the world for 1958–1968.

guano that had taken over 25 centuries to accumulate were removed in a few years. In 1909 the need to regulate the rate of removal and to protect the guano birds was officially recognized when Peru set up the Guano Administration. This administration attempted to restrict the amount removed per year to no more than the amount deposited. Given the magnitude of the annual fluctuations in the size of the bird populations, such a balance could be achieved only with the aid of the sort of highly sophisticated management information system which even today exists only for a very few natural resources. The Peruvian government also concentrated on protecting the birds and their eggs from human and animal predators. The bird sanctuary policy worked well and the numbers of birds began to increase after having been perilously low in the early 1900s.

AN OVERVIEW

The existing and rapidly increasing overcapacity of both the fish meal factories and the fishing fleets, and the inefficient use of labor and capital resulting from the use of extensive closed fishing periods to limit the size of the physical harvest, pose grave problems for Peru. She must devise and introduce an equitable means of reducing capacity that will neither encourage economic inefficiency nor foster corruption.

However, the first and most important objective of any regulatory scheme is to preserve the productivity of the resource. Other and almost equally important objectives include the following:

1   To harvest the correct amounts of the right parts of the population to maximize the physical yield of fishery products. At present these products are fish meal, oil, and guano but could include in the near future a product such as FPC (fish protein concentrate) for human consumption.

2   To minimize harvesting and processing costs in order to maximize economic return. This implies regulations that will force evolution of an efficient industry—technologically innovative in its harvesting methods and product utilization.

3   To set up a politically and socially acceptable scheme to distribute profits among fishermen, factory and fleet owners, and the people of Peru.

4   To achieve stability of employment and income from the fishery.

Traditionally fisheries management has taken a very narrow view of its mission and has not properly described to decisionmakers the total consequences of alternative courses of action. The calamitous economic performance of such fisheries as those for halibut and salmon on the Pacific coast of the United States can be attributed to an attitude that recognized only the biological aspects of conservation regulations. The anchoveta-based industries are so important to the total national economy of Peru that she cannot afford to run the risk of the disastrous economic and perhaps political consequences that could result from ignoring any significant aspects of the total problem of controlling exploitation of the anchoveta stocks. This chapter presents a broad view of the many background settings in which the regulatory problem is imbedded. Figure 2 is a diagrammatic overview of the current state of affairs and serves as an introduction to the rest of the paper.

In the sections that follow, we examine many of the factors—physical, biological, economic, and political—acting to shape the destinies of the Peru anchoveta stocks and the many industries and people dependent in one way or another upon maintaining the biological productivity of these stocks.

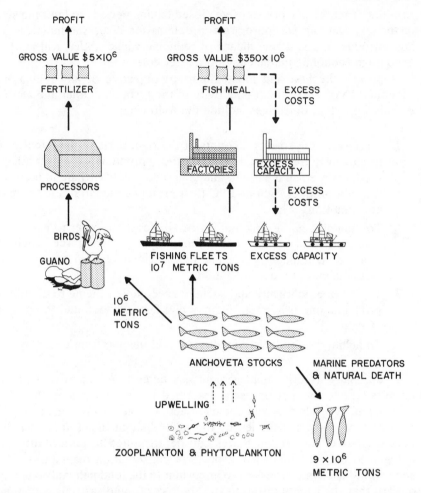

Figure 2 A pictorial systems diagram showing biological, economic, and social components involved in man's exploitation of the natural fertility of the Peru Current.

THE PHYSICAL SETTING

Peru's coastal strip is ideally suited for industrialization since the flat desert land provides spacious building sites, and huge inventories can be held with little or no protection from the weather in the stable arid climate. The 1400-mile long coast of Peru is about 100 miles longer than the Pacific coast of the United States. The narrow strip of coastal desert is usually less than 40 miles wide.

Peru's primary harbor is at Callao, 7 miles west of Lima and about halfway between the northern and southern boundaries. The extent to which Lima dominates the nation is expressed in an old saying "Peru is Lima and Lima is Peru." Lima and other cities of the Peruvian coastal plain are being invaded by poverty-stricken migrants from the high mountain areas. Lima itself is surrounded by an incredible array of makeshift housing. Most of the amenities of modern society, such as electricity, water, and sewers, will not be found in the *barriadas*—or slum areas. Many of Peru's fishermen and fish meal factory workers came from the *barriadas*.

Peru does not have the broad continental shelves that border many other countries. The western edge of the South American continent begins at the top of the Andes, falls precipitously to the flat and narrow coastal strip, and then plunges into the depths of the Peru-Chile Trench. Just off the coast from 22,205-ft Mount Huascaran, which lost part of its

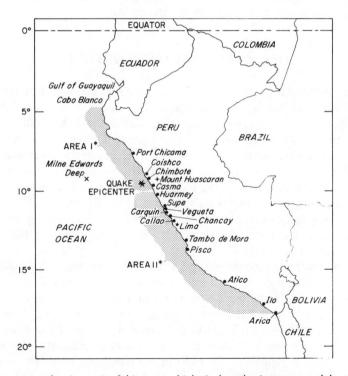

Figure 3 A map showing major fishing ports, biological production areas, and the epicenter of the May 31, 1970 earthquake. Areas I and II as defined by Cushing[2] for estimation of primary and secondary productivity.

peak in the May 31 earthquake, is the Milne-Edwards Deep, 20,394 feet below sea level (see Figure 3).

Beyond the 100-fathom or 200-m depth contour—a traditional measure of the outer edge of the continental shelf—the slope of the bottom increases abruptly. It is on the shallower shelf areas that exploitable minerals and petroleum resources are found. Not only is the narrowness of the shelf an obvious disadvantage, but the marine terrain of a narrow shelf makes economic exploitation extremely difficult. Thus Peru can expect little in the way of nonrenewable resources from her shelf, in comparison with countries with shelves that extend seaward as far as 200 miles below their adjacent waters. Lacking a broad continental shelf, Peru claimed as her own 200 miles of ocean adjacent to her shores. In one sense it is as if Peru regarded the Peru Current as being a substitute for the broad continental shelves claimed by other nations. The Peru Current is about 200 meters deep and approximately 200 miles wide, although these dimensions vary greatly along the coast. More by accident than by design, Peru's claim is a tidy extension of sovereignty in that it encompasses a single and unique marine regime.

OCEAN CURRENTS AND UPWELLING

From Valparaiso, Chile to the Gulf of Guayaquil in Ecuador, the coastal desert of South America is bathed by the cold northward flowing Peru Current. This current, sometimes called the Humboldt Current, swings to the west just south of the equator and joins the westward flowing South Equatorial Current. Its counterpart in the Northern Hemisphere, the California Current, flows south towards the equator where it swings to the west to join the Northern Equatorial Current.

A rough diagrammatic sketch of the basic dynamics of the swift western boundary currents, the slower and broader eastern boundary currents, and the huge slowly moving equatorial flows in the Pacific Ocean is given in Figure 4.

The interactive response of the winds and the ocean to differential heating between equatorial and polar regions combines with the earth's rotation to produce the Northern and Southern Hemisphere gyres illustrated. Excess equatorial heat is carried toward the poles along the western boundaries of the ocean and eastern shores of the continents. As this flow attempts to restore the heat balance of the earth, it is pushed by the spin of the earth to the right in the Northern Hemisphere and to the left in the Southern Hemisphere (Coriolis force).

The southeast trade winds are channeled by the Andes to blow parallel

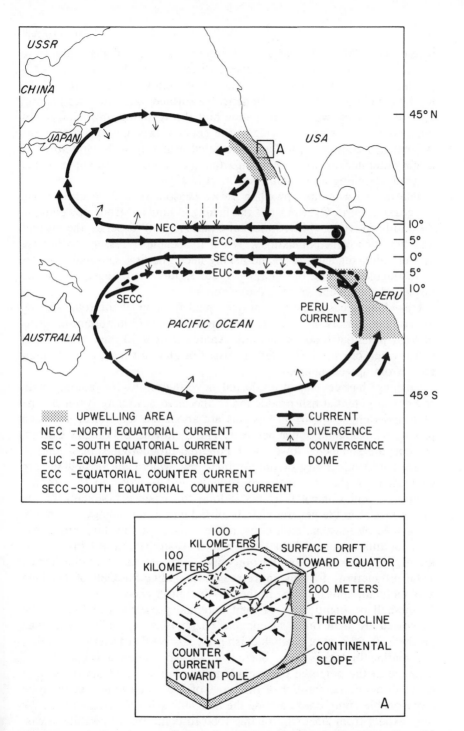

Figure 4 Eastern boundary upwellings and circulation patterns in Pacific Ocean. The solid line represents the upwelling and offshore drift. A secondary upwelling 100 km from shore is shown. Source: Cushing.[2]

to the coast. Mountain ranges in California act in a similar manner to direct prevailing winds in a southerly direction. As the Peru Current is pushed northward by the south wind, it is continually deflected to the left by the Coriolis force. As the deflected surface water moves off shore, it is replaced by water from below and the resulting upwelling forms huge eddies and spirals flowing in the general direction of the equator. A smaller secondary upwelling associated primarily with wind stress is often generated at about 100 km from the shore, where some of the original upwelling water sinks (see inset in Figure 4).

The intensity of the upwelling varies seasonally with location along the coast of Peru. Over most of the fishing grounds a strong upwelling in the southern winter (May, June, July, and August) disperses the current and the anchoveta populations. When it became necessary to impose conservation regulations on the fishing fleet, the first seasonal closures were imposed during the months of June, July, and August when the dispersed anchoveta were hard to catch anyway.

Upwelling areas constitute only a small fraction of the total surface area of the world's oceans, yet they are exceedingly important in terms of total potential food production. Approximately 50 mmt of the total world catch of 64 mmt in 1968 consisted of marine fishes. Over 15 mmt came from upwelling areas.

Cushing[2] has developed crude but nevertheless useful rankings with respect to potential fish production of the main upwelling systems. The Peru and Benguela Current systems are about the same size, and each produces 20–30 mmt of fish per year. The California and Canary Current systems are somewhat lower, with potential productivities of around 10 mmt. The productivity of the Arabian Sea, where the upwellings are generated by the seasonally shifting winds, is close to that of the California and Canary Currents. Along the eastern boundary of the Indian Ocean in the general vicinity of Indonesia, upwellings are greatly influenced by seasonal monsoons, and estimates of upwelling productivities for this region and for the Western Australia Current involve far more guesswork than for the other five more intensely studied areas. Total productivity for this entire region, including the Gulf of Thailand and Vietnam, appears to be in the 20–40 mmt/yr class.

One-half or slightly less of total potential production of fish can be harvested by man on a sustainable basis; thus, upwelling areas can produce 50–65 mmt of catchable fish per year. Neither fisheries scientists nor marine biologists have been able to agree upon a single "best" estimate of the potential harvestable production of fish from the sea. A figure of about 125 mmt/yr of the types of fish taken today seems to be a reasonable compromise among the available scientific conjectures. We may expect from 2/5 to 1/2 of this total to come from upwelling areas.

THE BIOLOGICAL SETTING

Upwelling systems may be compared to vertical chemostats. At the depths where the upwelling water originates, sparse populations of inoculating phytoplankton are present. Theoretically at least, as the water rises this seed population begins to grow as the plants and nutrients are slowly lifted along an increasing light gradient. The phytoplankton can multiply more rapidly than the herbivorous zooplankton that feed on them. According to this theory, the temporal sequence of events in a volume of upwelling water is similar to the temporal sequence of events during the spring bloom in temperate waters where the phytoplankton build up high densities before they are either overtaken by their predators or have their population growth limited by scarcity of critical nutrients.

Zooplankton, with generation lengths several times those of the phytoplankton, respond after an initial lag to increasing phytoplankton densities and build their own populations at the expense of the phytoplankton. High concentrations of zooplankton often are found far downstream of primary upwelling surface points.

Planktonic animals have difficulty in maintaining spawning populations in a moving habitat as dynamic as the Peru Current. To sustain populations in desirable locations, they must be able to withstand continuous washout from favorable zones or rely upon some mechanism such as a current/countercurrent system or a gyral to reseed the area with their sex products. How phytoplankton and zooplankton manage to continually inoculate the upwelling water with seed stock remains a nagging but most intriguing scientific problem.

A basic measure of biological productivity is the amount of carbon fixed per unit of surface area per unit time. Some of the highest values ever observed have been found in the Peru Current, where many investigators have observed values greater than 10 g carbon/$m^2 \cdot$ day. Estimated annual production figures in grams of carbon per square meter per day for two geographical divisions (Figure 3) of the Peru Current upwelling are given in Table 2. Production figures for the Peru Current may be contrasted with those for the California Current which, in the most productive of three primary upwelling areas, produces an average of 0.362 g carbon/m^2 and in the lowest, 0.234 g.

By extending this productivity to the next trophic level, which includes zooplankton and anchoveta, we can estimate total anchoveta production available from the Peru Current. Production in the upwelling farther to the south off the Coast of Chile should not be included in this calculation because of the considerable evidence that the Chilean component of the coastal circulation is separated from the Peruvian component by a belt of warm water at about 20°S.[3]

TABLE 2. Carbon Production in the Peru Current

	Seasonal Average, g Carbon/m²·day	Area km² (× 10³)	Season days	Tons, Carbon/yr (× 10⁶)
Area I				
Cape Blanco to 10°S	0.471	288	270	36.64
Area II				
10°S to Arica	1.479	191	270	76.27
Total				112.91

From Cushing.[2]

Zooplankton are usually represented in aquatic food-chain diagrams as a link between the plants or primary producers and the third stage carnivores. Most zooplankton in the Peru Current system consume phytoplankton as they should to fit into a neatly arranged food chain; the anchoveta are not so obliging. Adult anchoveta are far more herbivorous than the very young, which feed upon the larvae and adults of small copepods found in the Peru Current. By feeding directly on phytoplankton, the adults gain access to additional energy; 5–15 g phytoplankton are needed to produce every gram of zooplankton the anchoveta consume. By eliminating the zooplankton link, the anchoveta bypass a middleman who otherwise would remove a large portion of the available energy. Some zooplankton, especially the large shrimplike euphausids, which consume other zooplankton and organic debris as well as phytoplankton, occupy about the same position in the food web as the anchoveta.

A standard ecological accounting procedure is to equate some fraction of primary production with production at an artificial second level. Even though the figure of 10% has been widely used as the efficiency of energy transfer from one trophic level to another, in actuality such efficiencies are notoriously variable.

Cushing[2] computed secondary production in the Peru Current by two methods: (1) indirectly, using 10% of observed phytoplankton production; and (2) directly, converting average standing crop as measured by net catches of zooplankton by the number of generations produced per season. He assumed the bottom of the layer sampled by nets to be 300 m and arbitrarily extended generation times by 25% from observations made under ideal conditions to account for the intermittent character of upwellings. His figures (using unadjusted generation lengths) for secondary production of the Peru Current are given in Table 3.

The last column, trophic efficiency, compares the direct calculations of second-stage production to the measurements of primary productivity.

TABLE 3. Peru Current Secondary Production

Area	Net Catch, ml Displaced/ m² Surface Area	Carbon, g/m²	Generation Time, Days	Number of Generations per Season	Carbon, Tons/yr ($\times 10^6$)	Trophic Efficiency, %
Area I Cape Blanco to 10°S	68	3.78	52	5.18	7.50	20.47
Area II 10°S to Arica	60	3.36	52	5.18	4.43	5.81
Total					11.98	

From Cushing.[2]

Gerald J. Paulik

The low trophic efficiency of 5.81% for central Peru may be partly the result of competitive grazing of phytoplankton by anchoveta not included in the zooplankton standing crop measurements.

Recent proclamations on the potential production of fish in the Peru Current system by three eminent scientists provide a rare opportunity to examine the state of the art in estimating potential production of food from the sea. Gulland,[4] working primarily from an extrapolation of past fish catch statistics and assuming that about half the total mortality is caused by the fishery, estimates the total productivity of fish in the main Peruvian upwelling area to be on the order of 20 mmt. The accuracy of Gulland's estimate depends upon the implicit assumption that the fishery is currently exploiting all of the anchoveta stocks in the current system and there are no geographically isolated stocks either between major fishing ports or farther out to sea to be discovered in the future. Both possibilities seem unlikely, although the fish resources of the offshore regions of the Peru Current have not been adequately explored.

Two other scientists, Ryther[5] and Cushing,[2] used entirely different methods to calculate the productivity of the Peruvian upwelling system, and it is most revealing to compare their calculations. Both Ryther and Cushing estimate total production at about 20 mmt and both call attention to the agreement between their calculations and Gulland's. Ryther also cites Cushing's results as supporting the validity of his estimate. Although total production estimates turn out to be the same, the components used to construct these two totals differ shockingly. Below, the four factors mutiplied by Cushing and Ryther to compute production in the Peru Current system are laid out. (T − C = metric tons of carbon).

According to Cushing[2]

$(T − C/km^2 \cdot yr) \times (km^2)$
$\times$ (2-step ecological efficiency factor of 0.01)
$\times$ (carbon wet weight conversion factor)
= Metric tons fish
$(235.7) \times (479 \times 10^3) \times (0.01) \times (17.85)$
$= 20.15 \times 10^6$

According to Ryther[5]

$(T − C/km^2 \cdot yr) \times (km^2)$
$\times$ (1.5 step ecological efficiency factor of 0.12)
$\times$ (carbon wet weight conversion factor)
= Metric tons fish
$(300) \times (60 \times 10^3) \times (0.12) \times (10)$
$= 21.6 \times 10^6$

The agreement between the final products is not too reassuring when we compare the individual elements in the equations. If certain elements were interchanged, for example, if Cushing had used Ryther's 1.5 step ecological efficiency factor of 0.12, he would have obtained about 240 mmt rather than 20 mmt. On the other hand, if Ryther had used Cushing's estimate of the area of biological productivity, he would have obtained 172 mmt rather than 21.6 mmt. The discrepancies between the calculations of these two scientists illustrate existing inadequacies in our understanding of oceanic production. There is a great deal of important work waiting to be done.

DYNAMICS OF THE ANCHOVETA STOCKS

There are two vitally important reasons for understanding the population dynamics of the Peruvian anchoveta. The most obvious is that unless the population processes are understood and the rate of harvest adjusted to conform with the natural productivity of the stocks, the rapidly expanding industry could destroy the stocks by overfishing. A less obvious reason is that the growing importance to world fisheries of species similar to the anchoveta make it an extremely valuable prototype of an entire group of fisheries of the future. As man's ability to harvest the wild stocks of fish in the farthest reaches of the ocean grows, the larger, more valuable, easier to capture animals are placed in dire jeopardy. The political and practical difficulties of developing suitable international machinery for controlling the total take of high-level carnivores, such as cods, mackerels, tunas, salmons, flatfishes, and sharks, and of dividing the take between nations, may well prove insurmountable. Any future species succession in catches taken from a given region of the ocean will probably involve the replacement of higher trophic level animals with those from lower levels. It is also likely that any such change in species composition will increase the physical yield while reducing its economic worth.

There are indications that such replacements are beginning to occur. For example, the severe overharvesting and near extinction of several of the Antarctic baleen whales has led the Russians[6] to attempt to replace the whales as a harvester of krill. While the sustainable yield from whales was in the vicinity of 1–1.5 mmt/yr, these whales consumed about 50 mmt krill/yr. So far krill harvesting has not proven economically feasible on a large scale.

The surge of interest by the world's fishermen in anchovies and related species suitable for reduction to fish meal has been examined by W. M. Chapman:

> The overwhelmingly largest change in trend of usage over the years has been as raw material for fish meal and other undifferentiated protein production. In 1938, 8.1% of world production was used in this form and by 1968 this had reached 35.6%. . . . The reason for this is the rapid spread of modern animal husbandry practices (and particularly poultry production) into the rest of the world. This is really just beginning . . . by the year 2000 a larger proportion of this undifferentiated protein will be used for direct human consumption than at present.[7]

Thus anyone interested in food production from the sea must be concerned about the population dynamics of small, short-lived, anchovy-type species. The prospects for bypassing these animals and harvesting primary producers directly appear to be nil using any foreseeable technology. Until man can separate from the water the microscopic plants that form the basis of the food chain, he will continue to depend upon the herbivores to convert the plants into more available and palatable forms of food.

We examine the population biology of the Peru anchoveta as a representative type specimen in a taxonomy based upon the productivity characteristics of animals rather than upon their morphological characteristics. All members of the genus *Engraulis* are pelagic and typically are found in oceanic areas within 200 and 300 miles of a coastline or, in a few cases, in shallow inshore bays and estuaries. They spawn in huge schools or spawning aggregations and spawning is usually spread out over an extended period.

Fisheries scientists have developed a special brand of demographics in which a number of vital population parameters are first measured and then combined in a mathematical model to estimate the maximum sustainable yield of a fish stock. The critical factors include the relation between the stock and the production of young animals subsequently recruited to the population, the rate of natural mortality, and the rate of growth of individual animals. These factors may change with population density, and measurements taken on a virgin stock may be quite different from those for fully exploited stocks. These key processes for the Peru anchoveta will be examined in some detail.

Anchoveta spawn during the entire year in the Peru Current; however, the main spawning activity takes place around September. A secondary spawning peak, smaller and less distinct than the main peak, occurs in the southern fall, that is, sometime in April or May. The anchoveta are recruited to the fishery at about 5 months of age. Thus, the progeny of the first spawning peak enter the fishery in January or February and those of the second in July or August. A large percentage of anchoveta mature at an age of 1 year and fecundity is high; a 2-year old female may spawn over 20,000 eggs.

These pelagic eggs have an almost negligible store of nutrient material. Consequently, newly hatched anchoveta larvae require a high density of food in their immediate environment if they are to survive. They have only a few wiggles to locate and consume a food particle before their entire energy supply is exhausted and they die. In addition to the ever constant threat of starvation, the pelagic eggs and larvae are subjected to severe predation from a variety of plankton feeders ranging from carnivorous zooplankters to their own parents. The spawning aggregations intermix large numbers of filter-feeding adults with the drifting eggs and larvae, and field observations confirm that the numbers of eggs and larvae consumed by the spawners are sufficiently large that cannibalism during the period of larval drift could serve as a basic mechanism controlling population density and stabilizing the numbers of recruits produced.

By considering the population consequences of the reproductive behavior of the anchoveta, we can obtain insight into the manner in which the population evolved and how it is likely to respond to exploitation. Because of its short life span, the anchoveta is forced to adopt an entirely different reproductive strategy than that of a long-lived species such as cod, which may have upward of 20 year classes extant in a population at any one time. These long-lived species can survive in highly variable environments where conditions adverse to survival of eggs and larvae during the spawning season may occur for a series of years. Typically, spawing takes place in a very short period of time and, if not properly sequenced in a chain of events controlled by climatic conditions in a given year, will produce almost no recruitment to the population. The anchoveta, with 2 or 3 year classes in its population, cannot afford the luxury of gambling on the weather. Lack of spawning success in a single year could be disastrous. So the anchoveta hedges its bet by spawning over the entire year and increasing its chances of encountering favorable conditions during some part of the extended spawning season. We would expect the combination of this type of spawning behavior, a fairly stable environment, and cannibalism on the eggs and larvae to dampen annual variations in recruitment. Observations support this hypothesis. Recruitment has not varied by more than a factor of 3 for many years and in the last 4 to 5 years, by less than a factor of 2. This consistency is remarkable when compared to longer-lived species of fish.

Peruvian anchoveta suffer high mortalities. Estimates of the fraction of adult fish surviving per year at the level of fishing before the 1969–1970 season are in the vicinity of 8–15%. During the mid-1960s, mortality appeared to be fairly evenly split between the fishery and all other causes of death, including natural deaths from diseases or parasites and from predators such as birds, squid, bonita, mackerel, and hake. In the 1969–1970 season increased fishing probably increased man's share from 50 to 60% largely at the expense of new recruits that otherwise would have survived until the beginning in June of the 3-month closure.

The decrease in numbers of a hypothetical cohort entering the fishery at an age of 5 months is illustrated in Figure 5. Available data do not justify extension of the numbers-at-age diagram beyond 25 months. For all practical purposes, the maximum possible life span is 42 months.

Growth of anchoveta in length and in weight is also illustrated in Figure 5. Growth is rapid indeed, especially during the first year of life. From ages 5 through 8 months the weight added to a cohort by growth is much higher than the weight subtracted by natural mortality. It is apparent that the Peruvian stocks are now being exploited at a size and an age considerably before first maturity. This means that, in addition to the waste of potential growth and the waste involved with catching and reducing to meal and oil the low-yield *peladilla*, the present intensity of fishing on the young flirts with the danger of reducing the number of spawners below some unknown but nevertheless real level of abundance needed to sustain the population.

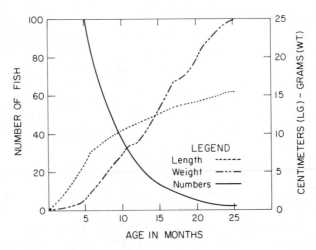

Figure 5 Growth in length and weight of Peruvian anchoveta; numbers surviving by age from a hypothetical cohort of 100 fish starting at 5 months.

One of the most important problems confronting fisheries biologists is recognizing and regulating self-sustaining population subunits within a population exploited by a common fishery. Such individual stocks may have unique recruitment, growth, natural mortality, and migratory characteristics and may be of quite different numerical abundances. Regulating a mixture of individual stock units on the basis of total allowable catch can be ruinous if the catch is not properly distributed over the stocks. This is precisely the type of management mistake that caused the destruction of several of the Antarctic baleen whale stocks. If the total quota of 16,000 blue whale units had been properly divided among the four main species, it is likely the whale stocks could have sustained this catch indefinitely. Instead the most valuable species was overfished until nearly destroyed and then the fishery concentrated on the next most valuable species. This pattern was repeated until desperate measures were required to save the whales from extinction.

Little is known about the genetic composition of the anchoveta populations in the Peru Current. Some indirect evidence on feeding behavior, intestine length, and number of gill rakers indicates the population off Chile and southern Peru may be genetically distinct from the fish off central and northern Peru. Characters such as vertebral number, size composition, and growth rates have been examined but do not provide definitive evidence for any further separation of the population and cast some doubt on the separateness of the southern from the central-northern stocks. Tagging studies that are just now beginning may be the only means of obtaining indisputable information on stock structure.

From the above description of the Peruvian anchoveta and from published studies[8,9,10] of other members of the genus *Engraulis,* an anchovy type of life history pattern begins to emerge. Anchovy are short-lived animals, with extremely high natural mortality even as adults and with maximum ages ranging from less than 3 to at most 6 or 7 years, depending on the species. Spawning activities are generally spread out over long periods and individual females produce large numbers of eggs. Anchovies are filter feeders and their population densities are closely related to the productivity of phytoplankton and zooplankton in particular marine habitats. They browse down their own eggs and by so doing help stabilize their numbers. Most of their growth has been completed by first maturity and ratios of length at first maturity to ultimate attainable length range from 0.6 to 0.8.

Growth during early life is extremely rapid. This type of life history is most productive at extremely high fishing intensities if recruitment to the fishery occurs at about the time of first maturity and if there is no direct relation between the number of spawning adults and subsequent recruitment. However, this is true only in terms of physical yield, not in terms of net economic yield. The marginal increase per unit of gear added at higher intensities decreases rapidly even though total physical yield may go up. If the natural order does not regulate the age at first entry into the fishery, man must assume the role or else face the prospect of trading a slight increase in production over a very few years for the continued existence of the population as an economically viable fishery resource.

GUANO BIRD STOCKS

Manufacture of the finest natural fertilizer known is the specialty of the colonies of guano birds found on the headlands and on the islands off Peru's coastal desert strip. Guano birds crowd every available bit of space on islands located where the ocean provides an especially abundant and available supply of anchoveta and other food. Interspersed with these island "fertilizer factories" along the coast are the shore-based fish meal factories with which they compete for raw material.

The guano birds, all members of the order Pelecaniformes, have an unmistakable pattern of foot webbing joining the hind toe to the front three. Virtually all the guano is produced by three species—the "Guanay" (cormorant), the "Piquero" (gannet or booby), and the "Alcatraz" (pelican). All three of these fish eaters have throat pouches, the most spectacular by far belonging to the pelican.

Guano birds are fairly uniformly distributed along the coast from about 8°S to 14°S, roughly from Chicama to Pisco (Figure 3). The distribution of prime fishing grounds coincides nicely with the distribution of birds. The purest and highest quality guano is produced by the cormorants, which comprised about 80% of the total bird population from the mid-1940s to the mid-1960s; about 18% were gannets and 2% were pelicans. These proportions have changed during the last 5 years and the present relative composition is closer to 60% cormorants, 35% gannets, and 5% pelicans.

According to Jordan[11] the cormorants abstain almost religiously from eating anything but anchoveta, which compose 95% of their diet. The gannets and pelicans, with 80% anchoveta diets, are only slightly less dependent upon a single prey species. The flocks of birds feed most actively at the break of dawn when the plankton and the feeding fish are near the surface. The cormorant, an accomplished diver and powerful underwater swimmer, pursues the anchoveta to depths exceeding 15 m. The gannet and pelican, on the other hand, dive at their prey from the air, hitting the water like dive bombers that failed to pull out. Although the gannet's effective feeding depth appears to approach the cormorant's the pelican feeds effectively only near the surface. The fish-catching birds, like their human colleagues, prefer to fish close to home and rarely venture more than 40 to 50 miles offshore.

The rates at which the birds consume anchoveta and produce guano have been the subject of many scientific studies.[11] The most reliable estimate of average per-day consumption of the cormorant weighing 2 kg appears to be about 430 g. From this meal of anchoveta the cormorant produces about 45 g of recoverable guano. The reader will note that the familiar and apparently inescapable ecological efficiency factor of 10% is applicable even to guano production.

The numerical abundances of guano birds and anchoveta catches during recent years are shown in Figure 6. The precipitous drops in abundance that occurred between 1957 and 1958, and also between 1965 and 1966, are associated with the El Niño phenomenon, so named because it occurs around Christmas time, the season of the small child.

When an El Niño occurs, a tongue of warm and low-salinity surface

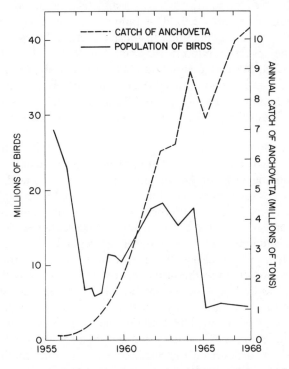

Figure 6 Guano bird population census figures and anchoveta catch by year for 1955–1968. Source: Redrawn from figure in Appendix 3 of reference 1.

water extends to the south over the Peru Current.[12] Heavy rains and north winds often accompany an El Niño and there may be an active surface flow to the south. The periodicity of El Niño is the subject of some dispute; a severe El Niño seems to occur about every 7 years, but there is considerable evidence that an El Niño is not an all or none phenomenon, but rather a condition whose degree and severity differ from year to year. The biological damage caused by the El Niño is dependent on its duration, magnitude, and the speed with which it invades the southern water.

In spite of the often-repeated statements that an El Niño destroys anchoveta, I believe it does just the opposite. The warm surface water of the El Niño serves as a protective blanket, forcing the anchoveta to sound and avoid the severe bird predation in the top 10 m. This behavior exerts a selective stress on the birds, discriminating against weak swimmers and divers unable to capture anchoveta or to switch to another type of prey. Census data[13] confirm that weak, immature birds are almost totally decimated by the El Niño.

Why the 1965–1966 El Niño should have such a prolonged effect is a mystery; the bird populations have completely failed to regain their previous abundance (Figure 6). This failure is in direct contrast to what occurred after severe reductions in the past. After being reduced to 6 million by the severe El Niño in 1957 and then to 5.5 million by a milder El Niño in 1958, the bird population climbed to about 11 million in 1959 and 1960 and by 1963 had reached over 16 million. When the population fell from 17 million in 1964–1965 to 4.3 million in 1965–1966, it did not recover and according to the most recent census in 1968–1969 was only 5.4 million.

We do not know what factors are acting to keep the population at such a low level. Two lines of speculation seem most plausible. One holds that the standing crop of anchoveta has been so reduced by intense fishing that the increased expenditure of energy needed to feed on a more diffuse prey does not leave the birds any excess for population growth. The other conjecture is that the Peru Current upwelling system may trap persistent pesticides just as the oceanographically similar California Current system does, and that these pollutants are concentrated as they move up the food chain to the birds, where they interfere with reproduction. In many other parts of the world, DDT has been shown to cause fish-eating birds to produce eggs so thin-shelled that the parents were unable to incubate them. The well-documented demise of the brown pelican in the California Current system is a disturbingly similar example.

Schaefer[14] has articulated the need to maintain some minimum population level of guano birds to preserve the species " . . . since this genetic material may be of great future value." The minimum abundance for survival of the species is unknown but it is well established that social birds, such as the guano birds, do require some minimum density for successful breeding.

THE INDUSTRIAL AND ECONOMIC SETTING

The same lusty excitement that must have swirled about the boom towns during the Gold Rush in the United States runs through the fishing towns along the Peruvian coast. The fish meal industry grew in a manner reminiscent of the building up during the nineteenth century in the United States of industrial empires based on natural resources such as timber, oil, and minerals. As large and as important as the industry is by absolute measures—total meal production during the 1969–1970 season was 1,920,000 tons, which was worth over 350 million dollars; about 20,000 full-time fisherman and 3,000 factory workers were employed; capital equipment consisted of 1,300 purse seiners and about 125 fish meal plants—such statisitcs do not even begin to convey the psychological impact of the industrial boom that has culminated in the largest fish meal industry the world has ever known.

To understand why the impressive economic and physical statistics are dwarfed by psychological factors, recall the circumstances under which this industry developed. During the past 15 years, the average population of Peru was about 10 million people. These 10 million people lived in a rigid and stratified society greatly influenced by semifeudal Spanish traditions. Although democratic political traditions have been much stronger in Peru than in some other South American countries, the social hierarchy, which correlates very closely with the economic hierarchy, was dominated by a small elite class of landowners and wealthy urbanites. The industrial explosion that made Peru the world's leading producer of fish meal should be pictured against a background in which social and economic rigidity were maintained by the sorts of institutions allied with the landed aristocracy, the church, and the military.

Opportunities to acquire wealth were heavily weighted in favor of members of a few large and socially prominent families who controlled much of the property in Peru. Because the fish meal industry had no precedent and developed so explosively, it spawned a small but important class of "nouveau riche." It also provided either directly or indirectly many new jobs for middleclass professionals such as lawyers, engineers, boat skippers, factory managers, and white collar workers. During the chaotic growth of the fish meal industry, men of action with brains and good luck became millionaires overnight.

One could hardly blame these industrial empire builders if they were skeptical of the idea of a fixed limit to the amount of fish that could be removed each year from the giant ocean upwellings so obviously teeming with life. Both the government and the industry are to be complimented on their recognition of the sobering possibility that such a finite limit might exist and that exceeding the limit could destroy the renewable resource upon which the industry depends. The fact that the government has been willing to impose limitations on the industry and that the industry has been willing to accept these limitations is reason for optimism. However, the tremendous excess capacity building in both the factories and in the fleets is severely straining any regulatory agency's ability to impose rational controls on harvesting.

It is interesting to compare Peru's performance to date with the exploitation of natural resources that occurred during the opening of the West in the United States. One great difference is the presence of teams of United Nations scientists and economists and the education of local counterparts. The primary mission of these consultants is to advise the government on the type of exploitation policy that will allow maximum production from the resource without destroying it. To men caught up in the grand enthusiasm of industrial empire building, these consultants must have sounded like prophets of doom. Yet it is abundantly clear that their presence and advice has served to cool down the industrial growth, which easily could have raced completely out of control.

In spite of the extreme scarcity of capital in Peru, the fishing industry has managed to acquire far more equipment than it needs to harvest the anchoveta available. Such scarcity generates pressure to employ available capital equipment to its maximum capacity. Many of the fish meal companies that sprang up overnight during the chaotic growth of the industry are models of industrial inefficiency, and their precarious financial state is becoming chronic. When a large number of marginal operators are combined with wildly fluctuating international market prices, the threat of overharvesting the resource becomes very great. The scientists and economists both from Peru and abroad who are responsible for monitoring the industry and for establishing harvesting policies are now engaged in one of the most delicate tightwire acts playing in the international resource circus.

THE FISHING FLEET

In the late 1950s, when it was becoming apparent that the fish meal industry was going to be substantial, tiny shipyards sprouted around Callao. These shipyards turned out wooden fishing vessels with hold capacities from 40 to 100 tons. As the fishery grew, so did the size of the boats, and the large numbers of small vessels were replaced by smaller numbers of much larger vessels. The increases in vessel size were accompanied by increases in the size of the purse seine nets used to surround the schools of anchoveta.

Most of the boats now under construction in Peru have hold capacities in excess of 300 tons and are made of steel; one venturous operator is beginning to build 350 tonners out of fiberglass. The vessels are around 100 feet long and powered by 500–600 horsepower engines. Basic equipment includes the hydraulic power block invented by Mario Puretic, which has revolutionized purse seining for salmon and herring as well as for anchoveta. The heavy manual labor formerly necessary to brail the batch from the net to the hold and to stack the net on the deck has been eliminated by power equipment; a pump sucks the anchoveta from the net into the hold and the Puretic power block hauls the net from the ocean. All Peruvian *bolicheras* are equipped with echo sounders and some have horizontally sweeping sonar or asdic fish detectors.

A 350-ton vessel requires a crew of only 12 to 14 men, as compared to the 10-man crews often needed to run a 100-ton vessel. Obviously this labor reduction per ton of capacity is a major source of efficiency associated with the larger vessels.

The net result of the decrease in numbers of vessels and the increases in average hold capacity and power and electronic equipment has been to increase total fishing power of the fleet far beyond that needed to harvest the anchoveta. More frightening, however, than the size of existing fleet capacity is the rapid rate at which that capacity is growing.

The fleet existing at the end of the 1969–1970 season could be reduced by 25% and still easily harvest the quota of 9.5 mmt in 8 months of fishing 5-day weeks. If the boats were allowed to fish as many days as they would like per week, the size of the fleet could be reduced by half or perhaps even 75%. If present construction trends continue in spite of the 5-day week and 4-month closure, capacity will be 50% greater than needed by 1971–1972. The irrationality of the present method of coping with excess capacity is worth examining in more detail.

When free entry is allowed but total removal must be restricted, the management agency finds itself on the horns of a dilemma. As more gear enters and the gear becomes more technologically efficient, the agency, in order to restrict the catch to some preset target quota, limits fishing time. While this regulation negates the effects on the stock of the new technology and the larger fleet, it wastes the economic potential of the resource. Fixed costs of vessel operation (normal depreciation, annual maintenance costs, insurance, and so on) are independent of the total number of days fished. They must be paid whether the vessel fishes 1 day a year or 365 days a year. For the Peruvian fishery fixed costs are about 47% of total costs. When the vessels are only allowed to fish 6.5 months as in the 1969–1970 season, fixed costs become increasingly important and greatly increase the price per ton of removing the fish from the ocean. Consider as an extreme case the difference in costs between catching some fixed quota with one vessel fishing 365 days a year and 365 vessels each fishing one day a year.

There is a limit to such artificial cost raising. The computers of the feed mixers of the world constantly compare the price of fish meal to the prices of competitive protein sources such as soybean, sunflower, bone, blood, ground nut, palm kernel, cotton seed, coconut, and poultry byproduct meal. If harvesting inefficiencies become excessive and the resultant price increases, it will cause fish meal to be replaced by other products in the world protein market.

Ownership and fleet size also affect efficiency. In the Peruvian anchoveta fishery and in other industries, private ownership of a piece of capital equipment promotes efficiency in its use. Other efficiencies result from the geographical mobility of the vessels. One of the most successful operators can move his fleet along Peru's 1400-mile long coast to help supply factories near concentrations of fish. Such mobility requires a certain scale of operation and wide deployment of processing equipment.

Economic considerations alone cannot be used to dictate a scheme for reducing capacity. Social and political factors are sure to play an important, if not dominant, role. For example, as vessels become larger, they require fewer fishermen per unit of harvesting capacity. Unemployed fishermen are notoriously difficult to retrain and relocate. Regardless of the scheme adopted, it is quite likely that some element of compulsion will be necessary to make it work and, if so, it will be extremely difficult to be just and equitable while maintaining efficiency and technological innovativity.

THE MEAL FACTORIES

The basic anatomy of a modern Peruvian fish meal plant is depicted in Figure 7a. A meal factory is a link transferring energy from solar radiation in the southern hemisphere to the plate of a citizen eating poultry, beef, or pork in an industrialized country in the northern hemisphere. The specific function of the factory is to dewater, crush, cook, and compact the anchoveta so that it can be easily shipped to the animal feed mixers. When the fish arrive at the factory, they have already

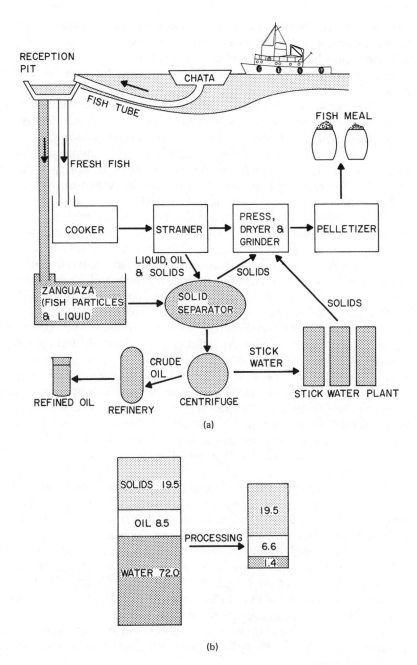

Figure 7 Components of a fish meal factory and flow of material during processing of anchoveta into meal. Source: Figure on pages 35–36 of Fisheries Supplement, *Peruvian Times*, July 24, 1970.

concentrated the nearly microscopic plants and tiny zooplankton from the Peru Current and transformed unpalatable materials such as cellulose and chitin into high-quality protein. By comparison the factory's job is simple; it must separate the solids and oils from the water in the fish flesh.

The bar-graph diagram shown in Figure 7b shows the amount of fat, solids, and water recovered from the raw fish in the oil and meal output of an efficient plant. The highly publicized fish protein concentrate (FPC) that some persons think will be a major source of animal protein for human consumption in underdeveloped countries is simply a refined version of fish meal with more of the oils and fats removed. Fishy odor and fishy flavor disappear with the oils and fats.

Figure 7 shows the *chata*, or unloading barge, which may be anchored as far as a half mile offshore from the factory. From the *chata* the fish are pumped into the reception pit, dewatered, and weighed. These weights are the official catch statistics. The drain material containing blood and flesh particles is known as *zanguaza* and is processed to separate the oil from solids recycled into the fish pulp. The raw fish are cooked and then pressed to squeeze out the oil and water. The press liquor is centrifuged to obtain crude oil for refinement and the residual liquid then pumped through a "stick-water" plant to extract soluble proteins and fine solids by a distillation process. As one might guess, the stick water contains the raw materials for glue manufacture. A stick-water evaporator is a sophisticated innovation and a fairly recent development in the Peruvian meal industry; it is found primarily in the large newer plants. Such equipment is one means of increasing output from a limited resource without increasing input. The dried fish pulp is then ground, cooled, and treated with antioxidants to make it less volatile and safer to ship and store.

Most complications shown in Figure 7 were developed to squeeze every last bit of energy from the anchoveta. Any small increase in efficiency of utilization is important for an industry handling 10 mmt of raw material. A 1% increase in efficiency salvages an extra 100,000 tons of meal. Removing soluble proteins from the stick water also avoids eutorphication caused by overfertilization of bays and harbors in the immediate localities of the fish plants. In the existing factories, efficiency of utilization varies from 15 to 21%. About 20% of the plants representing about 60% of the total production are equipped with stick-water evaporators. Although capital investment for stick-water evaporators is high, perhaps $250,000 for a fairly large plant, the costs may be recovered in less than 2 years.

According to reliable sources, the Peruvian government will soon make stick-water recovery equipment mandatory for all fish meal plants. Utilization of stick-water byproducts is not only economically and ecologically reasonable, but is also consonant with man's innate desire to avoid waste of food.

The processing capacity of the fish factories in Peru has grown far above that needed to reduce a catch of 9.5 mmt to meal. The most reliable estimate of processing capacity operating in the first half of 1970 was 8000 tons per hour. By running 350 days a year, 20 hours a day, these plants could process 56 mmt (or the total world catch of marine fish). Because of geographical and seasonal variations in the flow of the raw material and because of operational breakdowns, it is unfair to conclude that factory capacity is 5.6 times greater than needed to process 10 mmt. On the other hand, the immense overcapacity can be neither denied nor minimized. The same sorts of tradeoffs discussed for the fleet exist for the factories. A small number of factories can operate continuously during the year or a large number of factories can operate for a short time. Given a limited supply of fish, adding a new factory means all factories must reduce their operating time. Obviously, an infinite number of combinations of capacity and operating time exist. With a 4-month *veda* and a 5-day week, and providing a comfortable allowance for variability in supply and breakdowns, about 2700 tons per hour of capacity could be retired and the remaining factories could easily process the available catch.

Just why is excess capacity in fleet size and in the number of factories so bad? One answer is purely economic. Surplus specialized equipment ties up capital which could otherwise be employed to generate income for Peru. The country's development is retarded because of misallocation of capital. Even more serious, however, are the less tangible consequences of overcapacity in an industry based on a renewable natural resource. When the industry's harvesting and processing equipment outgrows its resource base, the industry will be in chronic economic difficulty and will be almost impossible to regulate on a rational scientific basis. An FAO-sponsored panel on economic effects of alternative regulatory measures in the Peruvian anchoveta fishery concluded:

> One of the greatest gains of limiting capacity to that capable of dealing economically with a catch of 9.5 million tons is the removal of the temptation to court disaster by significantly surpassing that catch. But, as no value can be put on keeping disaster safely at bay, this is not brought into the reckoning. . . . [15]

Capacity reduction can be done in such a manner as to reduce costs and at the same time increase the physical and economic yields from a fixed input of raw material. To see how one can have one's cake and eat it too, consider the consequences of eliminating 25% of the factories at the lower end of the efficiency scale. First, the amount of meal produced would increase by about 225,000 tons since raw material wasted by the inefficient plants would now be turned into meal by efficient plants. On the cost side of the coin, marginal costs of the more efficient plants are considerably less than the average costs of the less efficient, so transferring fish to them would significantly reduce the cost per ton of meal. Under the reduction proposed, about 6 million dollars worth of additional meal would be produced and 9 million dollars in costs saved. Peru would gain an annual income of about 15 million dollars in direct profit from its fish meal industry. In the short run, part of this money would be used to compensate owners of factories closed and to relocate their employees. Just and equitable compensations could be financed comfortably from the additional 15 million dollars of income generated.

MARKETING

While Peru's new industrialists were mastering the secrets of harvesting anchoveta from the giant swirls and eddies of the Peru Current, they were also learning the hard way about a different sort of wilderness — the international commodities market. It is said that much of the industry was financed with money and bank credits obtained from the future sales of fish not yet born that would be caught and processed by fleets and factories not yet constructed. In the early years of the fishery, Peruvian producers, unorganized and inexperienced in selling in a world market as complex and sensitive as that for commodities, saw much of the wealth generated by the fishery end up in the pockets of speculators and traders who served as middlemen, buying the meal from Peru and selling it to the animal feed manufacturers in Europe and the United States.

Market prices for meal are determined by the interaction of delicately sensitive market demand and market supply curves. Prices have fluctuated violently during the entire history of the fishery. For example, the average price of meal in 1964 was $108.50 per ton. An intrusion of warm water on the fishing grounds caused catches to drop in early 1965 and sent prices jumping to $207.50 per ton; prices then fell as catches increased. Prices during 1965 averaged $127 per ton. In 1968 meal sold for an average price of $100 per ton. The average price in 1969 climbed to $121 and during the first 4 months of 1970, prices were $180–$205 per ton.

Although Peru's fish oil is much less important on the world scene than her fish meal, oil prices also have exhibited great instability. Fish oil is used for industrial purposes and also for human consumption, primarily as an ingredient of margarine; its market price is controlled by soybean oil and sunflower oil prices. In March 1969 fish oil was selling for $119 per ton and in March of 1970, $228 per ton.

International traders who not only could follow and interpret events anywhere in the world affecting competitive protein sources, such as soybean meal, but also could monitor the catches and fish meal production of other nations and even knew what was happening in Peru before many of the Peruvian producers, enjoyed a considerable advantage in the market place. Accurate and timely information is exceedingly valuable in a touchy and enormously complicated market. Useful information includes knowledge of all physical and biological factors influencing the quantity, quality, and cost of the catch; industrial problems, such as strike rumors, port congestion, and the flow of sea traffic; and even national and international politics. When many sales are for future delivery, ability to forecast the demand by the poultry, swine, and cattle raising industries of several countries, and the fish meal output from Norway, Canada, South America, the United States, Angola, Denmark, Iceland, and Chile becomes most important.

In the first part of 1970 the government in Peru set up a state company, Empresa Publica de Comercializacion de Harina y Aceite de Pescado, EPCHAP, to exercise monopoly control over all fish meal and fish oil buying and marketing activities for Peruvian producers. The government took over the largest of the four industry marketing groups that had conducted most international sales. The other three were disbanded. This government takeover is an attempt to stabilize prices paid to the Peruvian producers primarily by selling directly to the feed mixers, bypassing the international brokers. The government also has assumed responsibility for providing the short-term financial assistance sometimes required by the industry. It is far too early to know if this government marketing monopoly will be successful.

THE GUANO INDUSTRY

The magnitude, value, and effect on the anchoveta stocks of the guano industry can be calculated indirectly from bird census, guano production per bird, market price per ton, and anchoveta consumption per day per bird figures. The total bird population is 5.4×10^6; each bird produces 45 g of guano per day; guano sells for $56 per ton; and each bird eats 430 g of anchoveta per day. Thus the birds consume 0.85 mmt of anchoveta to produce 90,000 tons of guano per year. Annual gross value of the guano is 5.05 million dollars.

Comparative gross values of a ton of anchoveta (live weight) are $5.94 in terms of guano and $30.20 in terms of fish meal (assuming 5.3 tons of anchoveta and a price of $160 per ton of fish meal). In the words of the FAO-sponsored special economic mission to Peru:

> If 50% of any fish denied to the birds were caught by Peruvian fishermen, then the loss of $5.94 of guano would result in the addition of $15.10 to the value of meal output. Assuming similar rates of return from the trade in guano and that in fish meal, if the industry's marginal costs of catching and processing one ton of fish into meal are less than the fish meal price to an extent which at least is as great as the cost of exterminating the bird population to the extent required to save two tons of fish (including an appropriate rate of return on the cost of extermination) then it is worthwhile to incur such costs regardless of the price of guano.
>
> . . . This approach to the matter ignores the threat it presents to the very existence of a bird population that has already reached perilously low levels. Furthermore, the ecological consequences of eliminating the bird populations are not understood. . . . Until more is known about the interactions between the bird and anchovy populations any type of predator control programme to reduce the birds is extremely hazardous.[15]

THE INTERNATIONAL SETTING

In September of 1945 the Truman Proclamation was issued. This document stated that the exploitable resources of the seabed adjacent to the continental United States belonged to the United States and that under certain conditions the United States could establish fish conservation zones on the high seas. Although the Proclamation was originally intended to satisfy certain needs of the petroleum and salmon power politics of the era, it opened a Pandora's box of diverse governmental and industrial activities inimical to traditional concepts of freedom of the seas. When the president of Peru, Jose Luis Bustamante, issued his Supreme Decree of August 1, 1947, declaring that Peru's jurisdiction over its "territorial" waters extended 200 miles off its coast, he cited as a precedent the Truman Proclamation. At present ex-President Bustamante is president of the International Court of Justice at The Hague. However, his interpretation of the Truman proclamation has been challenged by experts in the field of international law.[16]

In 1952 Peru, Chile, and Ecuador issued a tripartite declaration claiming unilateral jurisdiction by the coastal state over territorial seas of 200 miles. The CEP nations, as these three countries have become known, asserted that this extension of their territorial seas was purely for protection of natural resources and did not in any way involve questions of free navigation. Nations with high-seas distant-water fishing fleets and far-roaming navies such as the United States, Japan, and Russia became skeptical. Irrespective of the logic or legality of their territorial claims, the inescapable fact seems to be that the CEP countries are in a position to enforce their claims. Peru, for example, has a small fleet of modern British-built 100-foot gun boats capable of 30 knots and armed with Swiss Oerlikon antiaircarft cannon. U. S. tuna clippers are easy prey for such vessels and every tuna season brings with it the drama of high seas seizures followed by payment of fines and purchase of licenses by adventuresome clippers. The Peruvian Navy's greatest victory over a foreign fishing fleet occurred in 1954 when Aristotle Onassis' whaling fleet was caught in Peruvian waters with 3000 whales and fined 3 million dollars by a Peruvian court before the vessels were released.

Peruvian motives are not entirely obvious. One faction in Peru apparently favors using the 200-mile limit as a bargaining point to gain special exemption from the United States' 35% *ad valorem* tax on imported canned tuna, while another views extended jurisdiction as providing protection of fishery resources for future use by the coastal state. Another, perhaps more pragmatic faction, welcomes the income generated by selling fishing licenses and imposing various fines and fees on foreign vessels without licenses; there also seems to be some pressure to encourage foreign vessels to set up Peruvian-based operations. United States' response has been to reimburse fishermen the amount of fines levied by CEP nations while at the same time attempting to negotiate with representatives of the CEP nations some sort of settlement that does not recognize 200-mile territorial sea claims. While U.S. diplomats argue against foreign claims to extended jurisdictional zones, spokesmen for certain sectors of the U.S. domestic industry and sport fishing groups are asking that the United States unilaterally declare a 200-mile exclusive fishery zone around the continent and Alaska.

Other interests are maneuvering to set up a new international conference on the law of the sea. Although other powerful nations such as Russia have similar interests in restricting territorial seas to 12 miles, a new Geneva-type conference might very well provide the large number of developing nations with an opportunity to impose wider territorial limits on the more developed nations. Ironically, the U.S. flag fleet (excluding tuna and shrimp fishermen) would be a prime beneficiary of an internationally accepted 200-mile limit. The United States enjoys a broad continental shelf, a huge biomass of lightly or underexploited species, and the world's most important market for seafood. The real losers would be developing nations without seacoasts.

President Nixon, to the amazement and consternation of the petroleum industry, has proposed that the United States renounce any claims to resources of the seabed beyond the 200-meter depth contour and that an agency of the United Nations should manage exploitation of these resources for the benefit of mankind. Informed experts feel the likelihood that such a proposal will be adopted by the United Nations is near zero. With few exceptions, developing nations bordering on the sea have shown no inclination to share potential resources from the seabed.

Peru's extended jurisdictional claims have been one of its most stable policies. In spite of drastic changes in governments, the 200-mile claim has remained invariant. The current military regime in Peru has inaugurated many social reforms and there has been widespread concern since nationalization of the International Petroleum Corporation that similar actions might be taken against foreign-owned or foreign-controlled segments of the fishing industry. So far this has not happened, but more stringent requirements on the percentage of Peruvian interest in fishing enterprises were issued during the first part of 1970. It is perhaps notable that several foreign corporations have been quietly disposing of their holdings in the Peruvian industry.

GHOSTS OF FISHERIES PAST

Not too many years ago scientists and laymen alike believed there were no practical limits on the amount of fish that could be removed from the sea. In the early 1900s it was clear to any reasonable man that the puny catches taken from huge shoals of herring which sometimes stretched for miles in the North Sea could not possbily damage the populations. Indeed, even in recent years, the steady rise in tonnage of the total world catch supports this view. In 1950 the world's commercial fisheries took about 20 mmt of fish and other aquatic creatures and plants; in 1970 the total will approach 70 mmt.

Given this increase, why are many scientists expressing alarm over the future of the world's fisheries? Are there ominous signs portending trouble? Summary statistics can be misleading and pessimists point at the declining catch records of certain heavily exploited stocks and worry about how long total world tonnage can continue to increase. Fishermen of the world have adopted the motto "move on" as catches from one stock after another reached or exceeded maximum sustainable yields.

Without question, some stocks have been so decimated they are no longer worth fishing. Giant distant-water fishing fleets continue to explore new ground and to exploit previously unfished stocks. The consequences of a "move on" policy actively pursued in a finite area and coupled with changing environmental conditions can be seen in the Great Lakes. U.S. and Canadian fishermen overexploited local stock after local stock and watched long-lived economically valuable species disappear and be replaced by short-lived species for which no markets exist.

The major catch increases during the past 20 years have come largely from new areas in the southeast, southwest, and northeast Pacific and from the west coast of Africa and the Indian Ocean. It is most significant that no new major resources have been discovered in either the north Atlantic or the northwest Pacific. Fishermen are beginning to run out of unexplored parts of the ocean; within 20 years it appears all possibilities for discovering new major stocks exploitable with conventional gear will have been exhausted.

In 1949 a United Nations Scientific Conference on the Conservation and Utilization of Resources held at Lake Success in New York drew up a map which showed around 30 stocks of fish then believed to be underfished. When re-examined in 1968, at least half of these stocks were either being exploited to their maximum capacities or had already been overexploited.

Fisheries are most profitable in their early years as the expanding fishery harvests a standing stock that may have taken several years, or even decades, to accumulate. When the stock can no longer sustain any further increase in catch or is destroyed, the surplus of harvesting gear built during the expansion phase is available to fish other stocks. As the world's shipyards pour out more fishing vessels, the time between first exploitation and overexploitation is becoming shorter.

For many of the Peruvians who personally witnessed the growth of the anchovy fishery, the idea of a finite maximum sustainable yield is more than just an abstract theory developed by population dynamicists. The first meal factories in Peru were equipped with machinery from idle sardine-processing plants in California.

The California sardine story is well known. In the late 1930s and early 1940s, catches of over 600,000 tons were common. Catches then began to oscillate violently until the early 1950s, when in 3 years they dropped from over 350,000 tons to less than 150,000 tons. Exactly what caused the collapse of the California sardines still generates arguments among fishery scientists, who have not yet completed their autopsies. Recruitment failed as the fishery teetered before its final collapse. At the same time, the anchovy population in the California Current system increased tremendously. What caused the anchovy to replace the sardines is a matter of conjecture. However, we do know most of the older sardines had been caught and the entire fishery depended upon very few year classes, and that water temperature fell in the California Current system about the time the sardines began to disappear. What would have happened if there had been no fishery? Would the sardines still have disappeared? No one knows and the question seems beyond the capability of even the most sophisticated computerized crystal balls available.

Peruvian vessel and factory builders must have shuddered when they looked to the south and watched Chile dismantle fish meal plants and convert to other uses fishing vessels built to harvest an anchoveta resource much larger than actually exists there. The Chilean fishery has not collapsed; it simply never existed at a level matching up to the expectations of expansion-minded industrialists.

In spite of the half century it has taken for man to realize he is capable of fishing to economic extinction pelagic stocks, at the turn of the century it was clear marine mammals are highly susceptible to overexploitation. Stocks of seals, whales, and sea otters not protected by a national or international conservation agency have been overharvested into oblivion.

Fish with life patterns similar to mammals also have fared poorly. The spiny dogfish *(Squalus acanthias)* is such a species; it does not become mature until 11 to 13 years old and only then produces three pups. Litter sizes increase almost linearly and may reach 14–15 pups for sharks between 30 and 35 years old. Off the northwest corner of the United States, the vulnerability of this life pattern to fishing pressure was exposed with startling clarity. Within a few years, after a good market developed for shark livers during World War II, the waters of Puget Sound, the Strait of Juan de Fuca, and the Georgia Strait were swept almost clean of dogfish. The stocks began to rebuild only after the liver market and the fishery collapsed in 1946 because of synthetic vitamin A production.

The list of individual fish stocks in trouble at this writing is too long to tabulate here but includes representatives of the Gadiformes (cods), Pleuronectoidei (flounders), Salmonoidei (salmon and trout), and Clupeoidei (herrings).

Marine mammals and fish have no monopoly on susceptibility to exploitation. King crab catches by United States, Russian and Japanese fishermen in the Bering Sea and Gulf of Alaska dropped from 80 thousand metric tons in 1966 to 62 in 1967, 42 in 1968, and 27 in 1969. Stocks of early-maturing penaeid shrimps with short life spans, usually less than 2 years, and high fecundities have been able to withstand severe exploitation, while several stocks of pandalid shrimps, which have much longer life spans and delayed maturity caused by their habit of switching from males to females in their third or fourth year of life, have vanished when fished.

Since the characteristics of an animal's life pattern are so critical in determining ability to sustain a fishery, it is fair to ask if any of the ghosts of former fisheries have relevance for an "anchoveta-type" of life history.

Our collection of fishery ghosts does not include any defunct anchovy fisheries. Either the anchovies are an extremely hard clan able to withstand tremendous fishing pressure, or anchovy fisheries are such a recent phenomenon they simply have not been in business long enough. It would be tragic if this gap in the taxonomy of dead fisheries were to be occupied by the world's greatest fishery.

The herrings, sardines, and pilchards are a group of species closely related to the anchovies and with a long history of exploitation. Many of these stocks are fished by Peru's competitors on the world fish meal market.

One reason for the sharp rise in fish meal prices in late 1969 and early 1970 was the collapse under intensive fishing of many traditional sources of raw material. The recent experience of the Norwegians who fish the Atlanto-Scandian herring stocks is alarming.

Between 1964 and 1968 Norwegians invested almost a hundred million dollars building up their herring purse seine fleet. Unfortunately, the herring stocks were unable to increase their productivity to match the increase in harvesting capacity. In 1967 Norway took 1.2 mmt of herring; in 1968, 0.70; and in 1969, in spite of the massive harvesting power of the expanded fleet, the catch dropped to 0.10 mmt. Norway now owns a huge surplus of herring purse seiners with nothing to catch. The herring stocks appear to have dropped to less than 1/15 of their abundance in 1950. As might be guessed, Iceland's herring fishery, which harvests the same stocks, has experienced a similar disaster—catches from 1966 through 1969 were 0.77, 0.46, 0.14, and 0.05 mmt, respectively. Herring catches off the west coast of Canada, although much smaller than those in the western Atlantic, declined even more severely because of the necessity to impose emergency conservation regulations.

The catch of pilchards in the Benguela Current system off South Africa and southwest Africa appears to be holding up fairly well. However, recent increases in catches of anchovy from this system have caused some scientists to speculate we may be witnessing the type of replacement of a sardinelike species by an anchovy that occurred in the California Current system.

Gerald J. Paulik

THE PERUVIAN ANCHOVETA FUTURE

Thinking about unthinkable catastrophes is one way of preventing them. Destruction of the world's greatest single species fishery is unthinkable. Could it really occur?

Consider the following scenario: Heavy fishing pressure has so reduced the abundance of 1- and 2-year old anchoveta that the bulk of the catch is taken from recruits before their first birthday. Most of the catch is made during January through May. A moratorium on vessel construction has held the size of the fleet to 300,000 tons of hold capacity. In the year 197(?) adverse oceanographic conditions cause a near failure of the entering year class and low catches in the southern spring and early summer (September through December) cause fish meal prices to climb to $300 per ton. Weak upwelling currents in January, February, and March concentrate the residual population in a narrow band extending about 10 kilometers from the coast. The total catch during the first 4 months of the season is 1 mmt and the total biomass of the entire residual population plus the entering recruits is about 5 mmt. Although many danger signals are flashing and scientists warn the industry that fishing effort must be curtailed immediately, the industry's creditors are clamoring for payment of short-term loans, and many factories and fleets are unable to meet payrolls and pay operating expenses while fish meal prices are at an all time high.

The January catches total 0.5 mmt instead of the 2.0 mmt expected. The industry rationalizes that recruitment is later than usual and will occur during the February closure. When the season reopens in March, the entire fleet is poised and the residual population is vulnerable. Within 2 months the entire population of 4.5 mmt is caught and sold at record prices. The government imposes an emergency closure when the catch-per-unit drops to zero at the end of April. The fishery is closed during May but it is too late. The Peruvian anchoveta stocks have been fished into oblivion. Only small scattered schools of anchoveta remain; dead birds, rusting bolicheras, and idle fish meal factories will soon litter the beaches from Chicama to Arica. Just how unlikely is this apocalyptic vision?

During the 1968–1969 and 1969–1970 seasons, the fishery depended heavily on the entering year class for the bulk of the catch. The hold capacity of the fleet at the beginning of the 1970–1971 season is going to be near 225,000 tons. Fish meal prices are flirting with the $200 a ton level and the disappearance of other sources of meal fish may push them even higher. At the time of this writing, no definitive plan for restricting fleet growth and reducing present fleet capacity has been put forth.

During the 1969–1970 season, the fishermen were allowed the exceed substantially the recommended quota of 9.5 mmt to harvest 11.0 mmt. About 9.0 mmt of this total was caught during January through May when recruitment is high. Is the stage being set for a major ecological disaster?

The close fit of a logistic model with a maximum sustainable yield of 9.5 mmt to the catch history of the anchoveta fishery and the bird populations' failure to increase after the 1965 decline both indicate the stocks may be in some difficulty at present levels of exploitation.

In opposition to this line of reasoning, optimists point to the relative constancy of annual recruitments during the past decade, the short life span of the anchoveta, its high fecundity, early maturity, and extended spawning season.

Beyond the unthinkable lies the unknown. If the Peruvian anchoveta were to disappear, what, if anything, might replace it? Prophets of doom can assemble a long list of noxious or uncatchable marine creatures. If the successor species were a shrimplike euphasid, costs of catching and processing might be too high for the animal feed market; even a small component of poisonous animals in the replacement community could mean detoxification costs that would spell economic disaster. The importance of the huge schools of anchoveta in trapping and recirculating nutrients in the lighted surface layers of the Peru Current is not known. Conceivably if the stocks were removed, the basic productivity of the whole system could be a casualty.

Another possibility is that the anchoveta populations are composed of several functionally independent stocks and that the destruction of one stock (for example, the southern stock that Peru shares with Chile) might lead to major regulatory reforms in time to save the remaining stocks. The racial structure of the population is being investigated at the time of this writing by tagging experiments that should have been conducted long ago.

COMPUTER MODELING

Qualitiative descriptions of the important variables controlling the performance of the industries dependent upon the anchoveta stocks have occupied most of this chapter. Many of these variables have been incorporated in a computer simulation model by a team of biologists, economists, and computer scientists at the University of Washington. The logical structure of this model of the anchoveta stocks and the meal, oil, and guano industries is shown in Figure 8. This model is being used

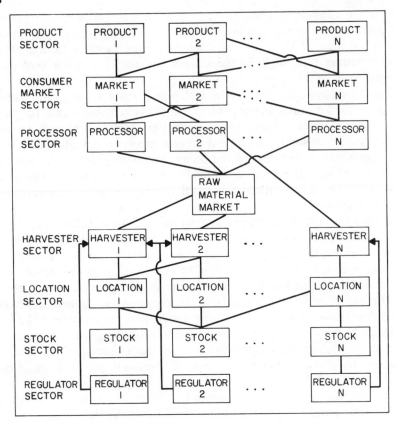

Figure 8 Logical structure of a generalized computer simulation model of the industrial exploitation of the anchoveta stocks off the coast of Peru.

to explore the sensitivity of the performance of the total system to modifications in parameters values and to changes in the functions used to describe various processes. Although the model is far too complicated to describe here and has not been tested sufficiently to determine if it will be accepted by, and can provide significant help to, the decision-makers in Peru, it represents a radical departure from earlier fisheries models. It has been specifically designed to incorporate the wealth of data colleted by the Instituto del Mar del Peru on the biology of the anchoveta, the catch and effort data accumulated as the fishery developed, and the extensive information on guano birds. A detailed economic sector of the program incorporates available information on harvesters, processors, and markets. The staff using this model plans to explore systematically the strengths and weaknesses of various regulatory strategies. The model can be run in one of two basic operational modes. It can be run as a conventional computer simulation model with built-in management decisionmaking algorithms specifying the behavior of factory owners, fleets, and regulators, or it can be run with human intervention at frequent intervals to allow intuitive and heuristic decisionmaking. Mixed modes are also possible. For example, a human decisionmaker may play the role of factory owner while the computer makes all decisions for the regulatory agency and fishermen.

The primary objective of the simulation exercise is to quantify biological and economic effects of alternative *regulatory measures* that might be imposed in the Peruvian anchoveta fishery. The output is also designed to help social scientists evaluate generalized welfare functions extremely difficult to quantify at the present time. Use of human participants is expected to be especially valuable in detecting regulatory and fiscal policies that encourage administrative corruption.

Ongoing research is investigating the effects upon regulators, harvesters, processors, marketers, and the natural population of the following types of regulations:

1 Input quotas for factories.
2 Output quotas for factories.
3 Quotas assigned to groups of vessels as either fixed absolute amounts or fixed percentages of a total national quota.
4 Catch taxes.
5 Harvester licensing schemes.
6 Auctioning of catch quotas for harvesters and for processors.
7 Limitations on number of units of gear.
8 Limitations on gear technology and size of individual units.
9 Limitations on lengths and seasonal placements of fishing periods.
10 Catch quotas by area and time.
11 Control of harvesting locations.
12 Combinations of two or more of the above types of regulations.

The number of combinations that could be run is truly formidable and present exploratory work is aimed at narrowing areas of ignorance. For the most promising methods it will be necessary to investigate further their performance in both deterministic and stochastic versions of the model and under a variety of plausible biological hypotheses. The interaction between regulatory policies and a variety of exogenous factors affecting markets and costs of the industry also must be investigated, as well as the value of various types of information to human participants.

One of the most surprising findings to date is the almost total inadequacy of present biological understanding of recruitment mechanisms in fish. Although the computer can help pinpoint such important gaps in current understanding, more field studies are indispensable in order to produce the additional input needed to fuel a realistic simulator.

REFERENCES

1   W. E. Ricker, Ed. and panel chairman 1970. Report of the panel of experts on the population dynamics of Peruvian anchoveta (unpublished). FAO Department Fisheries, Rome, Italy.

2   D. H. Cushing, "Upwelling and fish production," FAO Fishery Technical Paper 84, (1969).

3   W. S. Wooster, Eastern Boundary Currents in the South Pacific. In W. S. Wooster, Ed., Scientific Exploration of the South Pacific, National Academy of Sciences, Washington, D.C. 1970, pp. 60–68.

4   J. A. Gulland, "Population dynamics of the Peruvian anchoveta," FAO Fishery Technical Paper 72 (1968).

5   J. H. Ryther, "Photosynthesis and fish production from the sea." Science, 166, 72–80 (1969).

6   R. N. Burukovskii, ed., Antarkticheskii Krill, Biologiiai Promysel. Kaliningrad, Antarjucheskii Naukno-issledovatel 'skii, Institut Rybnogo Khoziaistua i Okeanografii (1965).

7   W. M. Chapman, Seafood Supply and World Famine—Positive Approach (unpublished). Prepared for delivery at the AAS Symposium "Food from the Sea," September 29, 1969.

8   R. J. H. Beverton, "Maturation, growth and mortality of clupeoid and engraulid stocks in relation to fishing," Rapp. P. - v. Reun. Cons. Perm. Int. Explor. Mer 154, 44–67 (1962)

9   R. J. H. Beverton and S. J. Holt, "A Review of the Lifespans and Mortality Rates of Fish in Nature, and their Relation to Growth and other Physiological Characteristics," in The Lifespan of Animals, CIBA Foundation Colloquia on Ageing, No. 5, 1959, pp. 142–177.

10   J. L. Baxter, Ed., Symposium on Anchovies, Genus Engraulis. California Cooperative Oceanic Fishery Investigation Report 11, 1967, pp. 28–139.

11   R. Jordan and J. Fuentes, Los Poblaciones de Aves Guaneras y su Situacion Actual, Instituto del Mar del Peru Report 10, 1966.

12   G. S. Posner, "The Peru current," Bulletin of the Bingham Oceanographic Collective, 16, 106–155 (1957).

13   H. Fuentes, Las Poblaciones de Aves Guaneras, Ser. Informes Especiales IM-54, Instituto del Mar—Callao (mimeo), 1969.

14   M. B. Schaefer, "Men, birds, and anchovies in the Peru current—dynamic interactions," *Transactions of the American Fishery Society,* **99** 461–467 (1970).

15   J. A. Gulland, A. Holmsen, A. Laing, G. J. Paulik, F. E. Popper, and J. Watzinger (members of the panel), Report of the Panel on Economic Effects of Alternative Regulator Measures in the Peruvian Anchoveta Industry (unpublished mimeo). FAO Department of Fisheries, Rome, Italy, 1970.

16   D. L. McKernan and W. Van Campen, "An approach to solving international fishery disputes," *Transactions of the American Fishery Society* **99,** 621–628 (1970).

CHAPTER 4

The Guano and
the Anchoveta Fishery*

ROBERT CUSHMAN MURPHY

Editor's Abstract

This chapter, *written in 1954,* discusses the development of the guano industry in Peru. He discusses the reassessment of priorities that the Guano Administration underwent at a time of near collapse of that industry because of overexploitation. As a result of that reassessment, the Guano Administration after 1909 began to view the guano as a renewable natural resource with Murphy commenting that the "increase of guano is their economic object, but removal of the crop does not exceed the limitations imposed by nature. The long view takes precedence over the immediate demand." The biological, meteorological, and oceanographic bases of the Peruvian living marine resources are presented. He discusses the linkages between guano production and the fluctuating anchoveta population on which the birds depend; all this at a time when entrepreneurs were about to develop the anchoveta fishery and the fish meal industry on a large scale. Murphy's report documents a struggle between two commercial ventures, one old and established (guano) and the other new and exciting (fish meal), the latter a new source of investment for entrepreneurs.

This report is a direct result of a reconnaissance along 2000 miles of the coast of Peru and northern Chile, made under the auspices of the Compañia Administradora del Guano in December 1953 and January 1954. It is based also on earlier experience of the writer in the same field, dating from the year 1919. Account has likewise been taken of the pertinent scientific literature in publications of the Guano Administration and other sources.

---

*Reprinted with permission of the American Museum of Natural History (New York).

GUANO IN RETROSPECT

The agriculture of coastal Peru, which is a normally rainless area, is dependent today, as it was in the time of the Incas, upon irrigation with river water and the application of fertilizer. Historically, as well as prehistorically, the principal fertilizer has been the famous guano of the islands. Formerly this substance existed in millions of tons and, if modern understanding of conservation had been a familiar concept 100 years ago, Peru might have possessed throughout all time an abundant supply of guano for its own use as well as for export.

Unfortunately, the resource was discovered as a commercial product in an era when frantic exploitation, rather than the principle of cropping, was the rule. The guano was "mined." It was viewed in the same light as coal or metals, rather than as a renewable natural resource. No thought was given to the welfare of the sea birds that produce it. The one aim was to dig it up, load it into vessels, export it to various seaports of the world, and convert it into cash. The fact that a small proportion was consigned for domestic needs was incidental. The function of the ancient guano beds, in the eyes of most Peruvians, was to eliminate taxation, to replace domestic imports of many or all kinds, and to supply funds with which to run the Government.

Between the years 1848 and 1875 more than 20 million tons of guano were exported from Peru to Europe and the United States. The value of such a quantity in the international market today would amount to 2000 million dollars.

THE REHABILITATION

The early history of guano reflects the usual stages of discovery; greed, erring faith in the inexhaustibility of natural treasure, and then ruin.

But in this instance the application of sound principles while there was still time at least forestalled an irrevocable tragedy. The restoration of the Peruvian industry or, rather, its conversion from a purely destructive exploitation into a lasting industry, represents one of the greatest examples of practical conservation that is known. It is perhaps more fully understood and admired in Europe and in the United States, and in even remoter parts of the world, than in the home country that has derived incalculable benefit from the accomplishment.

After the turn of the present century, Peruvian agriculture was threatened with disaster. A new and enlightened point of view was then needed to follow the spendthrift debauch known as the "Saturnalia," which most Peruvians would be glad to forget. If the heedless human animal could act so adversely upon the world of life, including his own interests, it was equally possible that by using his brains he might be able to recreate a profitable harmony with nature.

The *Compañia Administradora del Guano,* an institution devoted unequivocally to the thesis that guano is a renewable natural resource, and that the size of the increment is related to the intelligence and character put into management, was founded in 1909. After nearly half a century of consistent dedication, the system stands justified. Notwithstanding the ups and downs in the guano crop, which are correlated with climatic rhythms, the trend has been steadily upward. The more serious dips in the graph have coincided with such calamitous seasons as the so-called "Niño" years of 1925 and 1939, when a profound displacement of the atmospheric pressure field and a resultant shift in the distribution of southern-hemisphere winds caused a southward transgression of tropical surface water, cessation of upwelling, rising surface temperatures, disastrous rainfall, and the devastation of life in the sea, including microorganisms, fish, and birds. In each instance recovery followed rapidly upon the restoration of normal climatic conditions.

Since 1909 when sea fowl, the producers, were taken into partnership with man, the beneficiary, the place of the guano bird has been unchallenged. The bird has been recognized as the goose that lays the golden egg. Out of the wreck of the past, a great constructive regime has evolved. The guano islands are today bird sanctuaries, and by far the most impressive sanctuaries of their kind on the face of the globe. Increase of guano is their economic object, but removal of the crop does not exceed the limitations imposed by nature. The long view takes precedence over the immediate demand. It is realized that the guano output cannot be expected to be equal each year and that operations must be attuned to the climatic periods. A watchful eye is always kept upon exceptional circumstances affecting the wanderings and the reproductive capacity of the sea fowl. The bird itself—the producer—has at last become the determiner of what man shall undertake.

In 1953 the Guano Administration harvested 255,000 metric tons of guano, representing the largest annual increment since the early days of reckless extraction from beds that had been built up during many centuries. Future crops will probably be increasingly large. There are indications that the Executive Director of the Guano Administration, Ingeniero Carlos Llosa Belaúnde, and his informed and devoted associates, have already made discoveries that may ultimately carry guano production as far above its present yield as the latter surpasses the meager output of 45 years ago.

BASIS OF PERUVIAN MARINE RESOURCES

The conditions responsible for the accumulation of guano and the retention of its efficacy have become only gradually understood. They involve a series of interacting cosmic phenomena, including the wind systems of the west coast of South America, the circulation of water in the adjacent part of the Pacific Ocean, and the maintenance of extraordinarily low surface temperatures in tropical latitudes. A high degree of meteorological aridity allows guano to lie where it is deposited and prevents the reduction of its nitrate content.

Dependent also upon these briefly described conditions is the teeming microscopic plant life of the coastal waters. This in turn supports the higher and larger living organisms, all the way from crustaceans and small fish up to birds, seals, and whales. There is no other littoral district where birds can be found in such overwhelming profusion. There is likewise no area in which wild birds, requiring little human care other than to be left sufficiently alone, have been the source of such economic benefits to mankind. Guano is a commodity worth millions of dollars annually to those who collect and process it and those who use it for the enrichment of Peruvian soil.

No Peruvian needs to be informed that the coastal waters of his country are among the richest in the world. The explanation of the swarming marine life between central Chile and Point Pariñas lies in the upwelling of cool water from intermediate depths in the Humboldt or Peru coastal current. The rising water brings up to the zone of sunlight, where photosynthesis takes place, a constantly replenished supply of nutrients, which are absorbed osmotically by microscopic single-celled plants. Most important among the latter are the diatoms and naked green flagellates, which may properly be called the pasture of the sea.

The Bible tells us that "all flesh is grass." We readily understand this when we eat beef and mutton. We understand it also in the case of the jaguar that eats herbivorous mammals which, in their turn, live by grazing or browsing. Plants alone can manufacture food from inorganic material; animals must obtain theirs by piracy, namely by foraging on plants directly or by eating animals that have eaten plants. No matter how many links there may be in the chain, the ultimate source of all food is the green plants. This is true of the oceans no less than of the land, wherefore "all flesh is grass."

Now the pasture of the sea, or phytoplankton, supports larger organisms, which include free-swimming crustaceans and certain groups of fish. Many species of anchovies, for example, subsist by filtering out of the ocean water minute or invisible forms of life.

In the Peru coastal current the commonest species of fish that lives by such a process is the anchoveta *(Engraulis ringens)*, the most numerous of all vertebrate animals throughout the area. The anchoveta is the key organism in the food chain of this littoral zone. Its range is coextensive with that of the characteristically cool, upwelling water, from central Chile to a latitude near the Ecuadorian border. It comprises 95 percent of the food of the two principal species of guano birds (guanay and piquero). It is the main, or almost exclusive, food of the bonito, the skipjack, and the yellow-finned tuna and is therefore the basis of the leading fisheries of Peru. The anchoveta is eaten also by many other kinds of fishes, by whales and porpoises, squids, sea lions, and by vast numbers of sea birds native to Peru or present along the coast at certain seasons. The mountain gulls *(Larus serranus)*, which nest on the shores of lakes in the high Andes, descend to the ocean during the nonbreeding season and feed in part upon anchovetas until it is time for them to return to the mountains. The entire vast population of the pardela or Franklin's gull *(Larus pipixcan)*, which breeds in the interior of North America, migrates in the Northern Hemisphere autumn to the coastal waters of Peru and Chile. There its principal food is the anchoveta until the birds go north again in April and May. This statement is based upon long observation and the examination of stomachs of Franklin's gulls collected at several stations along the Peruvian coast.

The significance of the anchoveta as the key food resource in the relatively narrow belt of cool coastal waters is therefore apparent. It is the foundation of the fisheries and the guano industry alike. It interacts with the whole complex of surface and subsurface life. It is responsible for the favorable biological balance in the sea. It supports a permanently resident fauna, as well as the seasonal faunas that enter the coastal current from the outer Pacific, from lands toward the Antarctica, from the Galapagos archipelago, and from the Northern Hemisphere. If space permitted, examples of each of these components might be listed. Without the anchoveta, the littoral waters of Peru would be deprived of the most conspicuous and useful elements of their life.

THE NUMBERS OF THE ANCHOVETA

What is the population of this important little fish? The fisheries companies, which now use it and want to expand the catch, lay stress upon its enormous numbers. There is no reason to seek to minimize these numbers. The anchoveta population is inconceivably large, so that computation must deal in figures of what is called astronomical dimension. No calculation in the present state of our knowledge can be precise, but Dr. Erwin Schweigger, the conservatively minded oceanographer of the *Compañia Administradora del Guano,* has approached the problem by reasoned methods and has reached a quantitative conclusion that may serve until still better data can give us a closer approximation of the truth. Dr. Schweigger has taken into consideration the known individual consumption and the approximately known numbers of the Peruvian guano birds. He has also calculated the consumption of anchovetas by the bonito, and has assumed that the annual catch of the latter fish represents 10% of the total population in Peruvian coastal waters.

Upon these and related facts and tentative hypotheses, Dr. Schweigger infers that the population of the anchoveta in the waters off Peru, when the species is at the peak of its cycle, is of the order of 20 million metric tons. Since the average weight of one of these small fish is about 20 g, it is simple to compute that 50,000 of them would weigh 1 ton and that the individuals in 20 million tons would be represented by the figure 10^{12}—which is of the order of the number of known stars in our galaxy of the Milky Way.

PROBLEM OF FLUCTUATING POPULATIONS

Scientific forecast can proceed from information of many kinds, and it may be of interest to point out certain bearings of modern biological investigation upon our outlook in the sphere of conservation. Too often we have made our estimates upon a basis of certain continuing stable rates of plant and animal productivity. The return has in most cases fallen short of the expectation, even when liberal allowance has been made for improved methods of harvesting and the increased demand caused by an expanding human population.

The chief source of error has lain in the primary assumption. There is probably no such thing as stability in either numbers or productivity of living organisms, because all of them (except, evidently, modern man*) seem to share in the rhythmic fluctuations that are among the most familiar phenomena in our world, as well as in the universe outside it. General recognition of this truth, and the processes that stem from it, might go far toward enabling us to plan wisely for the future. It would not solve the whole problem, because mankind may become increasingly

---

*There is, perhaps, one other exception, namely, the beaver. This mammal feeds upon bark, rather than leaves, fruits, or seeds. It therefore depends upon "capital" rather than "interest."

faced with the quandary of "scant means in relation to given ends," but it would be likely at least to give us an ameliorating clue.

The changing populations of many kinds of animals and the alternating yield in the seed crop of plants are phenomena that have been observed since the dawn of history. In numerous instances such shifts are not marked by abrupt or startling breaks. Rather, the pendulum of numbers swings gently up and down, the increase and decrease being at comparable rates, neither of which is excessive, so that periodicity over a course of years becomes evident only after a large amount of quantitative data has been collected.

About such phenomena we urgently need to learn more. They will have to be included in the reckoning before we can begin to figure confidently about the future availability of many kinds of organic products for which we find a use or a dire need. This kind of foresight might protect us from making the slight overdrafts on natural resources that are different only in degree from the gross and obvious overdrafts. For whether we use many times too much or only a little too much of the material we can not create, the result in the long run will be the same, namely, depletion. We are already wise enough to discard as a complete fallacy the argument that a species is in no danger because it is still common.

Better known than the moderate and often obscure types of fluctuation in numbers and productivity are those in which organisms swing through such wide arcs that no observer can fail to be aware of them. These are the spectacular rhythms, in which increase to the point of overpopulation—in relation to new environmental conditions—is followed by a rapid and catastrophic die-off, so that for one or more subsequent seasons an area that had swarmed with a particular form of life may seem to be almost without a trace of it. But always, under conditions undisturbed by man, a sufficient nucleus remains to start the upswing of the cycle. It is a remarkable fact that although many animals approach extinction at regular intervals, they have rarely attained it since the last Ice Age except when man has added the weight of his hand to the falling beam of the scales. For, as Elton has stated, to have the right enemies is a biological advantage, but of all disasters, including predators, flood, drought, fire, avalanche and disease, the worst for the flora and fauna of primitive regions is the introduction of civilized man!

The rhythmic population growths, which are followed by a "boiling over," mass emigration, general mortality and other apparent abnormalities, are characteristic of a great many forms of life both on the land and in the sea. There is now an enormous scientific literature on this subject. The famous case of the lemmings, small rodents found on the land masses all around the northern pole, is known to nearly everybody. Closer to home, in Peru, is the occasional mass hysteria, exodus, and heavy mortality of the guano fowl. This is a particular liability of the northerly half of the Peruvian coast in such unfavorable Southern Hemisphere seasons as those of 1925 (when the writer was working in northern Peru). Such strange behavior by the birds is usually preceded by aberrations in the organisms of their food chain.

Through a period of years the numbers of the birds may expand as a result of the bountiful food in the upwelling bands of the coastal current. But eventually the oceanic circulation changes; warm surface water from the outlying ocean moves toward the continent, or upwelling of cool water may cease for other reasons; conditions become intolerable for the marine pasture, and the whole pyramidal structure begins to topple like a house of cards. Dead fish may litter the beaches and the stench carry miles inland. Or the anchovetas may migrate to other areas or to deeper levels. In either case the birds perish in quantities to be measured by scores of miles of their strewn corpses along the tide line. Their death is not always due directly to starvation but, rather, to some form of shock-disease that overtakes animals whose vitality has been sapped by the knocking from beneath their feet of the whole complex system upon which healthy existence depends.* The graph of production of Peruvian guano throughout a long course of years shows clearly the alternating peaks and depressions which are an index of the varying population of birds.

Between 1840 and the beginning of the present century, man aggravated the periodic reductions in guano bird population and hindered the recoveries. The birds were regarded not as inviolate capital but as an encumbrance and a nuisance on the thick and "inexhaustible" beds of fertilizer that their ancestors had laid down through the millennia. By the time the last cargo of ancient guano had been transported from the islands, the future looked dark, indeed, for a country that required a great quantity of such material to support its own essential agriculture.

Fishermen and the harvesters of guano are alike subject to the vagaries of ecological depressions. The argument that periodical fluctuations of the guano birds make the manufacture of fish meal all the more essential does not hold water. When the birds die in large numbers, high mortality of fish is likely to be concomitant, because the same environmental conditions influence the population cycles of both. Furthermore as will be discussed below, the problem of mitigating heavy mortality of the

---

*So little is yet known about the etiology of these "maladies of debilitation" that an intensive study is much needed.

guano birds and of maintaining population levels above 15,000,000 individuals may already have been largely solved. Finally, the constantly reiterated reference to "fish-meal fertilizer" is, in any case, academic and disingenuous. Its monetary value as cattle and poultry food is so high that little if any of the manufactured product has found its way into Peruvian soil.

STRENGTH AND WEAKNESS OF GREGARIOUSNESS

The reasons why certain animals have developed such a startling degree of gregariousness as the anchoveta are not yet fully understood. We may, however, point to the fact that an environment can support from a given total of energy (= food) a larger biomass of a closely associated organism than if the same organism were distributed either as smaller units or as scattered individuals. In this biophysical fact lies the strength of the anchoveta.

We know, on the other hand, that animals of extreme gregariousness also have a high extinction threshold, by which is meant that if they cease to live in large aggregations they may vanish altogether. The very condition of population density is essential to their survival. In this fact lies the weakness of the anchoveta and its precarious susceptibility to overfishing.

The history of extirpation demonstrates that when a notably schooling or flocking species suffers a reduction of half or more of its total population, the remaining part, even though very large, may not prove sufficient to win out over the normal causes of mortality. The same is sometimes true of dense but nongregarious populations, as has been learned through campaigns against the tsetse fly in parts of Africa. Economic entomologists engaged in the control find that if approximately 50 percent of the flies can be destroyed within a circumscribed area, most of the other half of the population will disappear from natural causes.

In the case of the North American passenger pigeon (*Ectopistes migratorius*), which was killed for food by "thousands of barrels" during the middle part of the 19th century, the rapid decline that ended in extinction began while millions of the birds were still occupying much of their old range. In other words, man did not wipe out the last of the passenger pigeons; he merely killed too great a proportion of the original population to enable the remainder to overcome normal hazards. The species became a victim of its own high extinction threshold.

A deficiency that affects understanding of the anchoveta is our ignorance of normal longevity in the species. Gregariousness and relative viability sometimes reveal unexpected correlations. Most cormorants, for example, are long-lived waterfowl. Ten years of life in aviaries have been frequently recorded, with maxima up to 23 years. Yet the uniquely gregarious guanay, a member of the cormorant family, seems to have a surprisingly brief average life. Recovery by the Guano Administration of several hundred numbered leg bands, from birds ringed as fledglings, indicates that most guanayes die within 2 years after hatching.

SHORTCUTS TO MORE FERTILIZER

At a conference held in the University of San Marcos on January 21, 1954, it was asserted that the needs of Peruvian agriculture for fertilizer amount today to a half million tons per annum. This means that guano production, despite its rapid gains, still falls short of satisfying the demands of agriculturalists within the Republic, and that no guano is now available, as it formerly was, for export. On the contrary, mineral fertilizer is imported, especially sodium and potassium nitrates and ammonium sulfate from Chile.

Under such circumstances it is natural that men's thoughts should have turned to other possible sources of fertilizer from the rich sea that bathes the shores of Peru. The search for a panacea commenced about 1935, and it was further stimulated by the ecological "crash" of the sea bird population of 1939–1940, and the consequent reduction of guano supply.

The cool waters of the coastal current abound in several sorts of schooling surface fish, of which the anchoveta is foremost in numbers and in consistent availability. Direct manufacture of fish meal fertilizer is carried out in other parts of the world, particularly in eastern North America where the extensive menhaden *(Brevoortia tyrannis)* fishery has developed from the use of the whole bodies of this herring as a crude fertilizer in the maize fields of the primitive Indians. It is obvious that the guano birds, despite their vast and constantly growing numbers, do not devour all the anchovetas in the sea. Why not, therefore, short-circuit the problem, and supplement the available fertilizer by a manmade product in accordance with tested industrial methods?

Such, at least, was the first conception in Peru. As it has been worked out in practice, the taking of anchovetas by organized fisheries has not increased the supply of fertilizer, because the fish meal is worth most as stock food. The practical result has been that the greater part of the product does not even enter the ports of Peru to enhance the resources of the country. Most of it that has been prepared to date has been exported to the United States or Europe, where its use has been chiefly that of cattle feed. The value of such a nutritious and palatable protein diet in carrying cattle through the winter season seems to have originated in Scandinavia and in Iceland, where in former times cows were fed during the dark months of the year upon the dried heads of cod and other fish. At any rate, the original intent in Peru has not fulfilled its purpose because it was sidetracked at the very beginning. No one has claimed that the export of cattle feed has been of benefit to Peruvian agriculture, which is as dependent as heretofore upon guano.

MAN, THE ONLY INSATIABLE PREDATOR

Why, it might fairly be asked, are there grounds for concern regarding an organized anchoveta fishery in Peru? In a statement covering the year 1952 one fishing company recorded a catch of 2000 tons, which is less by a wide margin than the quantity eaten by guano birds in a single day and on every day of the year. Even if the total human toll of anchovetas approached or equalled 100,000 metric tons, it might still appear to be a moderate drain upon the enormously rich resources of the current.

To answer such plausible assumptions, it is necessary to emphasize one of the oldest and most persistent of human errors, which has been aptly called "the fallacy of the inexhaustible." It is obvious that natural predators, such as fish or birds or beasts, can never seriously reduce the numbers of their prey, because their very existence is determined by the abundance, including a dynamic surplus, of the food organisms. Predator and prey are in equilibrium. Man, on the other hand, is capable of depleting any readily attacked natural resource. Unlike a guano bird, man has no automatic checks and balances upon his operations. He is not directly dependent upon the tissues of his prey for the energy with which he executes his exploitation. A guanay must eat anchovetas in order to catch more anchovetas, but fishermen hunt with energy from a totally different source, such as petroleum. Only the slow and disastrous consequences of exhaustion and financial failure can end their campaign.

Man, as a predator, can no longer be justly compared with any other organism because he is today not only the most numerous large animal in the world but also the only one whose numbers are constantly increasing. It would be an anachronism not to recognize these truths in the light of recent history, particularly of the 19th and 20th centuries. Lumber and other forest resources, game mammals and wild fowl, food fishes of the highest economic value (e.g., salmon), shellfish and crustaceans, and countless other recently abundant products have dwindled alarmingly before human demand and use. Improved facilities of transportation may for a time mask the actual depletion and loss. The ominous index of doubling and redoubling prices during the past half century remains for all to behold.

Such specific examples may seem to relate chiefly to the Northern Hemisphere, but Peru has also lost invaluable resources solely through lack of scientific foresight. Soil exhaustion and erosion on a huge scale are one instance. Easier to recognize are the complete disappearance of the once-profitable pearl fishery of Secura Bay and the extermination of the valuable fur seal that formerly lived in hordes along the Peruvian sea coast. Lesser, but possibly prophetic examples, may be found in the diminishing supply and the higher cost of such favorite seafood as scallops, camarones, langostas, etc. Any Peruvian whose memory carries back to an earlier generation can extend this list.

AN INTOLERABLE ABUSE

In 1919–1920 the writer of this report found dynamiting of fish to be an established and almost accepted practice along the greater part of the Peruvian coast. In 1924–1925 there was no evidence that it had diminished, despite the greater condemnation it was then receiving from the press. In 1954 the odious abuse was apparently as rife as ever.

Dynamiting is an incredibly antisocial procedure. It renders a proportion of each catch unfit for food. It kills an incalculable number of fish and other marine organisms that can never be recovered by the perpetrators. Nevertheless, it is now charged that in the neighborhood of hotels at isolated Peruvian bays a large part of the fish supply is taken with dynamite. In an editorial of January 21, 1954, a Lima newspaper even accused one or more commercial companies of employing dynamite in the anchoveta industry.

Statistical estimates of the damage wrought by dynamiting are difficult to reach, but several Peruvian authorities have not hesitated to state that "millions of tons" of fish are being wasted. There is no more sure path toward degradation of the whole beneficent complex of life in the Peru current.

The use of dynamite is not due to lack of law, but to lack of enforcement. In at least one Peruvian locality it ceased abruptly as soon as the Captain of the Port publicized the fact that he would make every effort to apply the penalties provided.

The intolerable evil of dynamiting has been widespread in the world, but in most countries it has long since been brought under control. In New Zealand, for instance, it has presented no problem since 1908, when an act imposed two months' imprisonment for the first offense, with the additional important provision that the burden of proving that dynamite in possession was not to be "used for the purpose of catching or destroying fish or whales shall lie upon the accused."

GUANO: FROM OBSERVATION TO CONTROL

The management of any renewable resource, such as guano, may be thought of as attaining four successive stages, namely:

1 Observation
2 Understanding
3 Prediction
4 Control

This succession is an accurate précis of the ideals and the scientific approach of the staff of the Guano Administration throughout nearly half a century of faithful stewardship, of which Peru may well be proud. In the course of the long program the best brains that could be mustered at home, and the most promising talent that could be brought from abroad, have worked in Peru with single-minded devotion for the welfare of the sea fowl, the increase of the guano yield, and the betterment of agriculture. An oceanographer has been brought from Germany to become a member of the permanent corps; a young Peruvian ornithologist has been sent to the University of Wisconsin, where he studied for 3 years under a world authority in the applied science of game management, the late professor Aldo Leopold; specialists in the fisheries and other aspects of marine science have been induced on several occasions to come from Great Britain, the United States, and elsewhere and to work under the auspices of, or in cooperation with, the Guano Administration for periods ranging from several months to as long as 3 years.

Notable scientific publications on the physics and chemistry of the coastal ocean, and on the ecology of marine organisms in the chain that leads from diatoms to birds, have appeared in the publications of the Guano Administration and have been favorably received in the technical journals throughout the world. Agricultural engineers, both within and outside Peru, have lent their counsel and their practical aid to projects fostered by the Guano Administration. Biological laboratories have been fruitfully operated at various strategic points along the coast, and a modern well-equipped chemical and agricultural plant at Callao conducts experimental research on the retention of fertilizing efficacy and the effects of guano upon various soils and crops. Today, thanks to such enlightened labors, the pulverized and stable form in which guano is distributed, and the techniques by which it is applied to tilled fields, represent a notable advance over the procedures of a generation ago.

Finally, the statistical analysis of quantitative data has been carried to a point of high effectiveness under the direction of Executive Director Señor Llosa Belaúnde. No longer is it necessary for conclusions and expectations to be based upon crude experience and the best possible guess. The guardians on the islands have been trained to record their daily observations in quantitative form. Information on the seasonal regimes of the birds and the constantly changing panorama of visible life in coastal waters is now correlated with the contemporary record of wind and cloud, the march of air and water temperatures, barometric pressure, rate of evaporation, the convergence and divergence lines and other phenomena of movement in the sea. The record of the recent past provides a trustworthy frame of reference and, as statistical information becomes more and more adequate, the prognosis of guano operations for the seasons ahead attains a constantly sounder validity.

Reverting now to the four progressive stages of management, i.e., observation, understanding, prediction and control, it is evident that operations of the Guano Administration have reached the stage of "prediction," and are apparently at the threshold of the final one, "control." This statement requires further elucidation.

Peruvian guano of high nitrogen content has been customarily known as "guano of the islands." This is because the islands have been recognized as the active centers of guano accumulation throughout the historic period. Mainland deposits have also, of course, been known and worked, but most of these have proved poor in nitrates though rich in phosphatic content. The beds of a few mainland localities, such as the famous *Pabellón de Pica,* included hundreds of thousands of tons of richly nitrogenous guano, but such sites were understood to be finished and static sources, places in which the guano had been laid down soon after the last Ice Age or in some other period of the remote past, and hence to be only once "mined" of their treasure. The dynamic and continuing sites at which guano could be regularly "cropped" have been confined to the islands.

Now along the southern half of the coast of Peru, where climatic conditions are ideal for the accumulation of nitrogenous guano, there are virtually no islands to offer suitable nesting grounds for the birds. The insular areas of exploitation lie, for the most part, between Santa Rosa Island, at Independencia Bay, and the Lobos Islands off northern Peru.

The range of the guano birds, however, particularly during the winter season, extends far to the southward of the island-studded coast. Since neither of the two most important species, the guanay and the piquero, spends the night on the ocean, it is necessary for these birds to roost upon projecting points and promontories of the southerly mainland. One result of this habit is that guano, sometimes in quantities worth extracting, has accumulated from time to time at the restricted sites of the nocturnal roosts.

Eight years ago (1946) Señor Llosa Belaúnde advanced the hypothesis that the preference of the guano birds for insular nesting grounds was due solely to the safety the islands offered from predatory mammals, such as the abundant *zorros (Dusicyon)* of the desert coast, rats, rayras, or their relatives, feral dogs, cats, and pigs, and, in some instances, man himself. He thereupon launched an experiment that has developed into an important demonstration of the soundness of his inspiration. Upon receiving reports that guanayes were roosting in significant numbers at a mainland locality, Señor Llosa Belaúnde ordered the construction of an animal-proof concrete wall, with a height of 2.5 meters, on the landward side of the aggregation of birds. The area was then put under the care of resident guardians whose duties were the same as those at the island stations.

The result has been an astonishing new development in the history of guano management. Due to security from disturbance provided by a wall, many roosting groups of guanayes, mingled with smaller numbers of pelicans and other species, have been converted into thriving breeding colonies on the mainland. At least one such has increased in size to a total exceeding two and a half million birds. Several successive seasons of guano extraction leave no doubt about the empirical success of a brilliant speculation. The system has since been copied by the officials of the *Sociedad Chilena de Fertilizantes,* who hope to profit mutually with Peru through an expansion of the guano bird population in the south.

Aside from providing a wholly new source of guano, the procedure just described holds an extraordinarily hopeful augury with reference to maintaining a high and consistent average in the total numbers of guano birds. It is well known, of course, that their fluctuations are correlated with meteorological cycles, especially along the northerly parts of the Peruvian coast. It is here that the transgression of tropical water, the inhibition of upwelling, the profound mortality of marine organisms, and the desertion of the islands by nesting guano birds are capable of causing, at more or less regular intervals, a decrease in the annual crop of guano. But through the induced process of greatly increasing the permanent guano birds in southern Peru and northern Chile, where they are least subject to the Niño or related cyclic disturbances, it is likely that the population and the guano crop may both be stabilized until a point of optimum saturation has been reached.

Aside from the successful establishment of mainland colonies of guanayes, it is likely that means will continue to be found of inducing the guano birds to occupy more territory on the large and lofty islands. Most of those were formerly devoid of colonies. Today the Isla Vioja, at Independencia Bay, has vast settlements of guanayes that extend from the edge of the cliffs uphill into the cloud zone, a thousand feet above sea level. A single guano crop of this island has attained 60,000 tons. San Lorenzo, off Callao, also has an important nucleus of guanayes covering steep slopes up to the easternmost summit of the island. Ultimately, it may prove possible to establish colonies on the great island of San Gallán.

Extension of bird colonies is the road to more guano, for it is admitted that food resources are still out of balance with the limited occupied nesting space.

In 1909–1910, the fresh guano crop amounted to 23,790 tons; in 1952 it exceeded 250,000 tons. Until 1914–1915 the quantity used in Peru was less than that exported. After that, exportation diminished until it ceased altogether in 1942. The average annual guano crops by decades have been:

1909–1918	76,000 tons
1919–1928	95,000 tons
1929–1938	142,000 tons
1939–1948	113,000 tons

The last figure was due to heavy bird mortality in 1939–1940, but the average for the decade beginning in 1949 bids fair to attain 200,000 tons or more. There is no apparent reason why the future annual guano crop should not increase in direct ratio with added nesting area, provided the producing birds can find 3,000,000 plus or minus tons of anchovetas available to them.

Such are the reasons for asserting that Peruvian guano management has already arrived at the level of "prediction" and that the final level of "control" is at least within sight. The Guano Administration is engaged not merely in taking or exploiting guano; rather, it is providing for increased supplies 10, 20, 50 years, or even a century, hence.

PERUVIAN FISHERIES

Turning now to organized fisheries in Peru, it should be noted that, unlike the guano industry, all the undertakings and establishments are relatively new. When the writer carried out his first investigations in 1919–1920, there were little but primitive Indian fisheries along the entire coast between the Chilean and Ecuadorian borders. In subsequent published consideration, he wrote as follows:

> Upon visiting so rich a seacoast, anyone with a knowledge of fisheries cannot fail to be impressed by the unrealized opportunities for the development of desiccating, canning and packing plants in Peru. . . . So far as I have learned, no seafood of any kind is canned . . . , the only familiar native marine products being dried or salted fish of many sorts, picked anchovies, and preserved fish eggs and sea weed. . . .
> Beyond doubt organized fishing in Peru will eventually be conducted with success. The resources are great and varied; the bays or protected roadstands of Independencia, Chilca, Paracas, Pisco, Callao, Ancón, Huarmey, Samanco, Ferrol, and others offer inviting sites for stations; native fuel oil is obtainable in quantity; adaptable labor is already on the ground; the climatic conditions of the coastal ocean are almost without parallel; the government is eager to see the country's natural resources developed. A rich potential market exists along both coasts of South America and in the interior, besides which it is likely that many prepared products might profitably be exported to the Orient and even to the northern hemisphere.

These paragraphs were published in 1925. Nearly two decades passed before exploitation of the opportunities began, but now (1954) the fishing industry has completed 10 years of modern undertakings. Between 1943 and 1952 production increased from 21,603 to 106,610 metric tons. The annual gain has been progressive with the single exception of 1947, which showed a slight falling off from the previous year's catch. Total annual value of the catch, on the other hand, has enjoyed uninterrupted advance. The 40 or more establishments in operation have a yearly capacity of one million boxes of 48 tins. Fish meal exports amount to 10,000 tons a year, and it is admitted that the current total catch of anchovetas is about 70,000 tons.

Pressure to enlarge the take of anchovetas is, for several reasons, active and constant. In the first place, the anchovetas are a convenient slack-season resource for companies primarily interested in food species, such as bonito and tuna. More important is the fact that the companies devoting all their energies to a purse-net fishery of the anchoveta are in a favorable position because of the presence of a constant supply of fish and the simplification of their factory operations. The firms that concentrate upon the reduction of anchovetas to fish meal—thereby eliminating the prepartion of human food—are commonly stated to achieve the best returns on their financial investment. Therefore a constant temptation exists to slide over from food fish to anchoveta meal. Herein lies the threat to the future of both guano production and food fisheries.

POINT OF VIEW OF THE FISHING COMPANIES

On the day after a public conference on the subject of this report, which was held at the Universidad Nacional Mayor de San Marcos on January 21, 1954, the Sociedad Nacional de Pesquería inserted the following notice in the newspapers of Lima:

In an effort to counteract the interposition of the Guano Administration's agents during the discourse which Mr. Robert Cushman Murphy, distinguished North American scientist, gave yesterday in the Faculty of Sciences of the Universidad Nacional Mayor de San Marcos, the National Fisheries Society finds itself obliged to make the following statement:

1. This organization is fully as interested as the Guano Administration in seeing the fulfillment of appropriate investigations that will determine up to what point it is prudent to continue exploiting the anchoveta; it therefore deplores any premature commentary.

2. At the present time the Fish Meal Industry takes a total quantity equivalent to only one per cent of the tonnage of anchovetas consumed by the guano birds; this figure can certainly be no cause for alarm.

3. It is important to note that at the present time two different industries, of equal importance and respectability, are based upon the anchoveta; that of Fish Meal, which provides a highly nutritious foodstuff for both local consumption and export for agriculture and cattle-raising; and the Guano of the Islands, which is an excellent fertilizer for agriculture.

Lima, January 22, 1954.
National Fisheries Society

The sentiments expressed are worthy of consideration, and it is no reflection upon the integrity of individuals engaged in fisheries if we attempt to show that their confidence in the harmony of two "equally important and respectable industries" is not necessarily justified.

We have already seen that the research conducted by the Guano Administration bears directly upon *increasing* the future supply of fertilizer. The reliability of this inference is assured not only by the nature of the program but even more by a *fait accompli*—namely the empirical test of a steadily growing output between 1909 and the present date.

What are the Peruvian fishing companies doing to provide greater natural supplies of fish 25 or 50 years in the future? It will be granted that they are improving their techniques for capturing larger quantities of fish from the supply provided by unassisted nature, but this tells us nothing about the ultimate effect of the one-way process. The universal consensus today regarding renewable natural resources is that they can be made to function permanently only when conducted in accordance with banking principles. This means that withdrawals must be continuously balanced by deposits, or else credit collapses. That is exactly the philosophy of the Peruvian Guano Administration.

SCIENTIST VERSUS "PRACTICAL FISHERMEN"

Most prophecies concerning the future supply of renewable natural resources have been based upon uncritical opinion and have proved correspondingly ill-founded. Fisheries in many parts of the world offer many examples. In Europe and the United States large-scale investigations aimed toward the maintenance of marine and fresh-water food resources began in the 19th century. In the face of failing supplies, they have been intensified in the present century, but effective action has all too often come tragically late, as in the case of the oyster fishery of Chesapeake Bay. The phenomena of "Saturnalia" are by no means confined to Peru.

Sixty-five years ago, Professor W. K. Brooks, of the Johns Hopkins University, foretold with uncanny accuracy the decline and fall of the vast "oyster empire" of the Chesapeake. The report he published in 1890 has ever since stood the severest tests of criticism. His researches had already revealed a 50% decline in the Maryland oyster population within the 3-year term of 1879–1882, and he boldly predicted what the subsequent course of events would be if the findings of science were ignored.

The fatal changes of the next half century represent one of the best validations of scientific forecast. The reasons behind the failure were that politicians rejected the sound foundations of the biologist. Again and again, successive Maryland legislatures sought to protect and encourage the fishery by consulting the so-called "practical oystermen." In other words, uninformed opinions of men whose business was that of taking oysters out of the water in the largest possible numbers were substituted for knowledge of the life history of the oyster based upon biological data and the application of the inductive method.

In North America and elsewhere, it would be easy to multiply similar examples from many fields. Lumbermen have hardly proved justified in their optimism concerning the quantity and price of marketable board-feet, nor sportsmen regarding the crop of waterfowl and other game, at least in relation to the increasing demand.

A tabulation of the coastal fisheries of New England shows that the annual catch of such former staples as sturgeon, salmon, shad, lobster, and edible mollusks has fallen off 70% or more, mostly during the past one to two generations in which the human population has doubled. The losses are partly masked by better facilities for transporting an equivalent product from a greater distance, but the deficit still stands against a final reckoning.

Knowledge and recognition grow, even though slowly. It would be possible, if space permitted, to point out happier examples of study and regulation of coastal fisheries in the United States. The main purpose of mentioning them, however, is merely to call attention to a fact that many businessmen find unwelcome. It is this: when matters take a turn for the better in a fishery, where do the ultimate springs of action prove to lie? With "practical" commercial fishermen? With the hard-headed substantial citizens whose money is invested in such ventures? Not at all! We can be thankful enough if these gentlemen merely fall in line and lend their cooperation before the time is too late. The call is nearly always sounded, years before action is taken, by the reputedly "impractical" man of science. It is very easy to demonstrate that the large industrial concerns have given far more opportunity and credit to research in pure science than have the business interests which tap the natural wealth of organic resources. Yet the benefits to be derived from the scientific approach are as sure in the one field as in the other. They have been superbly displayed in the last half century of guano management.

RESPONSIBILITY OF THE FISHERIES

The Peruvian fisheries operators are correct, perhaps to an extent beyond their own realization, in stating that they are equally interested with the Guano Administration in the future resources of the coastal current. They are right because, in the end, fisheries and guano will stand or fall together. But, while the Guano Administration is fulfilling its part toward creating that future, what is the simultaneous contribution of the fishing companies? Where are their biological laboratories along the coast? What studies of the plankton, the anchoveta, the bonito, have been issued under their auspices? Upon what criteria do they base their opinion as to the toll that the anchovetas can safely tolerate before the healthful state of the guano industry, and of the food fisheries as well, becomes endangered? What data have they acquired with reference to cyclic fluctuations in the populations of the food fishes they capture, and of the prey that supports both the food fishes and the guano birds?

POPULAR MISINFORMATION

The controversy relating to guano and fish meal belongs in an impersonal realm and it is unfortunate that it has generated so much more heat than light. The fault is mainly due to misinformation. The writer of this report has devoted approximately six months to close and *daily* observation of the Peruvian guano birds—on their islands and nesting peninsulas and at sea on the feeding grounds. As a professional naturalist of wide experience, he is prepared to defend the substantial accuracy of his observations. They have the advantage of general agreement with those of the trained scientists of the Guano Administration, who have spent not months but years in the same pursuit.

It is therefore disquieting to note how uncritical statements, expressed by individuals who have lacked opportunity to learn the facts, can be accepted as gospel truth by a part of both press and public. At the conference held in January at the University of San Marcos, a veteran Peruvian statesman, whose motives are beyond reproach, repeated the outdated belief that the guano birds must consume 32 tons of anchovetas in order to deposit 1 ton of guano, whereas in current manufacturing procedure the proportion of raw fish to fish meal is only five to one. In partial explanation of this alleged discrepancy, the speaker went on to say that about half the weight of fish devoured by the birds is subsequently lost by defecation into the sea water.

Let us consider the facts. Professor G. E. Hutchinson, of Yale University, author of the most important and exhaustive study of guano that has yet been published (1950), finds that the relation of food ingested by the birds to the guano laid down *on the nesting ground* is of the order of 9.7 to one (not 32 to one). Hutchinson's figure is probably conservative, because experiments conducted in the laboratory at La Puntilla, by Dr. Enrique Avila (1950) have shown that captive fledgling guanayes, fed for 90 consecutive days on an average of 350 g of fish, produced daily at least 55 g of guano. Other scrupulous computations indicate that wild adult guanayes and piqueros deposit on land throughout their lives a minimum dry weight of 1 kilogram of guano per month.

So on this point we may concede that factory methods of anchoveta reduction appear, by weight alone, to be about twice as efficient (not six times as efficient) as the digestive tracts of the birds. The birds, however, conduct their business according to a fool-proof system, which man does not. The birds convert no more fish into guano than is consistent with an indefinite persistence of all the parts of the ecosystem involved. No such margin of safety relates to fish-meal exploitation, which should never be permitted to replace the fortunate and admirable natural industry.

As regards qualitative, rather than quantitative comparison, it should be noted that the recent nitrogen content of Peruvian guano has been the highest in history (15.65%).

As to "defecation into the sea," this is a layman's notion based upon inference drawn from gulls and other familiar waterfowl. It possesses no meaning to naturalists who have lived among the guano-producing flocks. It so happens that the Peruvian members of the avian order Pelecaniformes, which includes the guanay, piquero, camanay, and alcatraz, ordinarily retain their excrement until it can be ejected on the nesting ground. This is known not only from direct observation but also from quantitative evidence. The food intake closely agrees with the calculated ratio of excreta at the nest. It is this characteristic, of course, that makes the Peruvian cormorant, boobies, and pelican *the* guano birds as contrasted with many other marine species of the region. One adaptive advantage of the curious specialization is that the surface water of their extensive feeding areas does not become opaque when the vast flocks are fishing. Another advantage is that, in a rainless and rocky environment, guano represents the only material for building countless thousands of the deep, strong, and walled nest bowls in which the slow-growing young birds are reared.

In other ways than defecation a certain proportion of guano is restored to the ocean. When flocks take wing in the wind from their breeding grounds, a cloud of guano dust rises and floats leeward. The birds, particularly the guanayes, become befouled with guano while brooding, and usually their first sortie of the morning is to carry out a spectacular flock-washing of their plumage in the ocean. All such guano that goes into solution serves to fertilize the water and to subserve plankton production. When we consider the extraordinarily minute proportion of stable manure required to affect fertility in pond-fish culture, as practiced by the French, the activation caused by small amounts of guano in sea water becomes comprehensible.* It fits directly into the cycle of marine

*According to a report submitted in 1952 by the Woods Hole Oceanographic Institution to the duck-rearing Towns of Brookhaven and Islip, Long Island, NY, the four million domestic ducks in the two townships produce 2.7 million pounds of nitrogen and 0.82 million pounds of phosphorus per year. The proportion of uric acid that reaches the water

equilibrium. In the fish meal industry there is no such return to the sea; there is only extraction.

Many other incorrect ideas about guano, in addition to those already cited, have received wide circulation through newspapers of Lima. There has been frequent comparison of Peru with "other countries possessing guano birds, such as South Africa." This is sheer absurdity. The South Africans, whose old guano was cleaned out like that of Peru in a former century, have recently been constructing "artificial islands" so as to increase the nesting space of their guano fowl. They look with longing eyes toward the supremely fortunate west coast of South America, because their own contemporary guano resources are not one-tenth of those derived annually by the Peruvian Guano Administration. Even the very young Chilean traffic in renewable guano greatly exceeds that of South Africa.

Peru is unique; it has neither rival nor peer. It is nonsense to attempt to belittle its eminence by referring to it as one of a number of guano-producing countries. The Peruvian guano bird population is one of the most amazing and permanently valuable climaxes in all nature.

It has also been asserted (*La Prensa,* March 21, 1954) that Peru has immense resources in chemicals that can provide a fertilizer much cheaper than guano. The proper time to exalt such a possibility will be when the product is made available to agriculturalists.

Much emphasis has been laid upon the quantity of food in the form of meat, milk, eggs, etc., resulting from the utilization of anchoveta meal. (This human sustenance, incidentally, is produced chiefly outside Peru.) The fact that guano is the first auxiliary of food production inside Peru for sugarcane, rice, and other grains, forage, vegetables, and fruit, not to mention the foremost textile crop, is overlooked. The price at which guano is sold in Peru—far below its value in the international market—is not its true monetary measure. The latter should rather be incorporated with the annual value of the total Peruvian agricultural crop, which in 1953 was 3,887,575,548.00 soles.* If we credit a modest 30% of this to guano, the latter would be worth well over a billion and a half soles.

It has likewise become fashionable among the detractors to refer patronizingly to the guano birds as an "archaic and romantic heritage from the Incas," hardly worthy of consideration in the present age of superefficient man. But the broad history of natural resources lends no support to such cocksureness. The Incas may well have been more hard-

---

of Moriches Bay and Great South Bay is quickly changed by bacteria into soluble nitrogenous compounds, which support a rich flowering of green algae. [Data from Mary Sears, *Deep-Sea Research,* **1,** 156–157 (1954).]

*Another estimate places the value in 1953 at S/4,500,000,000,000.

headed and realistic than the modern champions of machinery, which can process but which cannot produce.

Señor Llosa Belaúnde has written, "*La historia de la industria del guano en el Perú se pierde en los albores de los tiempos. Sólo tenemos noticias de que durante el período del Incanato se usó el guano de islas y que existieron reglamentaciones rígidas para impedir la interferencia del hombre en la vida de las aves marinas.* [*The history of the guano industry in Peru gets lost in the beginning of time. We have only records that during the Inca period guano of the Islands was used and that strict rules existed to prevent men's interference in the life of the marine birds.*]" Regulations as effective as those of the Incas, but adapted to modern circumstances, may assure for guano a future longer and more important than its past. It is still one of the most substantial and enduring riches upon which Peru can count.

THE COURSE OF WISDOM

No one has yet asserted that the useful purpose and financial success of Peruvian food fish exploitation is dependent upon freedom to net anchovetas. Most of the companies report that tinned or frozen fish are their fist concern, and that anchoveta meal is purely supplementary. It should be noted also that these firms are manufacturing fish meal from the heads, skin, fins, viscera, and other inedible parts of their food catch, as well as obtaining a residue of high-grade oil.

There is an element of truth in the claim that the annual toll of 50,000 or more tons of tuna, bonito, and barrilete corresponds to a "free gift" to the guano birds of the anchovetas that the captured fish would otherwise have eaten. It is possible that the increasing food fishery of the past decade is correlated with the simultaneous increase of the guano birds. The fish and birds are rivals with respect to a common prey—the anchoveta. By the same token, a massive anchoveta fishery would work to the detriment of both guano and fish production. There is thus no quarrel between the users of guano and the food fishery but, rather, a strong persuasion that the plankton-eating anchoveta had best be left undisturbed to supply food for the larger and more essential species of fish, as well as the birds.

The crux of the problem centers in the anchoveta fishery. Several companies are already engaged in it exclusively, and pressure for further franchises is persistent. The anchoveta is the source of the aroused hostility and the blunt expressions of opinion, many of which are far-fetched or grotesque. It has been seriously suggested, for example, that all restrictions against taking anchovetas be lifted, and that the Peruvians put into operation large fishing craft equipped with suction apparatus that would pump massed schools of anchovetas into the hold and assure an annual catch of 400,000 tons! This proposal ignores the fact that such machinery has not passed the experimental stage, that it would probably not pay for the cost of its operation, and that, if it were by any chance successful, it would wreak havoc with the ecology of the Peru current.

In the face of present indications, the part of wisdom would be to prevent extensive anchoveta fishery in the zone of the Pacific over which Peru exercises national control until very much more has been learned about the life history of the organism. What is the length of life of this fish, the prevailing direction and periods of its migrations, the dates of its ovulation, the rate of development of its eggs and of its larval stages? What is the normal replacement rate in the schools, and how many year classes do the vast aggregations comprise? For a sufficient term the appropriate Government department should compile a month-to-month and year-to-year census of the species, similar to the register that the State of California has long maintained with reference to the sardine, which has shown such wide amplitude in numbers off the western coast of the United States. Only by such measures will it be possible to learn the coefficient of reproduction necessary to hold the anchoveta in lasting equilibrium.

It must be remembered that men who are eager to tap natural resources are not inclined to be sympathetic toward an adequate weighing of the pros and cons. Their attitude is usually more vociferous than judicious. If they insist that overfishing has had no bearing upon the calamitous diminution of the California sardine (which may conceivably be true), it is, nevertheless, an assertion based upon inadequate data. Furthermore, the anchoveta is a different species in a distinct, more peculiarly specialized and delicately adjusted environment. The wishful thinking of fishermen should not become the criterion for action. To risk another Saturnalia would be a poor gamble.

Quite aside from the able corps of the Guano Administration, who might be regarded as special pleaders on one side of the controversy, Peru has plenty of citizens whose independent judgment would be extremely valuable. The members of the *Comité Nacional de Protección a la Naturaleza* are such a group. These gentlemen are disinterested, cosmopolitan, well informed, and closely in touch with ecological problems in various parts of the world. They are affiliated with a large and strong international organization. Their *Boletín,* now in its tenth volume, contains a record of sound opinion on conservational matters of many kinds, including industrial missteps that threaten public welfare.

Thus far, the organized anchoveta fishery in Peru may be no more than the cold nose of the camel inside the flap of the Arab's tent. It is the trend that creates the problem. *Mas vale pájaro en mano que ciento volando.* [*A bird in hand is worth more than a hundred in the air.*]

CHAPTER 5

The Collapse of the California Sardine Fishery

What Have We Learned?

JOHN RADOVICH

Abstract

For a number of years, Federal scientists, employed by an agency whose primary goal was to assist the development of the U.S. commercial fisheries, looked for causes, other than fishing, for the Pacific sardine's decline, while California State scientists, charged with the role of protector of the State's resources, sought reasons to support the premise that overfishing was having an effect. At the same time, scientists from Scripps Institution of Oceanography looked for fundamental generalizations in theory rather than the activities of man to explain changes in fish populations. For many years, California State personnel struggled without success to gain control over a burgeoning, and later declining sardine fishery.

Faced with the possibility that legislation might be enacted, giving the California Fish and Game Commission control over the sardine fishery, the California fishing industry sponsored the formation of the Marine Research Committee to collect and disburse funds and to coordinate and sponsor more "needed" research, thereby forestalling any action to allow management of the fishery to come under the authority of the California Fish and Game Commission. Subsequently, the California Cooperative Oceanic Fisheries Investigations (CalCOFI) was formed, under which cooperative research proceeded.

Oceanic conditions (temperature) was found to affect profoundly the distribution, year-class production, and yield of sardines. Nonintermingling or only partially intermingling stocks of sardines have been described. Considerable attention has been focused on the complementary role of sardines and anchovies as competing species acting as a single biomass while competing with each other as part of that biomass. Confirmation of this hypothesis was found to have been based on faulty interpretation of basic data. If such a relationship exists, it still needs to be demonstrated.

Density-controlling mechanisms, however, which may be of greater importance, include predation, cannibalism, and other behavioral characteristics. Schooling behavior, for instance, which has evolved through natural selection to decrease mortality from predation, may work toward destruction of the prey species when it is confronted by a fishery which evolves more rapidly than does the species defense against it. A model that is consistent with the results of all the previous studies on the sardine must bring one to the conclusion that the present scarcity of sardines off the coast of California, and their absence off the northwest, is an inescapable climax, given the characteristics and magnitude of the fishery and the behavior and life history of the species.

INTRODUCTION

At a symposium of the CalCOFI conference held in La Jolla, California, on December 5, 1975, on "The Anchovy Management Challenge," a paper was presented by W. G. Clark on "The Lessons of the Peruvian Anchoveta Fishery."[1] It is ironic that California's anchovy researchers felt compelled to learn lessons from the collapse of the Peruvian fishery which the Pacific sardine's failure could have provided to them as well as to the Peruvian anchoveta researchers. The observations which follow are presented with the hope that a discussion of some of the social, political, and biological factors associated with the decline of the California sardine fishery will have useful applications in interpreting events now taking place in our fisheries.

After 50 years of fishing for the Pacific sardine, *Sardinops sagax* (Jenyns), a moratorium on landings was imposed by the California Legislature in 1967, thus bringing to an end yet another act of one of the more emotionally charged fisheries exploitation-conservation controversies of the 20th century.

By the time the moratorium was imposed, however, the sardine fishery in southern California had already collapsed. The sardine fisheries in the northwest had long since ceased to exist with sardines last landed in British Columbia in the 1947–1948 season, in Oregon and Washington in the 1948–1949 season, and in San Francisco Bay in the 1951–1952 season (Table 1).

Even before the productivity and exploitation of the fishery peaked, researchers from the (then) California Division of Fish and Game issued warnings that the commercial exploitation of the fishery could not increase without limits, and recommended that an annual sardine quota be established to keep the population from being overfished.

Such recommendations were, of course, opposed by the fishing industry which was able to identify scientists who would state, officially or otherwise, that it was virtually impossible to overfish a pelagic species. This debate permeated the philosophies, research activities, and conclusions of the scientists working in this field at that time. The debate conformed to the basic charters (or *raisons d'être*) of each agency involved and persists today, long after the United States Pacific sardine fishery has ceased to exist. As a result of deep-rooted social and political feelings concerning the collapse of the Pacific sardine off California, many conflicting hypotheses have arisen, in spite of the completion of a vast amount of research.

For example, just recently, a prominent representative of a major oceanographic research institution asserted that it was a "false assumption that overfishing killed the former sardine fishery off Northern California. . . . The real cause of the disappearance of the California sardine was a climatic change."[2] The same official, addressing a group of scientists, stated that

> . . . the explanation of the disappearance [of the sardine] seems to be a change of climate that triggered a major biological upheaval. It was very quiet by our standards, we who live in the atmosphere, but it was violent in that several million tons of one species was replaced by another [anchovy].[3]

If this view is valid, one must ask why scientists of the California Department of Fish and Game supported a moratorium on fishing for sardines. Why did they recommend quotas of 250,000 to 300,000 tons of sardines as a measure to forestall the collapse which they had predicted would occur?[4] MacCall recently postulated that a safe estimate of the maximum sustainable yield (MSY) for the Pacific sardine, assuming it were to be rehabilitated, would be about 250,000 metric tons and that if the catch had been held to that limit the fishery would still be viable.[5]

Clearly, these views conflict. Why, and to what extent, do these conflicting views persist in scientific circles? Which concepts are in error? What seems to be the truth? The complete answers to these questions, particularly to the last one, are beyond the scope of this paper. However, it is time to recall a few pertinent events which may improve the historical perspective and provide better insights for the interpretation of the mass of ecological data already accumulated.

TABLE 1. Seasonal Catch (tons) of Sardines along the Pacific Coast (Each Season Includes June through the Following May[a])

Season	Pacific Northwest				California[b]							Baja[c] California	Grand Total
	British Columbia	Washington	Oregon	Total	Northern California				Southern California	Total California			
					Reduction Ships	San Francisco	Monterey	Total					
1916–1917	—	—	—	—	—	—	7,710	7,710	19,820	27,530		—	27,530
1917–1918	80	—	—	80	—	70	23,810	23,880	48,700	72,580		—	72,660
1918–1919	3,640	—	—	3,640	—	450	35,750	36,200	39,340	75,540		—	79,180
1919–1920	3,280	—	—	3,280	—	1,000	43,040	44,040	22,990	67,030		—	70,310
1920–1921	4,400	—	—	4,400	—	230	24,960	25,190	13,260	38,450		—	42,850
1921–1922	990	—	—	990	—	80	16,290	16,370	20,130	36,500		—	37,490
1922–1923	1,020	—	—	1,020	—	110	29,210	29,320	35,790	65,110		—	66,130
1923–1924	970	—	—	970	—	190	45,920	46,110	37,820	83,930		—	84,900
1924–1925	1,370	—	—	1,370	—	560	67,310	67,870	105,150	173,020		—	174,390
1925–1926	15,950	—	—	15,950	—	560	69,010	69,570	67,700	137,270		—	153,220
1926–1927	48,500	—	—	48,500	—	3,520	81,860	85,380	66,830	152,210		—	200,710
1927–1928	68,430	—	—	68,430	—	16,690	98,020	114,710	72,550	187,260		—	255,690
1928–1929	80,510	—	—	80,510	—	13,520	120,290	133,810	120,670	254,480		—	334,990
1929–1930	86,340	—	—	86,340	—	21,960	160,050	182,010	143,160	325,170		—	411,510
1930–1931	75,070	—	—	75,070	10,960	25,970	109,620	146,550	38,570	185,120		—	260,190
1931–1932	73,600	—	—	73,600	31,040	21,607	69,078	121,725	42,920	164,645		—	238,245
1932–1933	44,350	—	—	44,350	58,790	18,634	89,599	167,023	83,667	250,690		—	295,040
1933–1934	4,050	—	—	4,050	67,820	36,336	152,480	256,636	126,793	383,429		—	387,479
1934–1935	43,000	—	—	43,000	112,040	68,477	230,854	411,371	183,683	595,054		—	638,054
1935–1936	45,320	10	26,230	71,560	150,830	76,147	184,470	411,447	149,051	560,498		—	632,058
1936–1937	44,450	6,560	14,200	65,210	235,610	141,099	206,706	583,415	142,709	726,124		—	791,334
1937–1938	48,080	17,100	16,660	81,840	67,580	133,718	104,936	306,234	110,330	416,564		—	498,404
1938–1939	51,770	26,480	17,020	95,270	43,890	201,200	180,994	426,084	149,203	575,287		—	670,557
1939–1940	5,520	17,760	22,330	45,610	—	212,453	227,874	440,327	96,939	537,266		—	582,876
1940–1941	28,770	810	3,160	32,740	—	118,092	165,698	283,790	176,794	460,584		—	493,324
1941–1942	60,050	17,100	15,850	93,000	—	186,589	250,287	436,876	150,497	587,373		—	680,373

Season											
	580	1,950	68,410	—	115,884	184,399	300,283	204,378	504,661	—	573,071
1943–1944	88,740	1,820	101,000	—	126,512	213,616	340,128	138,001	478,129	—	579,129
1944–1945	59,120	—	59,140	—	136,598	237,246	373,844	181,061	554,905	—	614,045
1945–1946	34,300	90	36,700	—	84,103	145,519	229,622	174,061	403,683	—	440,383
1946–1947	3,990	3,960	14,090	—	2,869	31,391	34,260	199,542	233,802	—	247,892
1947–1948	490	6,930	8,780	—	94	17,630	17,724	103,617	121,341	—	130,121
1948–1949	—	5,320	5,370	—	112	47,862	47,974	135,752	183,726	—	189,096
1949–1950	—	—	—	—	17,442	131,769	149,211	189,714	338,925	—	338,925
1950–1951	—	—	—	—	12,727	33,699	46,426	306,662	353,088	—	353,088
1951–1952	—	—	—	—	82	15,897	15,979	113,125	129,104	16,184	145,288
1952–1953	—	—	—	—	—	49	49	5,662	5,711	9,162	14,873
1953–1954	—	—	—	—	—	58	58	4,434	4,492	14,306	18,798
1954–1955	—	—	—	—	—	856	856	67,609	68,465	12,440	80,905
1955–1956	—	—	—	—	—	518	518	73,943	74,461	4,207	78,668
1956–1957	—	—	—	—	—	63	63	33,580	33,643	13,655	47,298
1957–1958	—	—	—	—	—	17	17	22,255	22,272	9,924	32,196
1958–1959	—	—	—	—	—	24,701	24,701	79,270	103,971	22,334	126,305
1959–1960	—	—	—	—	—	16,109	16,109	21,147	37,256	21,446	58,702
1960–1961	—	—	—	—	—	2,340	2,340	26,538	28,878	19,899	48,777
1961–1962	—	—	—	—	—	2,231	2,231	23,297	25,528	21,270	46,798
1962–1963	—	—	—	—	—	1,211	1,211	2,961	4,172	14,620	18,792
1963–1964	—	—	—	—	—	1,015	1,015	1,927	2,942	18,384	21,326
1964–1965	—	—	—	—	—	308	308	5,795	6,103	27,120	33,223
1965–1966	—	—	—	—	—	151	151	568	719	22,247	22,966
1966–1967	—	—	—	—	—	23	23	321	344	19,531	19,875
1967–1968	—	—	—	—	—	—	—	71	71	27,657	27,728

[a]British Columbia data were supplied by the Canadian Bureau of Statistics and the province of British Columbia; Washington data by the Washington Department of Fisheries; and Oregon data by the Fish Commission of Oregon. Deliveries to reduction ships and data for Baja California were compiled by the United States Fish and Wildlife Service from records of companies receiving fish. California landings were derived from records of the California Department of Fish and Game.

[b]Prior to the 1931–1932 season, fish landed in Santa Barbara and San Luis Obispo Counties are included in southern California. Subsequent landings north of Point Arguello are included in Monterey and those south of Point Arguello are included in southern California.

[c]The amount of sardines landed in Baja California prior to the 1951–1952 season is not known.

HISTORICAL REVIEW

Differences in Agency Perspective

The California Fish and Game Commission began with the approval of an act of the California Legislature creating the Commissioners of Fisheries on April 2, 1870, by Governor Haight of California. The principal purpose of the Commission was embodied in the title of the legislation, "An act to provide for the restoration and preservation of fish in the waters of this State." While the objectives of the California Fish and Game Commission and its Department of Fish and Game have expanded since then, their role as protector of the State's fish and wildlife resources has remained paramount.

In 1871 the U.S. Bureau of Commercial Fisheries was created; the primary goal of the new Federal Bureau was to assist in the development and perpetuation of the United States fishing industries. This goal persists today, despite several agency name changes, even though the present National Marine Fisheries Service (NMFS) too has broadened its objectives somewhat in recent years.

For many years, federal personnel from the National Marine Fisheries Service debated vigorously with personnel from the California Department of Fish and Game on what was happening to the Pacific sardine. The Federal scientists, working for an agency whose fundamental charter was to assist the development and maintenance of U.S. commercial fisheries, looked for reasons other than fishing, for the sardine's declining condition, while the scientists employed by the State (whose basic role was protector of the State's resources) supported the premise that overfishing was having a detrimental effect on the standing stock. These were capable, competent scientists using the same data and coming up with different conclusions in part because they were employed by agencies whose fundamental goals were different.

Scientists are directly and indirectly influenced by the values of their society, their institutions, their academic disciplines, as well as by their personal political beliefs. Each scientific discipline is saturated with values imposed by its specific profession, and scientists are influenced by the agencies for which they work and to which they owe some allegiance. Thus, the definition of a problem becomes a biological one, a physical one, an economic one, a psychological one, a sociological one, even a philosophical one, depending on the researcher's discipline.

As another example, oceanographers frequently define their field as encompassing the ocean and all the sciences that are studied in relation to the ocean. This all-inclusive perspective relegates other sciences (biology, chemistry, physics, and geology) to the position of subdisciplines of oceanography. Such a disciplinary perspective tends to focus attention away from the effects of local human activities on various marine resources and to extend efforts, instead, toward the investigation of large-scale processes in search of fundamental generalizations to explain widespread phenomena. One might argue that an elitism tends to develop, where one finds, for the example given, at the top of the scale the physical oceanographer, and at the bottom, the biological oceanographer. Carrying this example one step further, perhaps because marine plankton is more dependent on currents, temperature, and other physical and chemical processes, phytoplanktonologists tend to be more influential than other biological oceanographers. Oceanographic institutions usually have an ichthyologist on their staff who may teach systematics and distribution of fishes, but other fisheries courses are not always taught in the largest oceanographic institutions. From the viewpoint of a school of oceanography, the solution to most fisheries problems invariably involves a large scale, multivesseled, physical and chemical assault on a large part of the world ocean, because that is how the problem is conceived—by definition, of course.

California State Biologists' Struggle for Fishery Control

A belief prevalent early this century was that the oceans were inexhaustible and that man could not affect the species in the sea. These concepts were expounded by McIntosh, who was impressed by the " . . . extraordinary powers of reproduction of animals and plants in the sea . . . and boldly asserted the inability of man to affect the species in the sea."[6] This general belief still exists (with some changes) at the present time. Others, however, felt that human activities could have a profound effect on living marine resources.[6]

Concerned with the protection of California's living marine resources, California state biologists consistently expressed concern about the rapidly growing exploitation of the sardine fishery. For example, as early as 1920, one such biologist, O. E. Sette, wrote about the sardine in Monterey Bay:

> The possibility of depletion cannot be much longer ignored. . . . we have definite clues to the answers . . . and it but remains . . . to . . . substantiate facts which we have concerning the age, rate of growth, migration and spawning. . . . It now remains for continuance of this study to solve all the problems concerned, and insure the perpetuity of our great resource, through the adoption of intelligent conservational measures.[7]

The difficulties and frustrations encountered by the California Fish and Game Commissioners and their staff in attempting to gain control over the burgeoning sardine fishery are well documented in the publications and Biennial Reports of the California Divison of Fish and Game (later called the California Department of Fish and Game). This early history of the sardine industry (its growth, economics, and legal regulation) has been summarized by Schaefer et al.[8]

After the states of Oregon and Washington approved the use of sardines for reduction to fish meal in the 1930s, there were essentially no restrictions on the quantity of landings or the use made of them in the Pacific Northwest. Inasmuch as the California Legislature never delegated full authority for regulating the sardine fishery to the Fish and Game Commission, the Commission was forced to attempt to control the fishery through the exercise of the only authority the Legislature had delegated to it, control over the reduction fishery. Schaefer et al.[8] and Ahlstrom and Radovich[9] have summarized the conflicts between canning and reduction interests and the desires of the Commission's biologists to protect the resource from overfishing and depletion.

During the 1930s, straight reductionists* bypassed State control over the reduction fishery by operating reduction ships outside the territorial sea limits, beyond the jurisdiction of the State of California.[9] (See Figures 1 and 2.) To stop the floating reduction plants, an initiative amendment to the California State Constitution was passed in November 1938 that prohibited any fishing vessel from operating in State waters if it delivered fish taken in the Pacific Ocean to points outside the State without authorization from the Commission. The enactment of this law, combined with lower fish oil prices and increased operating costs, ended reduction ship operations after 1938.[8]

As early as 1931, N. B. Scofield, the Chief of the Bureau of Commercial Fisheries of the California Division of Fish and Game, observed that

> the catch has not increased in proportion to the fishing effort expended, and there is every indication that the waters adjacent to the fishing ports have reached their limit of production and are already entering the first stages of depletion. The increase in the amount of sardines caught is the result of fishing farther from port with larger boats and improved fishing gear. . . .
> The Fish and Game Commission has consistently endeavored, through legislation and through cooperation with the canners, to restrict the amount

---

*Straight reductionists were those who reduced all fish received from the fishermen, while canners reduced only a part of the catch consisting mainly of heads and offal.

Figure 1 *Lake Miraflores,* the first reduction ship to operate outside the jurisdiction of the State of California, unloading sardines from a purse seiner in the early 1930s.

of sardines which canners are permitted to use in their reduction plants with the belief that the canning of sardines is the highest use to which they can be put and that the excessive use of these fish in reduction plants would, in time, result in depletion of the source of supply. The majority of the canners, on the other hand, have sought to get quick returns from sardine reduction and have made themselves believe there was no danger of depletion.[10]

In 1931, the State Division of Fish and Game advocated a seasonal limit of 200,000 tons on the amount of sardines that could be landed safely with little effect on the standing stock.[11] In 1934, N. B. Scofield[12] reiterated his view that the catch should be limited to 200,000 tons, indicating that this recommendation had been made 5 years earlier.

In 1938, W. L. Scofield[4] warned that overfishing was causing a collapse in the supply of sardines. He indicated that if the catch were cut to less than the amount replaced annually, the stock could rebuild back to its former productive level. He suggested 250,000 tons as the ideal level of catch. The 250,000 ton limit was also recommended by F. N. Clark,[13,14]

Figure 2 The reduction vessel, *Polarine*, unloading a purse seiner, with two other reduction ships anchored in the distance.

who suggested that the limit might be raised somewhat during limited periods of exceptional spawning survival.

Prior to the 1938 initiative amendment to the State Constitution, which happened to coincide with the discontinuance of the reduction ships, an attempt had been made in 1936 to pass federal legislation either making it unlawful to take sardines for reduction on the high seas, or making such operations subject to the laws of the adjacent state. While this legislation was not passed, the attempt to pass it gave impetus to pressures for the establishment of a federal research laboratory on the Pacific Coast. W. L. Scofield[15] wrote:

> The [reduction] ship operators, foreseeing future legislation, resorted to a plan (used before and since) by which anti-reduction legislation could be postponed by asking for a special study of the abundance of sardines and thus disregard the work of the St. F. Lab. [State Fisheries Laboratory] or at least throw doubt upon its findings. The ship operators (mid-1930s)

quietly promoted the plan of urging the legislatures of the 3 coastal states to ask Congress to have the U.S. Bureau [of Fisheries] make a study of sardine abundance. Wash. and Oregon complied but Calif. legislature refused to ask for Fed. [Federal] help. The U.S. Bur. was anxious to get a foothold in Calif. and sent out O. E. Sette (May 1937). This [was] a shrewd choice because Sette [was] a diplomat and personal friend of Calif. Lab. staff [actually he was a former state fisheries research biologist]. We told Sette he was not wanted in Calif. and asked him to go up to Wash. or Oregon who had asked for help.* Sette answered that he must work in Calif. because most of the sardines were here (not the real reason) and he pointed out that we could not afford to refuse our cooperation in a U.S. Bur. study of sardines. This was true and we had to grin and bear it. Sette started his sardine studies with a staff, housed at Stanford University. By 1938 a plan of cooperative sardine study for each agency was agreed upon.

From this beginning, the two agencies, one state and one federal, expended their efforts in different directions. On the one hand, the State's research biologists, with the responsibility for determining if and when "overfishing" was likely to occur and for making recommendations for appropriate management measures to prevent such overfishing, devoted their energies along those lines, even though their agency had not been delegated the authority to manage fully the commercial fishery. On the other hand, the Federal biologists, with no management responsibilities in (or obligations to) California, maintained a good rapport with the fishing industry, in that they were dedicated to assist the development and maintenance of a viable U.S. fishing industry, and looked for causes, other than fishing pressures, to explain the declines in the sardine fishery. This resulted in numerous debates at meetings and in conflicting scientific viewpoints in technical journals. The debates and conflicts were often based on the same data.

Despite warnings by State biologists that collapse of the sardine fishery was imminent, a large crop of young fish were produced in five successive years, 1936 to 1940 (Clark and Marr[16]). This gave rise to considerable speculation about the effect of environmental conditions on changes in the sardine population and to support for arguments by the fishing industry that nature, not the industry, caused much of the observed sardine population changes. The industry strongly supported the Federal

*W. L. Scofield related this incident to me, personally, as follows: "When Sette visited us after first contacting the major local fishing industry leaders, he asked how he could be of help to us. N. B. Scofield told him he could help us best by packing his bags and going back to Washington, D.C." O. E. Sette later personally confirmed this initial dialogue between the representatives of the two agencies.

biologists in their search for reasons, other than man, to explain the fluctuations in the sardine population.

The Marine Research Committee

After the large year-classes produced from 1936 through 1940 passed through the fishery, the sardine fishery collapsed to a low point in 1947 (Table 1). The fishing industry, concerned that legislation might be enacted to give the Commission control over the fishery, again resorted to the delaying tactic of advocating or sponsoring more research. In 1947, a meeting was held among representatives of the sardine fishing industry, United States Fish and Wildlife (later renamed U.S. Bureau of Fisheries), Scripps Institution of Oceanography, California Academy of Sciences, and California Division of Fish and Game. This group formulated a plan for a Marine Research Committee which would disburse funds collected from a tax on fish landings and would coordinate and sponsor research "to seek out the underlying principles that govern the Pacific sardine's behavior, availability, and total abundance."[17]

The Marine Research Committee was created by an act of the California Legislature in 1947, and was composed of nine members appointed by the Governor. Five members were specified to be selected from persons actively engaged in the canning or reduction industry, one member was the Chairman of the California Fish and Game Commission, one, the Executive Officer of the Division of Fish and Game, an additional member was taken from the Division of Fish and Game, and the ninth member was undesignated; the Director of the California Academy of Sciences was appointed to the undesignated position.

The work was to be carried out largely by Scripps Institution of Oceanography, U.S. Fish and Wildlife Service, California Division of Fish and Game, and the California Academy of Sciences, under the guidance of a technical committee representing the four agencies. Now there were two agencies: the Scripps Institution of Oceanography along with the Federal group, looking for reasons, other than fishing, to explain the sardine's decline; and State biologists also working, with mixed emotions, on a large-scale program. All of the above were somewhat under the auspices of a committee whose vote was controlled by the majority of five members from the sardine fishing industry.

The composition of the Marine Research Committee was changed in 1955 to consist of at least one member representing organized labor, at least one member representing organized sportsmen, two public members, and the same majority of five from the fishing industry.

The difference in perspectives of the biologists of the two fisheries agencies peaked in a joint paper by F. N. Clark (California Department of Fish and Game) and J. C. Marr (U.S. Fish and Wildlife Service)[16] in which the two authors drew different conclusions that were specifically identified from the same data. Also, the authors were careful to point out that the order of authorship was arranged alphabetically.

MORE RECENT RESEARCH EFFORTS

California Cooperative Oceanic Fisheries Investigations (CalCOFI)

Coordination of the efforts of the three principal agencies improved when the California Cooperative Oceanic Fisheries Investigations (CalCOFI) Committee was established in December 1957, with the working head of the unit in each of the three major agencies, Scripps Institution of Oceanography, the U.S. Bureau of Fisheries, and the California Department of Fish and Game, engaged in cooperative work. A fourth member, without voting power, was hired by the Marine Research Committee, and acted as Chairman.

Effects of Temperature on Population Size, Distribution, and Fishing Success

In 1957 dramatic changes in fish distribution revealed the close relationship of fish movements, fishing success, and local abundance of many marine species to seemingly subtle changes in average ocean temperatures.[18] Following these events and the World Sardine Conference that was convened in 1959 in Rome, in which the effects of fishing on the Pacific sardine were debated at length,[19] a change in attitudes of the two government agencies took place. California scientists became more aware of the effects of the environment, and Federal researchers began to appreciate that human activities could in fact adversely affect a pelagic marine resource.

At the 1959 Sardine Conference, Marr pointed out that a relationship existed between the average temperature from April of a given year to March of the following year and the sardine year-class size (Figure 3).[20] He also suggested that the northern anchovy, *Engraulis mordax* Girard, may prefer lower temperature optima than sardines. Radovich showed that up to the collapse of the fishery in the Pacific Northwest, ocean temperature correlated with an index of latitudinal distribution of young

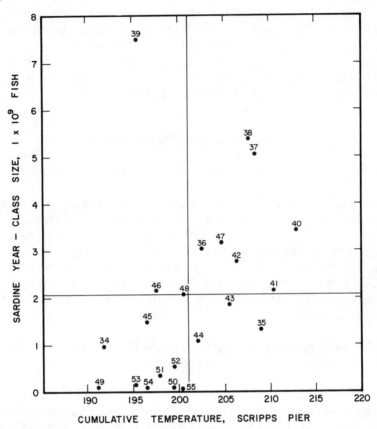

Figure 3 Relationship between year-class size and the sums of monthly mean sea temperature (April through March) at Scripps Pier. After Marr.[20]

sardines (Figure 4) and that year-classes of more northerly originating sardines tended to contribute more heavily to the fishery (Figure 5).[21] Inasmuch as year-class sizes were estimated from the catch, it was not clear, then, to what extent Marr's correlation with temperature was due to year-class size or to effect of temperature on the latitudinal distribution of the origin of the year-class and the effect of its early latitudinal distribution on its subsequent vulnerability to fishing.

Genetic Subpopulations

Another significant study resulted in delineating genetic strains of sardines by using erythrocyte antigens.[22,23] The studies agreed with

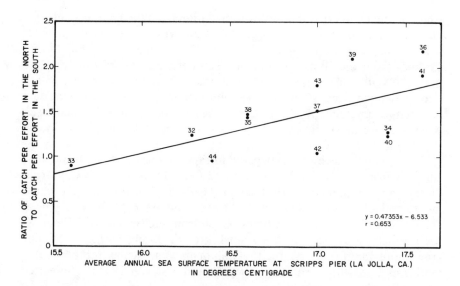

Figure 4 The relationship of sea surface temperatures at Scripps Pier to the index of north-south distribution of the Pacific sardine from the 1932–1933 to the 1944–1945 season. After Radovich.[21]

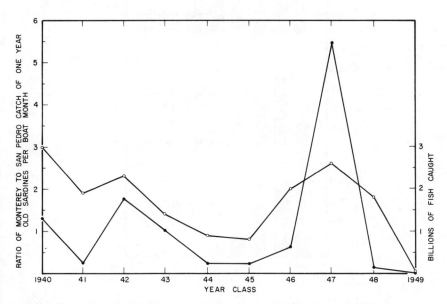

Figure 5 The ratio of the average lunar month catch at Monterey to the average lunar month catch at San Pedro of 1-year-old sardines (filled circles), and the cumulative total of each year-class of sardines taken in the fishery (open circles). After Radovich.[21]

121

Clark's conclusion that the sardine population from the Gulf of California and from the southern portion of Baja California were racially distinct from a single population to the north.[24] Vrooman[23] concluded that sardines from the Gulf of California, southern Baja California, and the northern California populations represented three distinct races, with a poorly defined (and somewhat variable) boundary separating the last two. Unfortunately, by the time that serological techniques had been developed for separating genetic stocks, sardines had disappeared from the Pacific Northwest.

There were, however, two good lines of evidence to indicate that sardines from the Pacific Northwest and those from southern California did not mix randomly: (1) sardines in the Pacific Northwest were much larger and older than those in California (Figure 6); (2) there was a significant difference in scale types of fish from the two areas.[25] Whereas the northern type had relatively small growth during its first year, but grew more rapidly afterward, the California type scales represented a more rapid growth during the first year and a slower growth thereafter

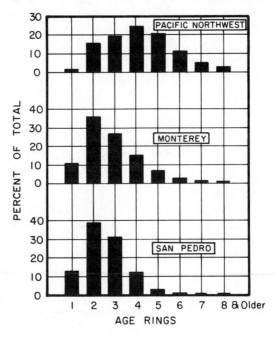

Figure 6 Average age composition of the Pacific sardine in different fishing areas during the 5-year period 1941–1942 through 1945–1946. After Clark and Marr.[16]

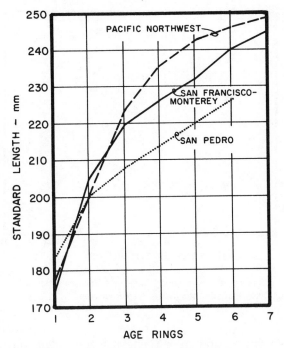

Figure 7   Observed growth curves (size on age) of sardines in several areas. After Clark and Marr.[16]

(Figure 7). Also, the northern type scales had well defined annuli (yearly rings) while the southern types had much fainter yearly rings.

Radovich, noting that the sardine temporarily restabilized at a much lower population regime following the decline in the Pacific Northwest (Figure 8), postulated that the stocks off the Pacific Northwest and off southern California were somewhat distinct, either genetically, or due to a strong tendency for fish to favor areas in which they were born (Figure 9).[26] He suggested that the fishery off the Northwest caught fish from the far northern stock during the summer, and that by winter much of this stock had moved off central California, where it was caught by California fishermen. The sizes and scale types of fish caught at these areas and seasons certainly suggested such a migration.[25] Scale types and sizes of fish also suggested that sardines caught in the fall off central California showed up off southern California in the winter the following year. J. L. McHugh (personal communication), in examining sardine samples from

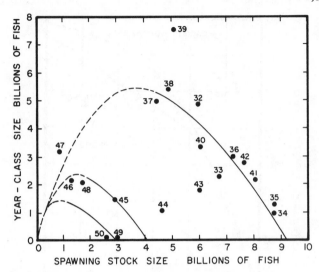

Figure 8 Hypothetical curves representing three probable regimes relating Pacific sardine year-class production to spawning stock size. After Radovich.[26]

the Pacific Northwest and from California concluded, on the basis of meristic and morphometric variations, that fish from the two areas remained somewhat distinct from each other and did not mix to any great degree. The results of this study were never published.

Murphy rejected the existence of a far northern sardine stock by saying " . . . it is not necessary to invoke a third race to explain the collapse of the fishery."[27] He concluded that " . . . the observed quantitative changes in the population offer a sufficient explanation of events without introducing the undocumented qualitative change in the population." In doing so, he ignored the considerable body of evidence that demonstrated the stocks were not uniform or randomly distributed.

Sardine-Anchovy Interspecific Competition

During the period of the CalCOFI expanded program, it had become apparent that the anchovy population was increasing in size,[20] giving cause to speculation that the sardine and the anchovy populations may be acting complementary to each other.[28] This speculation was based mainly on the increase in anchovy population in the 1950s, the co-occurrences and interrelationships of sardine and anchovy larvae in the

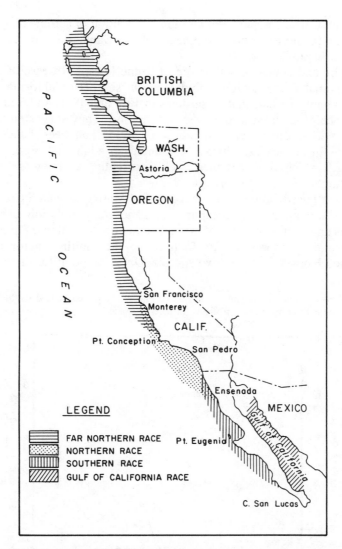

Figure 9 Diagramatic representation of four nonintermingling or partially intermingling stocks of Pacific sardines. The three stocks from the lower latitudes were delineated using arithrocyte antigens (Sprague and Vrooman,[22] Vrooman[23]); the far northern stock is suggested from studies of age and growth (Felin[25]) and population dynamics (Radovich[26]). Although the ranges are shown as overlapping, evidence suggests that the adjacent stocks did not generally occupy the same area at the same time. All stocks tended to range farther south during winters of cold years and farther north during summers of warm years.

California Current Region,[29,30] and the distribution of sardine and anchovy scales in the anaerobic sediments of the Santa Barbara Basin (off
Santa Barbara, California).[31]

Murphy and Isaacs in a 1964 report to the Marine Research Committee,[32] estimated the anchovy abundance in southern California at that
time at about one-half that of sardines in 1940 and 1941, and 6 times the
abundance of sardines in the 1950–1957 period. They suggested that the
decrease in sardines between the two periods had been balanced by
increases in anchovies. Murphy presented an additional report at that
meeting,[33] which also contained a table presenting anchovy and sardine
larval catches from 1951 through 1959.

Silliman[34] used data presented at the 1964 Marine Research Committee
meeting and an analog computer to generate population curves of
sardines and anchovies from Volterra competition equations (Figure 10).*
He assumed competition for food to be the limiting factor for the
combined biomasses of the two species. Only one point for the anchovy

*The Volterra competition equations are based on the logistic curve and mathematically
describe competition between organisms for food or space.

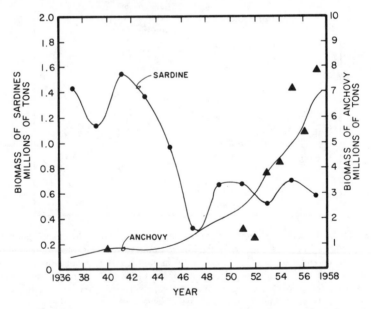

Figure 10 Annual biomass from 1936 to 1958 of the Pacific sardine and northern anchovy.
The anchovy curve was generated by analog computer from Volterra competition equations
and the sardine biomass was simulated using an analog computer method. After Silliman;[34]
data from Murphy[33] and Murphy and Isaacs.[32]

biomass prior to 1951 was used in this simulation, principally because the ichthyoplankton surveys of the CalCOFI program had not been fully implemented until 1951. The earlier point was in 1940 and resulted from the numbers of anchovy larvae taken in surveys made by Department of Fish and Game personnel then. Silliman's simulated curves have been cited in the literature as examples of competition and as substantiation of the Volterra competition equations.[35]

Smith indicated that the early cruises in 1940 and 1941 were conducted during the sardine spawning season and excluded an important portion of the anchovy spawning season.[36] In addition, the cruises only sampled 20% of the area that was later surveyed routinely. He derived total larva estimates for 1940 and 1941 for sardines and anchovies by comparing the 1940 and 1941 values with data obtained from analogous cruises in 1951 to 1960, conducted in the same season and covering the same area. His results (Table 2 and Figure 11) show his anchovy biomass values for 1940 and 1941 to be an order of magnitude higher than the value Silliman used. Smith concluded that both the anchovy and sardine populations declined between 1941 and 1951 and subsequently the anchovy population increased to over 5 million tons between 1962 and 1966. Smith's interpretation is the one commonly held at the present time by scientists working in the CalCOFI program, and is in direct contrast to the interpretation advanced by Silliman.

Murphy attributed the increase in the anchovy population to its use of the void left by the disappearance of the sardine.[27] He hypothesizes that food was the major resource for which the two species were competing and, in fact, this assumption was the basis for Silliman's simulation.

It was demonstrated by Soutar and Isaacs that the occurrence of sardine and anchovy scales are aggregated throughout the 1,850 year record in core samples of sediments from the anaerobic Santa Barbara Basin, and that sardine scales have appeared a number of times with a duration of between 20 and 150 years and with periods of absence between occurrences on an average of 80 years.[37] Northern anchovy scales were found to be more abundant throughout the time series. The hypothesis that the Pacific sardine and the northern anchovy are direct competitors is not supported by the less than significant positive correlation between the scale deposition of the two species in the Santa Barbara sediment.[38]

Iles concluded that, because the growth rates of the smaller year-classes were higher, the decline in the sardine population was not due to a reduction in its food supply resulting from environmental changes.[39] He reasoned that the increase in the length of sardines suggests the environment was not saturated with sardines, and hence food was not a

TABLE 2. Sardine and Anchovy Spawner Biomass Estimates by Ratio and Regression Methods

Year	Murphy Sardine Spawner Biomass (×10^6 T)	Regression Sardine Spawner Biomass (×10^6 T)	Sardine Larval Estimate (×10^12)	Anchovy Larval Estimate (×10^12)	Anchovy Sardine Ratio	Ratio Anchovy Spawner Biomass (×10^6 T)	Regression Anchovy Spawner Biomass (×10^6 T)
1940	1,296		1,634[a]	5,943[a]	3.64	2,359	
1941	2,001		2,476[a]	7,104[a]	2.87	2,871	
—							
1950	716		3,343	2,602	0.78	279	637
1951	570	553	2,685	6,504	2.42	690	797
1952	554	542	2,633	8,132	3.09	856	
1953	709	450	2,189 (3,442)[b]	13,632	6.23 (3.96)	2,209 (1,404)	1,335
1954	668	658	3,193	18,533	5.80	1,937	1,816
1955	425	404	1,959	17,100	8.73	1,855	1,676
1956	293	351	1,706	15,215	8.92	1,307	1,491
1957	212	234	1,137	20,040	17.63	1,869	1,964
1958	281	299	1,453	28,272	19.46	2,875	2,771
1959	190	117	570 (922)	23,463	41.16 (25.45)	3,910 (2,418)	2,299
1960		201	975	31,414	32.22		3,079
1961		132	642	32,538	50.68		3,189
1962		151	731	63,758	87.22		6,248
1963		78	379	61,533	162.36		6,030
1964		104	505	52,253	103.47		5,121
1965		226	1,098	79,292	72.21		7,771
1966		151	735				
—				52,200	71.02		5,116
1969		27[c]	132[c]	33,623[c]	254.72[c]		3,293[c]

[a] 1940, 1941—larval estimates seasonally adjusted.
[b] Parenthetic numbers for 1953 and 1959 assume larval numbers biased.
[c] 1969—larval counts 75% complete; adjusted for extra retention of small larvae.
After Smith.[36]

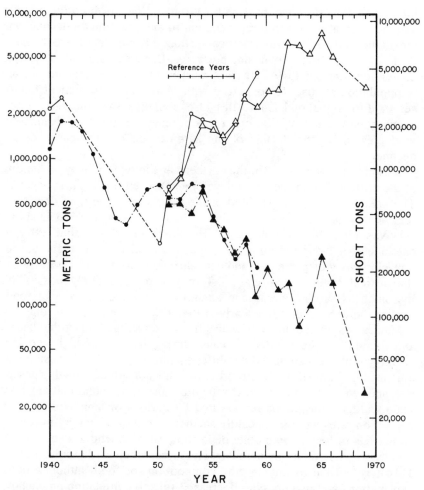

Figure 11 Sardine and anchovy biomass estimates from 1940 through 1969. The solid circle represents sardine biomass calculated from the fishery by Murphy.[27] The solid triangle represents sardine biomass derived from a regression estimate of the relationship between the Murphy biomass estimate and the annual total regional census estimate of sardine larvae during the reference years. The open circle represents the estimates of biomass derived from the ratio of anchovy larvae to sardine larvae and the Murphy sardine biomass estimate from 1940 to 1959. The open triangle represents anchovy biomass derived from a regression estimate of the relationship between the anchovy tonnage calculated from the anchovy to sardine larvae ratio and the annual total regional census estimate of anchovy larvae. Dashed lines represent interpolations between nonadjacent years. After Smith.[36] Compare with Figure 10.

limiting factor. He concurred with Murphy,[27] that fishing rates for the sardine population lowered reproduction to an extent that a decline was inevitable and that it was improbable that the population would have declined in the absence of fishing pressures. Iles disagrees with Murphy's contention that the 1949 year-class marked the most significant change in population status. He concurred with Marr[20] that recruitment failure set in during the mid-1940s. Iles also contends that the rise of the anchovy population off California was in response to the environmental void created by the decline of the sardine and not the cause of the decline.

Lasker has shown that, for the northern anchovy, food may be limiting at the critical time in the larval development when feeding first begins.[40] The presence of the proper food of the right size at the right density in the vicinity of the larva determines whether or not a larval anchovy survives past this critical stage. He has pointed out that extreme patchiness in the distribution of proper food exists in time and space. Upwelling may disperse the food concentrations so that, if it occurred at the critical time, a year-class failure would be a likely possibility. Inasmuch as sardine larvae and anchovy larvae feed on similar food, sardines may be affected similarly at the critical time of first feeding.

MacCall examined anchovy scales from sediments of anaerobic basins and found that the widths of scales during periods of high and low anchovy scale deposition did not differ significantly.[41] This suggested that intraspecific competition for food either did not affect growth rates or was masked by other factors. On the other hand, he found that anchovy scale widths from groups represented by periods of high sardine scale deposition rates were significantly smaller than those from groups based on periods of low sardine scale deposition rates. Anchovies seem to be smaller when sardines are abundant and larger when sardines are scarce. This may be due to interspecific competition for food, although other explanations are also possible. Increased selective predation on anchovies could result in a higher mortality and a smaller average length of the anchovy population. MacCall also pointed out that the record of the past century's abundance of sardine scales does not reveal a period of low abundance comparable to the present one and suggested that the present depletion was, therefore, not a natural one.[41]

Density-Controlling Mechanisms

If food does not appear to be the limiting factor related to poor sardine year-classes, except perhaps at the critical stage of first feeding after the yolk sac has been absorbed, then one should look for other density-controlling mechanisms for sardines and anchovies, such as predation and cannibalism. Hunter has found cannibalism of eggs by anchovies can account for about 50% of the total egg mortality.[42] The percentage would vary depending on the density of anchovies and of other food. Such a relationship would constitute a strong density-limiting force, and may well be the principal interaction between the two species. Sardines eat sardine eggs and larvae and anchovy eggs and larvae.[43]

Radovich suggested that, because man follows aggregations of schools and uses communication techniques to concentrate fishing effort on school groups, each nominal unit of fishing effort expended will take an increasingly larger portion of a declining pelagic fish population.[44-46] The catchability coefficient, then, is a variable function of the population.*

Radovich suggested that behavioral characteristics, such as schooling behavior, which have evolved through natural selection to decrease mortality from predation, may work toward the destruction of the prey species when it is suddenly confronted by a fishery which evolves more rapidly than does the fishes' defense against it.[51]

THE END OF THE MARINE RESEARCH COMMITTEE

With the passage of the Fishery Conservation and Management Act of 1976, the United States established a conservation zone between 3 miles

---

*MacCall used a power function to approximate the catchability coefficient[47]:

$$Q = \alpha N^\beta$$

where Q is analogous to the catchability coefficient, q, N is the mean population size, and α and β are constants.

$$C/f = QN$$

where C is the catch in number and f is the number of nominal effort units. If we assume the two previous equations, it follows that

$$C/f = \alpha N^{\beta+1}$$

At $\beta = 0$, the catchability coefficient is a constant and a linear relationship exists between catch-per-effort and population. At $\beta = -1.0$, C/f is a constant, and at all other values of β, C/f bears a curvilinear relationship to population size.

Fox calculated a β of -0.3 for the Pacific sardine fishery from 1932 to 1954.[48] MacCall estimated a β of -0.724 for the sardine.[47] With a β of these values, if effort is increased beyond a critical point, a population collapse is inevitable (Figure 12) instead of reaching some equilibrium as predicted by Schaefer's model.[49,50]

and 200 miles off the coast within which the United States has managment authority over fishery resources excepting tuna. The original utility to the fishing industry of the Marine Research Committee, that of forestalling managment of the resources, was somewhat removed.

Therefore, at the request of the California fishing industry, at the end of 1978, the Marine Research Committee was dissolved by an act of the California Legislature; however, by mutual agreement, the University of California, the National Marine Fisheries Service and the California Department of Fish and Game are continuing the California Cooperative Oceanic Fisheries Investigations as a viable cooperative research unit, beginning in 1979.

DISCUSSION

From the foregoing examination of only a small portion of the work which has been done on the Pacific sardine and the northern anchovy, it is apparent that most simplified generalizations are probably incorrect.

Any model attempting to describe these populations must be consistent with the results of all the studies on these species. Following is a brief summary of the major points in this paper, all of which must be considered in any modeling attempt.

A model for sardines must account for a population heterogeneity of sardines that does not randomly mix throughout its geographic range. The evidence suggests the Pacific sardine consists of a clinal distribution of intraspecific populations in which there is limited intermingling and a

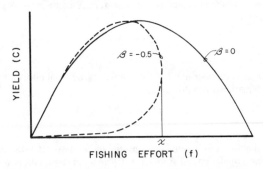

Figure 12 Effect of negative values of β on the equilibrium catch curve. At $\beta = 0$, the catchability coefficient becomes a constant and the usual parabolic relationship depicted by Schaefer[50] holds. If $\beta = -0.5$, as effort is increased above a critical point, x, the yield curve becomes unstable and the population collapses. Such an event appears to have happened with the Pacific sardine. After Fox.[48]

series of variable overlapping coastal migrations of more than one stock.[25] Sardines in the Pacific Northwest were distributed farther north in the summer months when the fishery in that area operated.[52] During the winter months, many of these fish migrated south and were caught off San Francisco as winter fish.[14] Similarly, sardines located off Monterey in the fall supported the winter fishing off southern California. Following the collapse of the fishery off the Pacific Northwest, the winter fishery failed off central California. With the failure of the central California stocks, the southern California stock migrated into Mexican waters and became unavailable soon after the fall season began.[53]

The model should include the higher vulnerability of the northern stocks and the proper sequence of the stocks' decline, with the northern stocks declining first. Wisner was able to find only the southern (central Baja California) "racial" types in the southern California fishery by the period 1950–1959, as indicated by vertebral number.[54]

The model must be consistent with variable and somewhat independent spawning success for the different areas along the coast,[18,21,25] and with different vulnerability of the different stocks resulting from sardine movements from one fishing area to another during each fishing season.

The model must be able to handle major single-season population shifts such as occurred in 1954 and 1958.[21] It must be consistent with higher average spawning success and less variability in large populations, and with a more concentrated inshore distribution of spawn in lower populations.[21]

The model must consider the intraspecific density-dependent relationship that seems to have existed for the various sardine stocks.[26] It could speculate on the effect of sardine population size on anchovy growth rates, but there is no evidence of the effect of anchovy populations on sardine growth rates. It should include cannibalism, as a population limiting mechanism. It should relate ocean temperature to the distribution of spawn success, and should be able to explain the success of the far northern 1939 year-class and its exceptional contribution to the fishery.

The model must consider the effect of a variable catchability coefficient, which increases as the population declines,[47,48,51] or as a population becomes more available to the fishery.[21] It must also consider the effect on the population of any major change in the abundance of a strong predator.[51] It must consider the variability in the temporal and aerial distribution of proper feed in relation to larvae at the time of first feeding and, finally, it must be consistent with results of all the many studies that have been conducted to date. A detailed model should contain a number of generalizations, many of which complement each other, and some of which do not.

I have presented the beginning of a conceptual framework which, I believe, makes a strong case that the present scarcity of sardines throughout their range and their complete absence off the Northwest is not a natural condition but, instead, is an inescapable climax, given the characteristics and the magnitude of the fishery and the behavior and life history of the species.

REFERENCES

1 W. G. Clark, "The lessons of the Peruvian anchoveta fishery," California Cooperative Oceanic Fisheries Investigations Reports, No. 19, 57–63 (1977).

2 C. Smith, *San Diego Union,* January 1979, B-2 (Col. 1), B-5 (Col. 4).

3 W. A. Nierenberg, "Physics and oceanography," Richtmyer Memorial Lecture, New York, January 30, 1979.

4 W. L. Scofield, "Sardine oil and our troubled waters," *California Fish and Game,* 24, 210–223 (1938).

5 A. D. MacCall, "Population estimates for the waning years of the Pacific sardine fishery," California Cooperative Oceanic Fisheries Investigations Reports, No. 20, 72–82 (1979).

6 W. F. Thompson, "The scientific investigation of marine fisheries, as related to the work of the Fish and Game Commission in southern California," *Fish Bulletin,* Calif. Div. Fish Game, No. 2, 1919.

7 O. E. Sette, "The sardine problem in the Monterey Bay district," *California Fish and Game,* 6, 181, 182 (1920).

8 M. B. Schaefer, O. E. Sette, and J. C. Marr, "Growth of Pacific coast pilchard fishery to 1942," U.S. Fish and Wild. Serv., Research Report No. 29, 1951.

9 E. H. Ahlstrom and J. Radovich, "Management of the Pacific Sardine," in N. G. Benson, Ed., *A Century of Fisheries in North America,* Spec. Pub. No. 7, Amer. Fish. Soc., Washington, D.C., 1970, pp. 183–193.

10 N. B. Scofield, "Report of the Bureau of Commercial Fisheries," *Thirty-First Biennial Report for the Years 1928–1930,* Calif. Div. Fish Game, 1931.

11 N. B. Scofield, "Report of the Bureau of Commercial Fisheries," *Thirty-Second Biennial Report for the Years 1930–1932,* Calif. Div. Fish Game, 1932.

12 N. B. Scofield, "Report of the Bureau of Commercial Fisheries," *Thirty-Third Biennial Report for the Years 1932–1934,* Calif. Div. Fish Game, 1934.

13 F. N. Clark, "Can the supply of sardines be maintained in California waters?" *California Fish and Game,* 25, 172–176 (1939).

14 F. N. Clark, "Measures of abundance of the sardine, *Sardinops caerulea,* in California waters," *Fish Bulletin,* Calif. Div. Fish Game, No. 53, 1939.

15 W. L. Scofield, "Marine fisheries dates," unpublished report, on file at California State Fisheries Lab., Long Beach, CA, 1957.

16 F. N. Clark and J. C. Marr, Population dynamics of the Pacific sardine, Progress Report, Calif. Coop. Ocean. Invest., 1 July 1953–30 March 1955.

17 California Cooperative Sardine Research Program, Introduction, Progress Report, 1950.

18 J. Radovich, "Relationship of some marine organisms of the Northeast Pacific to

water temperatures particularly during 1957 through 1959," *Fish Bulletin*, Calif. Dept. Fish Game, No. 112, 1961.

19 H. C. Davis, "World sardine conference report No. 2—does fishing really affect abundance of sardines?" *Pacific Fisherman* (1959).

20 J. C. Marr, "The causes of major variations in the catch of the Pacific sardine, *Sardinops caerulea* (Girard)," World Scientific Meeting on the Biology of Sardines and Related Species, Proceedings, 3, 667–791, FAO, Rome, 1960.

21 J. Radovich, "Some causes of fluctuations in catches of the Pacific sardine *(Sardinops caerulea,* Girard)," World Scientific Meeting on the Biology of Sardines and Related Species, Proceedings, 3, 1081–1093, FAO, Rome, 1960.

22 L. M. Sprague and A. M. Vrooman, "A racial analysis of the Pacific sardine *(Sardinops caerulea)*, based on studies of erythrocyte antigens," *Annals of the New York Academy of Science,* 97, 131–138 (1962).

23 A. M. Vrooman, "Serologically differentiated subpopulations of the Pacific sardine, *Sardinops caerulea,*" *Journal of Fisheries Research Board of Canada,* 21, 691–701 (1964).

24 F. N. Clark, "Analysis of populations of the Pacific sardine on the basis of vertebral counts," *Fish Bulletin,* Calif. Div. Fish Game, No. 54, 1947.

25 F. E. Felin, "Population heterogeneity in the Pacific pilchard," *Fishery Bulletin, U.S. Fish and Wild. Serv.,* 54, 201–225 (1954).

26 J. Radovich, "Effects of sardine spawning stock size and environment on year-class production," *California Fish and Game,* 48, 123–140 (1962).

27 G. I. Murphy, "Population biology of the Pacific sardine *(Sardinops caerulea),*" *Proceedings of the California Academy of Sciences,* 34, 1–84 (1966).

28 J. L. Baxter, J. D. Isaacs, A. R. Longhurst, and P. M. Roedel, "Partial review of and proposed program for research toward utilization of the California current fishery resources," California Cooperative Oceanic Fisheries Investigations Reports, No. 12, 5–9 (1968).

29 J. D. Isaacs, "Larval sardine and anchovy interrelationships," California Cooperative Oceanic Fisheries Investigations Reports, No. 10, 102–140 (1965).

30 E. H. Ahlstrom, "Co-occurrences of sardine and anchovy larvae in the California current region off California and Baja California," California Cooperative Oceanic Fisheries Investigations Reports, No. 11, 117–139 (1967).

31 A. Soutar, "The accumulation of fish debris in certain California coastal sediments," California Cooperative Oceanic Fisheries Investigations Reports, No. 11, 136–139 (1967).

32 G. I. Murphy and J. D. Isaacs, "Species replacement in marine ecosystems with reference to the California current," *Minutes of Meeting Marine Research Committee,* 7, 1–8, March 6, 1964.

33 G. I. Murphy, "Notes on the harvest of anchovies," *Minutes of Meeting Marine Research Committee,* 11, 1–6, March 6, 1964.

34 R. P. Silliman, "Population models and test populations as research tools," *BioScience,* 19, 524–528 (1969).

35 E. P. Odum, *Fundamentals of Ecology,* 3rd ed., W. B. Saunders, Philadelphia, 1971, p. 226.

36 P. E. Smith, "The increase in spawning biomass of northern anchovy, *Engraulis mordax,*" *Fishery Bulletin,* U.S. Nat. Mar. Fish. Serv., 70, 849–874 (1972).

37 A. Soutar and J. D. Isaacs, "History of fish populations inferred from fish scales in anaerobic sediments off California," California Cooperative Oceanic Fisheries Investigations Reports, No. 13, 63–70 (1969).

38 A. Soutar and J. D. Isaacs, "Abundance of pelagic fish during the 19th and 20th centuries as recorded in anaerobic sediment off the Californias," *Fishery Bulletin,* U.S. Nat. Mar. Fish. Serv., **72**, 257–273 (1974).

39 T. D. Iles, "The Interaction of Environment and Parent Stock Size in Determining Recruitment in the Pacific Sardine, as Revealed by Analysis of Density-Dependent O-Group Growth," in B. B. Parrish, Ed., *Fish Stocks and Recruitment,* Rapports Et. Procès-Verbaux Des Réunions, Cons. Inter. Explor. Mer, Vol. 164, 1973, pp. 228–240.

40 R. Lasker, "The relation between oceanographic conditions and larval anchovy food in the California current: Identification of factors contributing to recruitment failure, Rapports Et. Procès-Verbaux Des Réunions, Cons. Inter. Explor. Mer, **173**, 212–230 (1978).

41 A. D. MacCall, "Recent anaerobic varved sediments of the Santa Barbara and Guaymas basins with special reference to occurrence of fish scales," Southwest Fisheries Center, U.S. Nat. Mar. Fish. Serv., La Jolla, CA, Admin. Report No. LJ-79-24 (1979).

42 J. R. Hunter, "Egg cannibalism in the northern anchovy," unpublished manuscript, Southwest Fisheries Center, U.S. Nat. Mar. Fish. Serv., La Jolla, CA.

43 C. H. Hand and L. D. Berner, "Food of the Pacific sardine *(Sardinops caerulea),*" *Fishery Bulletin,* U.S. Fish and Wild. Serv., **60**, 175–184 (1959).

44 J. Radovich, "A challenge of accepted concepts involving catch-per-unit-of-effort data," *Cal-Neva Wildlife Transactions,* 64–67, 1973.

45 J. Radovich, "An Application of Optimum Sustainable Yield Theory and Marine Fisheries," in P. M. Roedel, ed., *Optimum Sustainable Yield as a Concept in Fisheries Management,* Amer. Fish. Soc., Spec. Pub. No. 9, 1975, pp. 21–28.

46 J. Radovich, "Catch-per-unit-of-effort: Fact, fiction or dogma," California Cooperative Oceanic Fisheries Investigations Reports, No. 18, 31–33 (1976).

47 A. D. MacCall, "Density dependence of catchability coefficient in the California Pacific sardine, *Sardinops sagax caerulea,* purse-seine fishery," California Cooperative Oceanic Fisheries Investigations Reports, No. 18, 136–148 (1976).

48 W. W. Fox, Jr., "An overview of production modeling," Southwest Fisheries Center, U.S. Nat. Mar. Fish. Serv., La Jolla, CA, Admin. Report No. 10, 1–27 (1974).

49 M. B. Schaefer, "Some aspects of the dynamics of populations important to the management of the commercial marine fisheries," *Inter-American Tropical Tuna Commission Bulletin,* **1**, 26–56 (1954).

50 M. B. Schaefer, "A study of the dynamics of the fishery for yellowfin tuna in the eastern tropical Pacific Ocean," *Inter-American Tropical Tuna Commission Bulletin,* **2**, 247–285 (1957).

51 J. Radovich, "Managing Pelagic Schooling Prey Species," in H. Clepper, Ed., *Predator-Prey Systems in Fisheries Management,* Sport. Fish. Inst., Washington, D.C., 1979, pp. 365–375.

52 F. E. Felin and J. B. Phillips, "Age and length composition of the sardine catch off the Pacific coast of the United States and Canada, 1941–42 through 1946–47," *Fish Bulletin,* Calif. Div. Fish Game, No. 69, 1948.

53 California Cooperative Oceanic Fisheries Investigations, *The Sardine Fishery in 1956–57,* Progress Report, 1958.

54 R. L. Wisner, "Evidence of a northward movement of stocks of the Pacific sardine based on the number of vertebrae," California Cooperative Oceanic Fisheries Investigations Reports, **8**, 75–82 (1961).

CHAPTER 6

Hidden Elements in
The Development and Implementation of
Marine Resource Conservation Policy

The Case of the South West Africa/Namibian Fisheries

DAVID CRAM

Editor's Abstract

This chapter presents the history of the development of the South West African/Namibian pilchard fishery. The author discusses the complex processes involved in the management of that fishery resource. He draws attention to administrative aspects such as the structural deficiencies of the Fisheries Advisory Council in which the scientific representation would be constantly overshadowed by several political and economic representatives. He refers to other alleged illegal practices such as dumping, scale manipulation, and mislabeling kinds of fish caught. The author notes that the principal problems of the fisheries began with increases in quotas and concessions to the fishing industry (such as allowing South African factory ships to operate off the South West African coast).

He concludes that "despite strenuous work and scientific initiatives, the hidden political, administrative, legal, and economic factors outweighed scientific arguments and led to the decline of the stock and the fishery."

INTRODUCTION

Scientific research into fish populations and their environments can produce valid comments on the behavior of fisheries. Furthermore, this research seemingly can be used to frame a management policy for the regulation of a fishery. As a consequence, however, when the fishery suffers a decline due either to a response to fishing or to environmental change (or a combination of both), it is frequently the scientific policy that is regarded as being faulty. The process of resource management is complex, involving elements other than fisheries research, and it is the objective of this chapter to use the example of a collapsed fishery to shed some light upon these "hidden" elements.

In a previous review of the history and management of Southeast Atlantic fisheries,[1] it was noted that the pilchard fishery based mostly on Walvis Bay (South West Africa/Namibia) had been tightly controlled since 1951 and was effectively studied between 1970 and 1974. Despite this strong background of legislation and research, the pilchard population did not ultimately benefit from the conservation measures enforced following the decline from 1968 to 1970 (Figure 1). The recovery experienced between 1971 and 1974 ended in a sudden collapse between 1975 and 1976. What went wrong? Or, to phrase the question more precisely, what elements of the conservation policy were missing?

There are two good reviews of the pilchard fishery, both giving purely scientific explanations for the 1975–1976 decline.[2,3] However, the management process involved in the decline of the fishery was not only influenced by science, but by powerful economic, administrative, legal, and political pressures as well. The formulation of conservation policy was always rendered particularly difficult by the conflicts of interest involved. It will be shown that these pressures tended to nullify the value of scientific research results.

The moral of this chapter is that success in the management of living resources is much more likely with a well-coordinated approach than without; for in the final analysis it was the lack of a well-coordinated

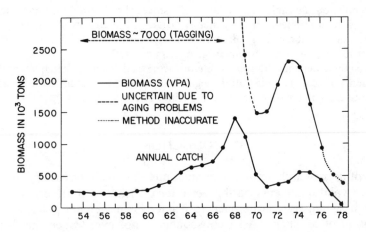

Figure 1   Annual catch 1953–1978 and biomass 1970–1976.[3]

scientific, economic, administrative, legal and political policy that brought about the latest decline in the pilchard population.

MANAGEMENT OF PELAGIC RESOURCES

The Sea Fisheries Branch of the South African Department of Industries has the statutory responsibility for conducting research into marine fisheries with the objective of presenting scientific information to the Department, which coordinates this information for submission to the Minister of Economic Affairs. The Minister has control of all aspects of the fishery through the wide-ranging Sea Fisheries Act (1973), which allows him to regulate any aspect of the industry by proclamation.

In addition, other legislation created the South West Africa Fisheries Advisory Council, a body with powers to make recommendations on the management of resources to the Minister directly. Composed mainly of industry representatives, the council met at least once per year to set quotas and settle other business before the opening of the fishing seasons. After 1970, it became the habit of the Branch to brief comprehensively the Council on the state of the stocks and to suggest courses of action.

Following the proclamation of regulations, it becomes the responsibility of the Branch to enforce them by means of its Inspectorate. The Sea Fisheries Act (1973) requires, for example, that all fish landed be weighed on a specified type of scale and that samples be taken from each catch for species identification to enforce quota regulations.

This, then, is the cycle of events: research accomplished by the Branch; recommendations made to the Department of Industries; recommendations incorporated with those from other sources into policy proposals for the Minister. In addition, a separate line of recommendations comes from the Advisory Council. After the Minister's decision, the regulations are promulgated in the Government Gazette, and the Branch comes into the picture again with the task of enforcement. It is within this chain that the economic, administrative, legal, and political pressures influenced the acceptance of scientific advice.

POPULATION HISTORY

There is no complete time-series of reliable data on the size of the pilchard population since the beginning of the fishery in 1947.[1] The longest time-series is that of the catch-per-unit-effort (CPUE) index of abundance, "tons landed/fleet hold capacity," which is available for 1957 to 1970. Unfortunately, it is a crude measure which underestimates effort, because it does not include increasing sophistication in equipment and methods; at least it has value as a minimum condition.

The annual "tons landed/fleet hold capacity" value is influenced by that year's quota. As surplus effort was available, the "tons landed/fleet hold capacity" value rose as quotas were progressively raised. However, after 1963 this CPUE index fell sharply, indicating that the increased quotas were no longer the main factor determining the size of the catch. A small temporary increase in CPUE from 1967–1968 was due to the effect of the very large quota increases in those years. The population was probably stable from 1957–1962 but rapidly declined thereafter (Figure 2). With the introduction of very restricted individual vessel quotas in 1971, this index of abundance ceased to have much meaning. Another catch-per-unit-effort index of abundance, "tons landed/fuel issued," shows that the decline continued until 1971, then recovered until 1973, but afterward declined until 1978.

Virtual population analysis (VPA) is a method of calculating population size from the total number of fish in each age class caught each year, a value for the coefficient of natural mortality, and an independently estimated value of fishing mortality for 1 year. It is unfortunate that age data collected before 1970 are inadequate, and prevent the successful application of this technique.[3] From 1970 to 1973, virtual population analysis showed that the population increased in size from 1.5 million

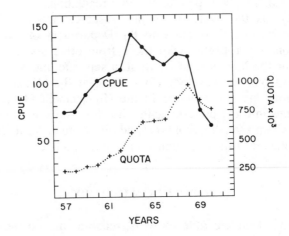

Figure 2 Catch-per-unit-effort (CPUE) and quotas 1957–1970. The CPUE index shown here is "tons of fish landed/gross fleet tonnage."

tons to 2.4 million tons, due mostly to the good 1972 year-class, and thereafter declined well below the 1.0 million ton mark (Figure 3).

The history of the population can be summarized as one of stability in the 1950s and early 1960s while quotas were low, followed by an accelerating decline as quotas were raised. During the 1960s, the catch represented an "eating away" of the resource rather than a harvest of the stock's production and, as such, left the reserves depressed from the level required to produce the maximum sustainable yield (MSY) and left the industry with an inflated idea of the stock's potential harvest. The process was halted in 1971 with a very reduced quota assisting a temporary recovery, before the decline continued in 1975. The 1972 year-class was a mixed blessing, for without it there would have been no recovery, but its presence added to the false optimism over the population's strength generated by the large catches prior to 1970.

One would imagine that such perturbations in a fishery would be noticed, and that remedial measures would be applied without delay. A recent review of a number of pelagic fisheries contained an interesting overview of this fishery which implied that its management was goal-

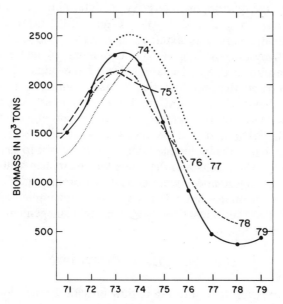

Figure 3 Virtual population analysis biomass estimates from 1971–1979 and projection as calculable at the start of successive years from 1974–1979.[3]

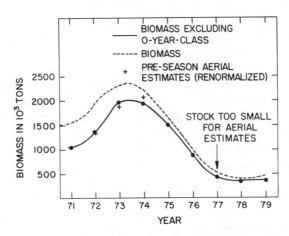

Figure 4 Comparison of virtual population analysis of the pilchard stock (excluding the 0-year-class) with pre-season aerial/acoustic survey estimates (renormalized to the VPA average over the 1972–1974 period).[3]

oriented from the start.[2] Unfortunately, this was not the case: until 1970 the management system was conservative but almost purely arbitrary, as it lacked the scientific input necessary to determine the effects of any action (see reference 1, pp. 36–38). From 1970 onward, a complex scientific infrastructure was available that allowed assessments of the stock to be accomplished prior to the opening of successive fishing seasons. Even so, a detailed examination of the pilchard fishery from 1971–1978 showed that catch-statistics-based indicators (VPA, CPUE, adult modal length) only confirmed the severe stock decline from 1975–1976 with absolute certainty by the beginning of 1978. Some preliminary indications of the decline were visible in 1976, but by that time the stock had already declined considerably. Prior to the 1972, 1973, 1974, and 1977 fishing seasons, the stock was examined with substantial surveys whose performances were considered promising.[3] It is therefore particularly unfortunate that no surveys were accomplished during the critical 1975–1976 period, when their input to management would have been invaluable (Figure 4).

Growth and Decline 1947–1970

There is not much to add to the review previously mentioned,[1] except to elaborate upon the way in which the fishery became unstable in the early 1960s. Despite the absence of a scientific assessment of the stock, it is recorded that the South West African Administration "associated itself gradually with the line of thought" that greater catches of pilchard could be made.[4] Quotas were increased and more effective fishing gear introduced. In 1965, the South African Government granted concessions for two factory ships to operate outside the 12-mile limit. It is also recorded that these concessions were "inordinately generous" and "set in motion an unhealthy trend in the fishery."[4] In 1966, the first factory ship operated off South West Africa and was joined in 1967 by the second. It was frequently alleged that the catching vessels fished inside the 12-mile limit on the stocks exploited by the land-based fleet. Patrols were not extensive, but at least one arrest was made and conviction secured.[5]

These concessions were made against strenuous opposition from the South West Africa Administration, which knew that, because of the presence of these vessels, they would be unable to resist an argument for more quota based upon the premise "if we don't catch the fish, they will." These quotas were granted (as well as the eventually extremely useful "research quota") before the factories were withdrawn in 1970 and 1971, respectively. But by then the damage had been done.

It is possible that the basis for the decision to overrule the South West Africa Administration's objections and to grant the concessions lay in a rapid decline of the South African pilchard fishery after a peak catch in 1963. This industry must have needed a substitute fishery, and the seemingly stable expansion of catches in South West Africa would have been attractive. The way in which the concessions were granted has been cautiously criticized by the Commission of Enquiry[4] and more openly in Parliament, where the cooperation between fishing industry personnel and Government was criticized as corrupt. It has been alleged that members of Government, the governing Party, and public officials arranged for concessions to be awarded to companies in which they had financial interests or from which they stood to gain financially.[6]

If the above interpretation is correct, the additional concessions and quota increases originated in political pressures that overcame vigorous opposition, and that were not supported by any scientific assessment of the ability of the stock to sustain those extra catches.

The Situation 1970–1978

In retrospect, it is not easy to get into perspective all the reasons for the decline after 1974. One review correctly notes that the management methods were empirical, but did not stress the importance of the surveys conducted prior to the 1971, 1972, 1973, 1974, and 1977 seasons.[2] It was realized at the time that short-term prognoses depended largely upon a knowledge of the current biomass, and once the surveys showed that the 1968–1970 decline had halted, the management policy was clear. In the absence of an MSY value, quotas would be allowed to rise by small annual increments as long as annual surveys showed the stock to be increasing substantially in size.[7] It has been pointed out that the yearly quotas were too large a proportion of current biomass to allow recovery of the stock.[3] The empirical approach to quota setting would have led to a reduction in this proportion, had the surveys continued through 1975 and 1976.

It was unfortunate that a reliable time-series of catch statistics did not exist in the early 1970s, and it was also unfortunate that doubt existed on the veracity of those that were collected. Population studies were always hampered by the difficulties of determining the age of the pilchard and of interpreting CPUE data in the multispecies fishery, where the distribution of fish and information exchange by radio made it difficult to distinguish reliably the effects of pilchard availability and abundance. In any case, it has been demonstrated that catch-statistics-based indicators of abundance did not succeed in identifying the 1975–1976 decline until too late.

Why then was the survey series discontinued in 1974? Why, in contradiction to stated management policy, were quotas raised, when there was no annual survey with which to assess biomass changes? What were the pressures on quotas? Why were catch statistics considered to be only partly reliable? How was the stock managed after 1974? How, in spite of everything, did the industry remain intact?

ADMINISTRATION

Administration and Research

Up to 1974, a comprehensive assessment of the stocks and a description of the research program were prepared for the Advisory Council prior to the season, with a version published in a local fisheries journal.[7–9] A large new program such as this required personnel, but the Department and its Branch did not have the administrative flexibility to adjust staffing to suit requirements. While the crisis was visible, an irregular arrangement of staff was tolerated. Once the crisis was believed to have passed, the administrative ad hoc measures were progressively removed until, in 1974, the staff available to maintain a satisfactory survey program became insufficient.

This inflexibility with personnel seems even more extraordinary when viewed against the great flexibility in funding. Due to the foresight of the

South West Africa Administration, a research fund had been established through a levy of $9 per ton (in addition to the normal 10 cent/ton) upon the special research quota awarded in 1967. This fund had grown by 1971 to more than $1 million, and could be used at Departmental discretion with exemption from certain Treasury and Tender Board regulations. This unparalleled flexibility provided the impetus for the program, but the monies could not be used to pay staff.

These personnel problems were quite visible, and there was criticism of "ill-informed" administrators making decisions without scientific input.[10] Criticism became increasingly sharp as time passed, particularly when research was directly linked to the future of an industry that supported 24,000 people.[11,12]

Social and Political Aspects

These problems were partly created when the Department took over the South West Africa Administration Marine Research Laboratory in 1969. It was a condition that the laboratory in Walvis Bay be maintained, but this town was unattractive to many people and the rate of staff loss was unacceptably high. In 1974, moves were initiated to transfer staff to Cape Town, but this was vigorously opposed by the Walvis Bay Municipality on local social and political grounds, and it was only in late 1977 that the consolidation in Cape Town finally occurred.[13] In parallel, some reorganization occurred at the Branch in late 1976 which allowed the aerial/acoustic survey program to begin again in early 1977, prior to the 1977 fishing season, but it was already too late.[14]

Administration and Management

One of the problems involved in arriving at a pre-season decision on management methods was the way in which policy was formulated during the South West Africa Fisheries Advisory Council meetings. First, the representation was unbalanced in favor of the industry with only one scientist, and one representative each from the Department of Industries, Fishermen's Association, Private Boat Owners' Association, a central marketing organization (observer status), and three representatives of the factory's senior management. The Chairman was the Secretary for Industries. Second, the Advisory Council did not meet sufficiently often, with the result that when they did meet, some confrontation was inevitable.

The decisions on policy were developed by consensus, which meant that concessions had to be made. Negotiating positions were established beforehand, but here scientists were at a disadvantage: they could not add on a factor to their stock assessments that could be "negotiated away" to arrive at their real objective. This was partly solved by extensive lobbying of industry leaders and members of other groups, to try to ensure that the audience at the Advisory Council would be receptive to the scientific arguments. The process worked well enough while the assessments were favorable to the industry, but when they were not (or they were incomplete) the lack of scientific influence at the meetings was very noticeable.

Another substantial problem was the minimal effectiveness of the Boat Limitation Committee in limiting the effort of the fleet. Although boat limitation (i.e., limited entry) was established in 1952, despite some stability in fleet size, total effort had steadily increased since then, possibly more slowly than it would otherwise have done. Between 1957 and 1970 the fleet hold capacity trebled, while numerous innovations occurred, such as the use of small mesh nets, sonar, fish pumps, and power-blocks. Between 1970 and 1975, the capacity increased by about 50% (9,263 to 14,280 tons),[15] despite contrary advice from the Branch and the quota-enforced reductions in total landings. Only in 1976 did the industry publicly admit that the fleet was underutilized. At that time, a 50% increase in landings was possible because of excess fleet capacity,[16] but prevented by quotas.

ECONOMICS

The major conservation measure imposed in 1970 was the splitting of quotas between pilchard and "other species," with the provision that the season closed when the pilchard portion of the quota was filled, regardless of the state of the "other species" quota. The ratio was initially 33⅓ : 66⅔ (pilchard : "other"), increasing ultimately to 60 : 40 before the final decline. The immediate effect of the imposition of "split quotas" was to reduce the 1971 landings to about 50% of the 1970 level.

To cope with this spare capacity (as well as the steadily increasing fleet hold capacity), the factories introduced a "tally" system of trip quotas on pilchard, designed to spread pilchard catches throughout the season, thus maximizing landings of other species. The tally system was also planned to eliminate gluts resulting in the conversion of potentially cannable fish into fish meal. Because of the infrequently reviewed fixed wet-fish price, the split quota and the tally system, the income of the fishermen and the private boat owners (about 50% of the fleet) fell in proportion to the quota and was spread over the whole season.

Although the fishing community was very cohesive, threats of strikes became common in support of demands for increased earnings.[17,18] Tallies worked against the interests of the more efficient vessels, particularly when other species were scarce.

The canneries also had problems with the reduced pilchard quotas from 1970 onward, and they were forced to increase efficiency. This was done by eliminating waste in the canneries and by introducing the tally system. The reduced quota, the provisions of the split quota, and consequent reduction in incomes had a series of unfortunate side effects.

Legal Matters

The "dumping" of unwanted catches is illegal, but appears to have been common in the past. The Commission of Enquiry into the Fishing Industry noted evidence that 10–50% of the overall catch was dumped,[4] and it has been alleged that dumping was common in the 1970s. Because of the trip tallies for pilchard, the fishermen strove to increase their income by concentrating on catching other species, chiefly anchovy. However, it is often difficult to distinguish anchovy from pilchard prior to catching, and it was widely alleged that any pilchard catches in excess of the tally were dumped.

Dumping had to be stopped, and it was thought early on that better identification of shoals before catching would help; the use of fishspotting aircraft was recommended.[8] Persuasion was tried before the 1973 and 1974 seasons, because dumping affected the quality of management advice by biasing the data, thus working against the best interests of the industry.[8,9] In 1974 it was realized that the "effectiveness of the split quota had been reduced over the years by illegal practices" and that a different system had to be devised that permitted single-species management.[7] This system would have been based on effort limitation rather than quotas, but was not devised because of the disruption of the research team mentioned above.

Prior to the 1977 season, allegations were made of dumping in 1976 when many sectors of the industry were worried about the future. Dumping has never been observed by authorities, nor have sworn affidavits been received from first-hand observers, but allegations were continuous and usually made in private for fear of repercussions.

The same Commission noted that the landing statistics provided by factories were always less than the amount of fish supplied.[4] Many factories already had automatic scales, but they were made compulsory before the 1971 season. A considerable expansion of the Inspectorate was made to supervise landings by recording scale readings before and after off loading and to take samples of the catch for species identification (cf., split quota regulations). Almost immediately, allegations occurred that pilchards were being "misidentified on landing" and recorded as other species to defeat quota limits. Because of the requirements of inspection, these allegations implied inefficiency or corruption on the part of the Inspectorate. Knowledge of this alleged practice was so widespread that efforts were made to stop it during each season,[8,9] but by 1974 it was plain that persuasion had failed and an alternate strategy would be required.[7]

In addition, there were allegations of abuse of the automatic scales and, at least until 1974–1975, it was apparently not appreciated by authorities that the scales could be made to record almost any value desired while operating at maximum throughput. The scales were assized once per year by a contractor, but the potential for underrecording lay in manipulating the response-time of the scale or temporarily applying a counterweight within the mechanism, rather than by affecting its absolute accuracy. As the scale printed "before and after" totals on a slip, any substantial underrecording could only occur with the agreement of the fishermen concerned.

The seriousness of these allegations was such that official investigations were made that demonstrated the potential for irregularities, but did not confirm any instance of corruption. Nevertheless, in 1976 all factories were required to totally enclose their scale houses (including the roof and floor), leaving only a sealed window to observe the scale and a locked and sealed door for access. The seal was to be broken only by an inspector, when the factory workers required access.[19]

The occurrence or extent of these alleged irregularities is difficult to demonstrate, and one can only look for indirect confirmation of their existence. Prior to 1970, it was estimated that 10–50% of the total catch was dumped. Perhaps stung by suggestions that landings up to 100% in excess of the quota were possible, one commentator obliged by stating that "a more realistic but still doubtful assessment (of underrecording at the scales) was 20%."[19] In addition, there were "hints that some pilchards were incorrectly classified to conserve the more valuable quota" (i.e., pilchards).[19] Reversed thinking, one suspects.

Another pointer, which can only be regarded as tenuous, is the record of production. From 1970 to 1976 fish meal production was approximately constant, but, for the same landings, the 1975 cannery production was 300% higher than in 1970. The proportion of fish canned is reported to have risen from about 25% to roughly 45% but mostly with the introduction of refrigerated seawater vessels after 1976.[17] Whether this apparent inconsistently high cannery production reflected undeclared pilchard landings is unknown.

Economics and Business

Why was there such pressure on the quota? The Government was committed to the conservation of the cannery industry because of its value[8] and, before the 1973 season, the choice seemed clear: "the industry had to decide what it wanted: restoration, or the stagnation of a reduced and fragile pilchard stock. It could not have increased quotas and [long term] growth."[9]

Until 1974, the Branch was able to hold quotas at approximately the desired levels because of the comprehensive survey data available, and the general agreement on increasing abundance. After 1974, when no more surveys were accomplished, the industry based its proposals for increases on unrealistic appraisals of the situation. It was apparently assumed that the population was increasing at the 1972–1974 rate: the future was regarded as "assured" in 1976.[19] In 1977, the industry was unwilling to accept that overexploitation had (again) occurred, and was suggesting that quotas could be raised after 1979.[17] It was predicted in 1978 that cannery production would rise in 1979 to double the 1975–1976 level. There were arguments of increased earnings based upon these predictions because export earnings increased by 89% from 1973 to 1974,[20] and canned fish exports increased again in 1975.[21] Extrapolations were highly appealing. The companies had to maintain growth and generally they did so: profits were generally maintained until 1977, with a drop experienced in 1978.

The repeated scientific view was for caution, especially after the surveys were stopped. Indeed, it was observed in 1976 that while fishermen were aware of the fast-dropping CPUE from daily realities, from the factory side things still "looked good."[19] The high catches of the middle sixties, while the stock was being invisibly but rapidly decreased in size, had produced a deep-seated optimism, which now prevailed again. Contrary to the stated policy, quota increases were demanded, and awarded, in the absence of data showing that the stock was still substantially increasing, due to the pressures caused by these pseudo-economic appraisals.

Fishing is regarded as a high-risk investment in South Africa, and these incorrect optimistic analyses should have increased that risk. In fact, they did not, because the optimism was long term and the risks were possibly minimized in the short term by landings in excess of the quota through the alleged irregularities at the scales and the alleged misidentification of species on landing. The long-term risks were also minimized by most companies through diversification out of fishing.

The proportion of after-tax profits generated by the whole industry in activities other than fishing in Southern African waters rose slowly from approximately 25 to 32% between 1975 and 1978.[20-24] Also, following the example of associated South African companies (but ignoring the fate of the South African pilchard) the multispecies nature of the fishery was seen as a safety net. When one species was no longer "available," another "would take its place." This did occur to an extent, although doubt has been expressed over the correctness of the tonnages reported for "other species" landings.

Economics and Resource Utilization

The wet-fish purse-seine industry developed on fish caught close to the canneries at Walvis Bay. When the search area began to expand rapidly after 1964, pilchards would often be delivered in an unsuitable condition for canning. Because of the persistently over-optimistic outlook of management, the observed need for preserving pilchards at sea was not met until early 1977 by the charter of a foreign purse-seiner with a refrigerated seawater system (RSW).[22] The advantages were immediately obvious and six local vessels were fitted out for ice-chilled seawater (CSW). Cannery ratios were reported to increase from 25 to 42%.[17] Ultimately, nine vessels out of a fleet of 72 were fitted out for refrigerated seawater and twelve for ice-chilled seawater.[25] A significant later development was the charter of a 600-ton RSW carrier vessel.[26]

In 1978, the maximum rate of utilization of raw fish by the canneries was 52%, with the average from RSW being 48% and from CSW slightly lower. In comparison with revenue gained, the costs of conversion were reasonable: up to $400,000 for converting steel boats to RSW (when the vessel was lengthened at the same time); up to $12,000 for converting wooden and glass-reinforced polyester boats for CSW.[27] In 1978, the pilchard catches were nearly all made with RSW and CSW at very great distances from the canneries, up to 36 hours away.

THE FINAL PHASE

The 1976 season started uneasily: fish were scarce. The quota was initially 560,000 tons, but the lack of availability of fish led to a "fishermen's strike" and an Advisory Council meeting, after which the quota was further reduced to 475,000 tons. In mid-1976, the lack of pilchard was apparent, and a survey organized by the industry failed to locate commercial quantities of fish. The quota remained at 475,000 tons, even though it was not quite filled.

Prior to the 1977 season, a large aerial/acoustic survey was mounted but too few pilchard shoals were located to use the method.[13] The pilchard quota was reduced to 250,000 tons for the season, and later reduced to 200,000 tons. By 1978 the collapse was quite clear and an initial quota of 125,000 tons was awarded, but the fishery was closed with pilchard landings at 46,000 tons. For the first time, quotas were submitted to the International Commission for Southeast Atlantic Fisheries (ICSEAF), with foreign quotas being insignificant.[28]

In January 1978 the control of the fishery was taken over by the new Administration of South West Africa/Namibia, which repealed the Sea Fisheries Act (1973), established its own Fisheries Advisory Council, and then ceded its powers back to the South African Department of Industries.[29] It is of interest that the new Advisory Council has the following representatives: one from fisheries science, eight from the fishing industry, two from the Pilchard Boat Owners Association, two from the Fishermen's Association, the Minister of Works from Kavango, a university economist, an official from the South African Bureau of Standards, four from the general business sector and one other scientist.

The following comment is from the annual report for 1977 of the SA Fisheries Development Corporation: "The crises (in a number of fisheries including SWA) underlined the necessity for decisive action, not only to contain foreign fleets hungry for fish, but to discipline domestic fishermen and industrialists insensitive to the necessity to conserve for the morrow. The high capital outlay involved in modern vessels evokes in fleet owners a desperation to keep fishing which often appears to cloud their judgment and results in unhealthy maneuvering to achieve their purpose."[30]

LESSONS FROM THE SOUTH WEST AFRICAN FISHERY

Catch-statistics-based methods of detecting biomass changes did not detect adverse change, with certainty, until too late. Part of the problem with fishery-dependent methods was the doubtful veracity of the data and their time-consuming nature: enormous backlogs of unprocessed data can build up, drastically reducing their value. Particularly damaging was the backlog of age material and the problems of correct interpretation. Such backlogs must be avoided, and the judicious allocation of contracts to outside organizations seems to help. An improvement in the reliability of catch statistics must be achieved, but has proven difficult in practice. Survey methods that are independent of fishery data did detect changes in stock size contemporaneously and showed the greatest promise as a management tool. They should be used whenever possible.

The great flexibility provided by the use of Research Levy Fund monies under the vital exemptions demonstrated what could be achieved by responsible action with the Public (i.e., government) Service. That the survey program ceased despite this was because of a lack of flexibility in personnel management. If similar flexibility could have been introduced into this aspect of administration, then the program could have succeeded. Ideally, fisheries research should be accomplished by a semiautonomous state body, by its nature free of the administrative burden of Public Service methods, and able to react on time-scales relevant to the management of very short-lived resources. This was one of the main recommendations of the Commission of Enquiry in 1970.[4]

The role of the Advisory Councils in formulating policy is valuable, and they can be useful forums, but they must have balanced representation including "protectionists" (usually of fish predators), resource users, and with many more than one fisheries scientist occupying a middle ground. A consequence of the Public Service jealously guarding its position in fisheries research was that there were no other experts who could be invited to participate. Involvement of universities in advanced or academic studies is essential, so that viewpoints independent of the Public Service can be heard. Any success of the Advisory Council resulted from good communications, usually lobbying, when conservation was involved. However, it was not enough, as ultimately there was no counter to the quasieconomic arguments for expansion.

The use of exclusive scientific, administrative, economic, and sociopolitical data sets led to the production of exclusive results that were contradictory when incorporated into policy. Examples are fishermen's earnings versus split-quota versus fixed wet-fish price; increasing cannery exports versus the state of the stock; management information needs versus inflexibility of funding. What is required is a total fisheries information system, incorporating the usual scientific data, as well as the administration, economic, and sociopolitical information essential to answer complex queries on the whole fishery. It is not implied that fisheries scientists should manipulate such a data base by themselves, but that it be genuinely shared between each user, whether scientist, economist, or administrator, so that each may view those sectors of information not normally available to his specialist field. This is an ambitious objective, but one that will be reached as the need is already apparent. Although existing population models are very imprecise and although it would be difficult to incorporate them into more sophisticated economics models, a start could have been made with simpler things like modeling the quota levels against the improvement in cannery ratio expected from RSW and CSW, and the costs of modifications of vessels, to maximize the value of cannery products for minimal quota. Hindsight is no help, but it is fair to comment that such a study would have brought a faster move into RSW/CSW, to the earlier benefit of the fishery.

Understanding the need for such a system is the greatest lesson from the dismal performance of the South West African pelagic fishery.

CONCLUSIONS

The fisheries' principal problems commenced with the granting of additional quotas and concessions under the questionable political circumstances surrounding the activities of the South African factory ships off the South West African coast. Even when they were removed, they left a legacy of increased quotas and too many vessels for the fishery to support. Attempts at conservation through strict quotas were difficult, as the regulations were easy to evade. Reducing catches by effort limitation instead of quotas was strongly resisted, but this does seem to be the only effective way of enforcing a reduction in catch. The decline of the stock was swift from 1968 to 1970.

The stated objective of management after 1970 was to conserve the pilchard and its valuable cannery industry. This policy nearly failed. The industry's economic appreciations and optimistic assessments of the stocks' condition led it to take the long term risk of pressing hard and continuously for increased quotas while making the necessary large investments, and at the same time minimizing those risks in the short term through allegedly illegal practices. Because of overwhelming evidence, the industry seemingly accepted a strict conservation regime between 1970 and 1974, but after 1974 it was difficult to resist quasieconomic arguments and the quota was raised slightly before the collapse began in 1976.

The effect of any alleged illegal fishing mortality through dumping, irregularities at the scales, and incorrect identification on landing, would be to slow down the rate of population increase and speed up the rate of decrease, as well as to introduce bias into conventional methods. The lack of surveys in 1975 and 1976 was possibly *the* decisive factor in the failure to detect the decline at an early date.

When administrative inflexibility made the aerial/acoustic and egg and larvae surveys too difficult to carry out prior to the 1975 and 1976 seasons, there were no effective means with which to counter the pseudoeconomic expansionist arguments put forward by the industry, so the pilchard portion of the split quota rose for a short while before its reduction, too little and too late, in mid-1976. Economic arguments used to support requests for increased quotas were weak. Increased export earnings were important, but should have been made on raw material derived from a reduction in post-harvest losses, rather than an increase in pilchard quota.

One of the risks taken by the industry was that "other species" could be exploited, if the pilchard declined. To a certain extent, landings of anchovy and maasbanker did help the industry through difficult periods, despite their much lower revenue potential as fish meal. The argument for retaining a 50% excess of effort in the fleet was that advantage could be taken of any good anchovy year-classes. It was a forlorn hope in the long run, with the unfortunate short-term effect of having too many vessels chasing too little quota. A fairly quick reduction of the fleet size in the early 1970s, plus a swifter change to RSW/CSW, would have seen the industry on a sounder basis than its own strategy ultimately allowed.

All discussion occurred in an unbalanced form. One scientist in an Advisory Council could not oppose persuasive arguments from different sources simultaneously. Furthermore, the scientist cannot "haggle:" assessments are made and contain no padding. In addition, the structure of the Advisory Council cast the scientist in the role of a protectionist, when his real role is one of promoting rational utilization of the resource. This was unfavorable polarization of viewpoints, particularly after 1974.

The fragmentary approach to the management of the resource, seeing each element of the policy as a nearly exclusive element, was necessary because the diverse scientific, economic, and other information was not integrated into a multidisciplinary data base. Few fisheries, if any, have advanced data base management systems operating now, and those planned may well be primarily for scientific purposes. It must be remembered that scientific information alone lacks essential influence and credibility in the decisionmaking process, and efforts must be made to broaden new data bases into all related disciplines.

The public view of fisheries management is that it is a purely scientific exercise, and that conservation policy is derived in an orderly way on a scientific basis. It has been demonstrated that this was not the case for the South West African pilchard fishery: while argument centered on scientific matters, other elements, normally hidden from view, dominated decisions and events. The result was that, despite strenuous work and scientific initiatives, the hidden political, administrative, legal, and economic factors outweighed scientific argument and led to the decline of the stock and the fishery.

This was not an unqualified disaster: the effect of scientific advice was to create the environment needed by the industry to increase its efficiency, and to provide the time necessary for the introduction of these improved methods. Consequently, the industry has survived, although substantially rationalized, albeit only in bearable economic health but well able to survive the waiting period until the pilchard population increases again or until anchovy and maasbanker make their regular appearance in cans on the world market. In the meantime, South Africa will continue to be an importer of canned fish, instead of an exporter.

REFERENCES

1 D. L. Cram, "Research and management in Southeast Atlantic fisheries," California Co-operative Oceanic Fishery Investigations (CalCOFI) Report 19, October 1977, pp. 33–56.

2 J. P. Troadec, W. G. Clarke, and J. A. Gulland, "A review of some pelagic fisheries in other areas," ICES Symposium, Biological Basis of Pelagic Fish Stock Management, Document 10, 1977.

3 D. S. Butterworth, "The value of catch statistics based management techniques for heavily fished pelagic stocks, with special reference to the recent decline of the SWA pilchard stock," Preprints Vol. 1, NATO Symposium. Operations Research in Fishing, Marine Technology Center, Trondheim, Norway, August 14–17, 1979.

4 Report of the Commission of Enquiry into the Fishing Industry. The Government Printer, Pretoria, South Africa, RP 47/1972, 1971.

5 Anon., "Fished inside 12 mile limit, fined R20,000," *Shipp. News Fish. Ind. Rev. S. Afr.,* **26**(3), 57 (1971).

6 J. P. Wiley, Proceedings of the House of Assembly of South Africa, 15 February, 1971, Vol. 1034.

7 D. L. Cram, "SWA pilchard stock continues to recover," *Shipp. News Fish. Ind. Rev. S. Afr.,* **29**(2), 74–75 (1974).

8 D. L. Cram and G. A. Visser, "Cape Cross Research Program—Phase 2," *Shipp. News Fish. Ind. Rev. S. Afr.,* **27**(2), 20–43 (1972).

9 D. L. Cram and G. A. Visser, "SWA pilchard stocks show first sign of recovery," *Shipp. News Fish. Ind. Rev. S. Afr.,* **28**(3), 56–63 (1973).

10 Editorial, "Untroubled conscience?" *Shipp. News Fish. Ind. Rev. S. Afr.,* **30**(4), 3 (1975).

11 Editorial, "That two-year gap," *Shipp. News Fish. Ind. Rev. S. Afr.,* **32**(5), 37 (1977).

12 Editorial, "The case for Cape Cross," *Shipp. News Fish. Ind. Rev. S. Afr.,* **30**(9), 3 (1975).

13 D. L. Cram, "Comprehensive research into SWA fisheries," *Shipp. News Fish. Ind. Rev. S. Afr.,* **33**(2), 36 (1978).

14 D. L. Cram, "Poor prospects for pilchards: need to diversify SWA fishery" *Shipp. News Fish. Ind. Rev. S. Afr.,* **32**(4), 38 (1977).

15 G. G. Newman, "The living resources of the Southeast Atlantic," FAO Fisheries Technical Paper 176, FAO Rome, 33 (1977).

16 Anon., "Fishing enters era of sophistication," *Shipp. News Fish. Ind. Rev. S. Afr.,* **31**(1), 73 (1976).

17 Anon., "SWA fishery faces uncertain future," *Shipp. News Fish. Ind. Rev. S. Afr.,* **32**(8), 41 (1977).

18 Anon., "Fishermen may strike," *Shipp. News Fish. Ind. Rev. S. Afr.,* **32**(12), 37 (1977).

19 Anon., "Satisfactory season despite poor availability," *Shipp. News Fish. Ind. Rev. S. Afr.,* **31**(9), 44 (1976).

20 Anon., "Fishing companies maintain profits, but face cost/market squeeze," *Shipp. News Fish. Ind. Rev. S. Afr.,* **30**(7) 57 (1975).

21 Anon., "Markets regain bouyancy, SWA tax concession steadies earnings," *Shipp. News Fish. Ind. Rev. S. Afr.,* **31**(7), 46 (1976).

22 Anon., "Fish companies forecast reduced profits," *Shipp. News Fish. Ind. Rev. S. Afr.,* **33**(7) 45 (1978).

23 Anon., "Fish companies call for 'realistic prices' as production declines," *Shipp. News Fish. Ind. Rev. S. Afr.,* **32**(7), 45 (1977).

24 Anon., "SA pelagic season looks good. RSW pilchards at Walvis Bay," *Shipp. News Fish. Ind. Rev. S. Afr.,* **32**(5), 40 (1977).

25 Anon., "Walvis Bay: Where the fishing had to stop," *Shipp. News Fish. Ind. Rev. S. Afr.,* **33**(9), 35 (1978).

26 Anon., "Pilchards slump, but industry invests for higher yields," *Shipp. News Fish. Ind. Rev. S. Afr.,* **32**(10), 43 (1977).

27 Ibid., p. 37.

28 Editorial, "ICSEAF Quotas," *Shipp. News Fish. Ind. Rev. S. Afr.,* **33**(1), 43 (1978).

29 Anon., "AG acts on SWA fishery management, but not on 200 mi. zone," *Shipp. News Fish. Ind. Rev. S. Afr.,* **33**(3), 37 (1978).

30 Annual Report for 1977, Fisheries Development Corp. of South Africa, Orange St., Cape Town.

PART II
SCIENTIFIC ASPECTS OF EL NIÑO

CHAPTER 7

Ocean Models of El Niño

JAMES J. O'BRIEN
ANTONIO BUSALACCHI
JOHN KINDLE

Abstract

Wyrtki suggested that El Niño is created by a weakening of the southeast trades in the central Pacific which excites an internal Kelvin wave that propagates into the eastern Pacific. The downwelling wave suppresses the thermocline along the coast of South America and creates poleward currents that transport water of equatorial origin all along the Ecuador and Peru coasts. In this paper we attempt to describe the physics of coastal upwelling, equatorial upwelling, free Kelvin and Rossby waves, and some numerical scenarios that produce El Niño-type events. We do not attempt to explain the most esoteric physics of each phenomena but instead present the simplest mathematical model that elucidates the basic physics.

A model of the seasonal circulation of the equatorial Pacific is described. This model is driven by the mean, annual, and semiannual observed east-west winds. It simulates the east-west slope of the equatorial thermocline and the 40 m seasonal excursion of the thermocline in the eastern Pacific. This circulation is used as the background flow for the subsequent El Niño experiments.

INTRODUCTION

El Niño is an anomalous oceanic and meteorological event character-ized by the sudden appearance of abnormally warm surface water on the scale of a thousand kilometers off the coasts of Peru and Ecuador. The financial consequences of this event are catastrophic for the local fisheries, and the economic repercussions are felt worldwide. Other chapters in this book discuss the economics. In this chapter we wish to explain some of the mathematical and physical ideas that scientists are presently using to develop an understanding of the physical oceanographic aspects of El Niño.

It is generally agreed that the causes of the warm water off western South America are linked to very large-scale atmosphere-ocean interactions.[1] Meteorologists have concentrated on describing the anomalous atmospheric circulation in the tropics and middle latitudes preceding and during El Niño events, while oceanographers have described the oceanic events. For each discipline the data base is sparse and not conclusive. After all, El Niño is not predictable at this time or periodic and, thus, observational expeditions cannot be preplanned easily.

Professor Klaus Wyrtki of the University of Hawaii presented in 1975 the basic idea which we will use to explain the present models of El Niño. The following is the abstract for Wyrtki's paper:

El Niño is the occasional appearance of warm water off the coast of Peru; its presence results in catastrophic consequences in the fishing industry. A new theory for the occurrence of El Niño is presented. It is shown that El Niño is not due to a weakening of the southeast trades over the waters off Peru, but that during the two years preceding El Niño, excessively strong southeast trades are present in the central Pacific. These strong southeast trades intensify the subtropical gyre of the South Pacific, strengthen the South Equatorial Current, and increase the east-west slope of sea level by building up water in the western equatorial Pacific. As soon as the wind stress in the central Pacific relaxes, the accumulated water flows eastward, probably in the form of an internal equatorial Kelvin wave. The wave leads to the accumulation of warm water off Ecuador and Peru and to a depression of the usually shallow thermocline. In total, El Niño is the result of the response of the equatorial Pacific Ocean to atmospheric forcing by the trade winds.

There have been many papers trying to discover a clue to the atmospheric signal that triggers El Niño. In 1974, Quinn found that El Niño is preceded by large peaks in the 12 month running means of the Southern Oscillation Index.[2] This index is simply the atmospheric pressure difference between Darwin, Australia and Easter Island. When the index is high, the trades blow strongly toward the equator; when it is low, the trades are weak. Wyrtki's hypothesis is that the most intense El Niño events occur when these strong southeasterly trades relax.

Hurlburt et al. have addressed some of these topics through a simple numerical simulation of the onset and early development of an El Niño event.[3] The major results of this simulation reveal the importance of internal Kelvin and Rossby waves to the dynamics of the equatorial and eastern boundary regions as well as to the communication between these regions during a model El Niño event. Also of great significance is the result that the nonlinear attributes of these waves play an important role in the dynamics of the model El Niño. The theory has been supported by additional observational evidence[4,5] and advanced statistical analysis.[6-8] Wyrtki has analyzed in detail the 1972 El Niño as detected by sea level to support his hypothesis.[9]

A more satisfying theoretical examination of El Niño was carried out by McCreary in which he used an analytical, linear, reduced-gravity, primitive equation model.[10] He examined primarily the forced interior and Rossby wave response to various changes in the meridional profile of the zonal and meridional wind, and applied the result to El Niño. He found that the time-dependent equatorial surface flow is very sensitive to changes in the zonal wind stress within a few degrees of the equator but not to those outside. It was also found that reasonable asymmetries in the changes of the meridional profile of the zonal wind were found to produce significant asymmetries in the western and equatorial ocean response, including some cross-equatorial flow. However, the symmetric part of the response was dominant.

Already we are introducing complicated ideas. In this paper, we want to present these and other ocean modeling ideas in a simple framework. We shall not always draw from the most esoteric theory or mathematics to illustrate our point. Instead we seek to provide a framework for appreciating new ideas as they unfold in the future.

We have defined that El Niño is a massive inundation of abnormally warm water into the coastal regions of Peru and Ecuador. Figure 1 contrasts the normal (1971) and El Niño conditions (1972), showing the warmer eastern tropical Pacific Ocean. In Figure 2, the temperature anomalies at coastal stations along Ecuador and Peru are shown, with the appearance of warm water clearly distinguishable. Figure 3 shows the principal surface currents in the Pacific. There is a well-defined, narrow, northward-setting current along the west coast of South America called the Peru or Humboldt current. We have known for hundreds of years that the surface waters close to Peru are colder than what is expected at these latitudes (Figure 4). One might suspect that the icy waters of Antarctica are being swept northward. However, as early as 1844, de Tessan recognized that there is an upwelling (from 100 to 300 meters) of deep water along the coast. Upwelling has been defined as ascending motion by which water from the subsurface layers is brought into the surface layer and is removed from the area of upwelling by horizontal flow.

SEA SURFACE TEMPERATURE ANOMALIES

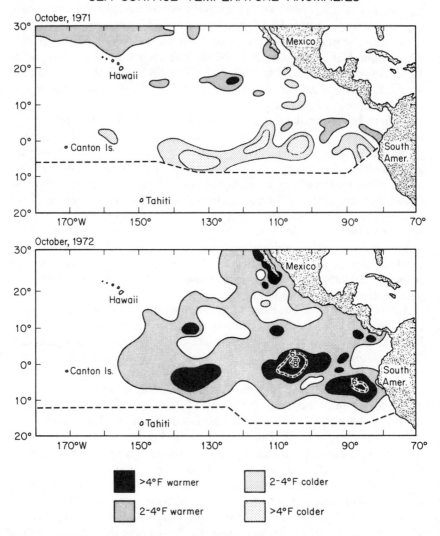

Figure 1 The sea-surface temperature anomaly is the observed temperature departure from the long-term climatic mean for the month. The year 1971 was considered "normal"; 1972, an El Niño year. Warm pools of water with anomalies of greater than 4°C were observed at the equator (110°W) and off the coast of Peru. (Courtesy NOAA.)

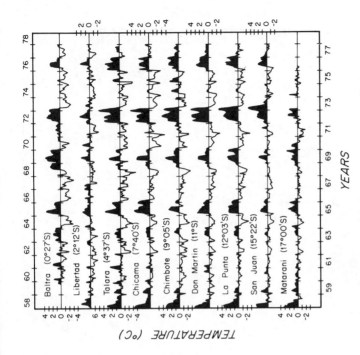

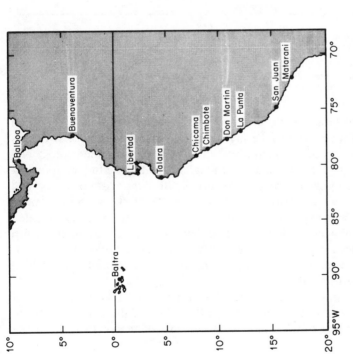

Figure 2 The map indicates coastal stations where ocean temperatures have been recorded. The monthly temperature anomalies are the deviation of the average monthly temperature from the long-term average. Therefore, the annual march of summer (warm) and winter (colder) temperatures has been removed. The El Niño years of 1958, 1965, 1972, and 1976 show warm water along South America. Scientists disagree on whether 1969 was an El Niño year. (Courtesy Dave Enfield and John Allen of Oregon State University.)

163

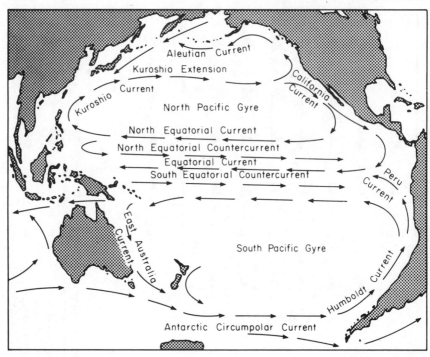

Figure 3 The current pattern in the equatorial region of the Pacific is more complex than that of the Atlantic, although to north and south similar enclosed circulatory gyres are formed. Recent research designates three west-flowing equatorial currents separated by two east-flowing equatorial countercurrents. (Courtesy Rand McNally Atlas of the Ocean.)

Coastal upwelling results when prevailing winds produce a condition of offshore flow in the surface layer near a coastal boundary. It occurs off the west coasts of continents and affects both the marine ecology and the climate of the adjacent land. Although coastal upwelling regions (for example, off California-Oregon, western South America, northwest and southwest Africa, and Portugal) cover only 1% of the world's oceans, John Ryther of the Woods Hole Oceanographic Institution has calculated that such areas account for more than half the world's commercial fish stocks. Lowered sea-surface temperatures in these areas suppress atmospheric convection and evaporation, producing a stable, but humid desertlike environment.

When the wind blows toward the equator off the west coast of continents, it sets up currents. Due to the effect of the earth's rotation, called the Coriolis force, there is a transport of water away from the

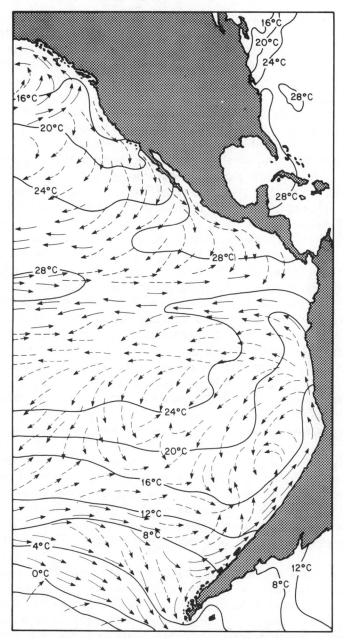

Figure 4 Mean temperatures in the eastern Pacific with indications of surface currents. Observe that the 16, 20, and 24°C isotherms are far northward along the Chile/Peru coastline. (Courtesy Rand McNally Atlas of the Oceans.)

coast. (This drift is known as Ekman flow after V. W. Ekman, who in 1903 studied why icebergs drift to the right of the wind.) Since the mass of water must be conserved, the water within 20 kilometers of the coast rises at 2 to 10 m per day. The water comes up from 100- to 300-m depths and is rich in dissolved nutrients. At these depths, there is not enough light for phytoplankton to grow, but at the surface, copious blooms of the tiny plants occur. Zooplankton, the animal forms of plankton, eat the plants and multiply. Fish are the next link in the food chain to benefit from the upwelling.

In middle latitudes, coastal upwelling is seasonal; off Oregon it lasts from April to October. Nearer the equator, off Baja or Peru, it occurs almost the entire year. As a result, the fish harvest can be bountiful. In 1970, Peru accounted for more than a fifth of the total world fish protein. The collapse of the fishery in 1972 may have been partly the result of overfishing, but experts agree that El Niño was a major factor.

Since El Niño may be linked to the wind field in the Pacific Ocean, we need to have some knowledge of the mean flow. The mean currents in Figure 3 are driven by these winds. Since El Niño only occurs rarely, it is more important to look at anomalous winds. We prepared monthly wind charts for each month of the 10-year period January 1961–December 1970. For each $2 \times 2°$ grid we separated the energy spectrum for east-west wind stress into a low-frequency part and a high-frequency part. In Figure 5 we have plotted all the power for periods 18 months or longer. This is a relative indication of the region of large variability versus low variability on time scales of 18 months or longer. Three regions show strong variability. In the Southern Hemisphere, we observe that from Chile to Australia there is a large amount of long period signal. This is the region of the Southern Oscillation studied by Quinn and others.

Around 10°N and particularly in the central Pacific another latitudinal belt of energy occurs. The region near the dateline is the most probable key area for El Niño events to start. However, the region in the western Pacific in the Asian Monsoon area may also be responsible (see Chapter 15). The last high variance area at 25–30°N in the central Pacific is indicating winters when deep cyclones penetrate further south than normal.

The most striking feature of this analysis is the very low variability found all along South and North America for these long-time periods. We can conclude that there is little long-period variability in the winds in these areas. Thus, the winds in these areas would not be responsible for climatic variations over a few years in the ocean. Enfield addresses this point in further detail (see Chapter 8). These data are further proof that, since El Niño is probably created by wind changes and since El

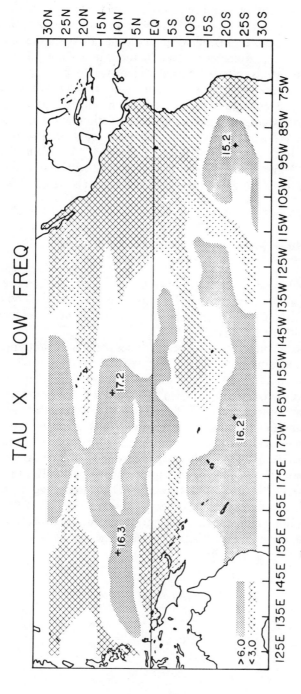

Figure 5 The relative kinetic energy per unit frequency in the eastward wind stress for time periods 18 months to 10 years for the equatorial Pacific. The region along the Americas is shaded because it is low. The other shading is for regions of high wind energy in the low-frequency bands.

Niño is aperiodic, the causes are remote from Ecuador and Peru. The Wyrtki idea says that the roots of El Niño are in the western Pacific.

This paper investigates a series of phenomena. First simple models of coastal upwelling are explained; then equatorial upwelling. These represent the normal events. Since internal Kelvin waves play an important role in the El Niño event, an ideal ocean with only Kelvin waves and their allies, the reflected Rossby waves, is displayed. It will be important to have some knowledge of the typical seasonal circulation due to winds at the equator. We will not solve the entire seasonal ocean circulation, instead an idealized circulation will be presented. Finally, in the computer, some El Niño events will be created and we will demonstrate how wind events in the western and central Pacific can create El Niño.

A SIMPLE MODEL OF COASTAL UPWELLING

Coastal upwelling of cold, nutrient-rich water occurs when prevailing winds produce a condition of surface divergence near a coastal boundary. Off the west coasts of continents, this condition exists when the longshore wind blows toward the equator.

Two-layer ocean models have been a useful tool for studying many ocean circulation situations. When the ratio of vertical to horizontal length scales is small compared to unity, the hydrostatic approximation may be utilized to construct so-called "reduced-gravity" models. These are appropriate for coastal upwelling. However, numerical models of two-layer oceans are not easily understood. In this paper, we present some simple analytical solutions for two-layer models which approximate more complicated numerical solutions. These closed-form solutions explain the essential physics of coastal upwelling. In particular, the important role of the north-south pressure gradient in producing a poleward undercurrent is demonstrated.

Let us consider an inviscid, rotating, stratified, hydrostatic, linear, two-layer fluid on a flat continental shelf. The equations have been derived by O'Brien and Hurlburt.[11] They are

$$u_{1t} = fv_1 - g(h_1 + h_2)_x + \frac{\tau^{sx}}{\rho H_1} \tag{1}$$

$$v_{1t} = -fu_1 - g(h_1 + h_2)_y + \frac{\tau^{sy}}{\rho H_1} \tag{2}$$

$$h_{1t} = -H_1 u_{1x} \tag{3}$$

$$u_{2t} = fv_2 - g(h_1 + h_2)_x + g'h_{1x} \tag{4}$$

$$v_{2t} = -fu_2 - g(h_1 + h_2)_y + g'h_{1y} \tag{5}$$

$$h_{2t} = -H_2 u_{2x} \tag{6}$$

where $g' = g(\rho_2 - \rho_1)/\rho_2$.

The velocities, u_1, u_2, v_1, v_2, are assumed to be independent of the north-south coordinate, y. The flow is inviscid; horizontal, interior, and bottom stresses have been neglected; H_1 and H_2 are the undisturbed values of h_1, h_2. H_1 can be considered the mean depth of the warm mixed layer.

If the Coriolis parameter $f = f_0 + \beta y$, then the north-south pressure gradient must be retained.[12] If $f = f_0$ then the north-south pressure gradient is at most a constant, since Eq. (1) may be written

$$\beta v_1 = g(h_1 + h_2)_{xy} \tag{7}$$

Let us assume the forcing is simply,

$$\tau^{sy} = \text{(a constant)} \qquad t \geq 0$$

$$\tau^{sy} = 0 \qquad\qquad\qquad t < 0 \tag{8}$$

$$\tau^{sx} = 0$$

This impulsive forcing will excite external and internal inertia-gravity waves as well as coastal upwelling. The gravity waves are excited on a time scale of f^{-1}, whereas the onset of coastal upwelling will occur on a longer time scale (say, several days). We can filter the former from the equations by requiring the longshore flow to be geostrophic. If $\beta = 0$, we retain the essential physics of the numerical model of O'Brien and Hurlburt[11]; if $\beta = 0$, we have the essence of the Hurlburt and Thompson model.[12]

Following Garvine[13] and Hurlburt and Thompson,[12] we impose a constant N-S sea surface slope such that, away from the coast at $x = -\infty$, the integrated mass transport is zero.

$$H_1 u_1 + H_2 u_2 = 0 \tag{9}$$

If the flow is geostrophic, then at $x = -\infty$,

$$u_1 = -\frac{g}{f}(h_1 + h_2)_y + \frac{\tau}{\rho H_1 f}$$

$$u_2 = -\frac{g}{f}(h_1 + h_2)_y$$

and Eq. (9) requires

$$\frac{g}{f}(h_1 + h_2)_y = U_E \frac{H_1}{H_1 + H_2}$$

where $U_E = \tau/\rho H f_0$ is the Ekman velocity. We define reduced zonal velocities, u_1' and u_2'

$$u_1' = u_1 - U_E - \frac{g}{f}(h_1 + h_2)_y = u_1 - \frac{H_2}{H_1 + H_2} U_E$$

$$u_2' = u_2 - \frac{g}{f}(h_1 + h_2)_y = u_2 - \frac{H_1}{H_1 + H_2} U_E$$

(10)

The equations to be solved are

$$fv_1 = g(h_1 + h_2)_x$$
$$v_{1t} = -f u_1'$$
$$h_{1t} = -H_1 u_{1x}'$$
$$fv_2 = g(h_1 + h_2)_x - g' h_{1x}$$
$$v_{2t} = -f u_2'$$
$$h_{2t} = -H_2 u_{2x}'$$

(11)

The boundary conditions are

$$v_1 = v_2 = 0 \qquad \text{at } t = 0$$

$$h_1 = H_1 \qquad h_2 = H_2 \qquad \text{at } t = 0$$

$$u_1' = -\frac{H_2}{H_1 + H_2} U_E \qquad \text{at } x = 0$$

$$y_2' = -\frac{H_1}{H_1 + H_2} U_E \qquad \text{at } x = 0$$

$$u_1' = u_2' = 0 \qquad \text{at } x = -\infty$$

We need to define

$$\gamma_I^2 \equiv \frac{g'H_1}{f^2}\left(\frac{1}{1 + H_1/H_2}\right) \tag{12}$$

$$\gamma_E^2 \equiv \frac{g(H_1 + H_2)}{f^2} \tag{13}$$

The length γ_I is the baroclinic Rossby radius of deformation for a two-layer fluid. The length γ_E is the barotropic Rossby radius of deformation. By eliminating in favor of u_1', Eq. (11) may be written

$$\gamma_I^2\gamma_E^2\, u_{1xxxx}' - \gamma_E^2\, u_{1xx}' + u_1' = 0 \ . \tag{14}$$

This may be solved by boundary layer techniques if we recognize that

$$\left(\frac{\gamma_I}{\gamma_E}\right)^2 \ll 1 \tag{15}$$

This ratio is *always* small for the real ocean. This simply states that the barotropic radius of deformation is always much larger than the baroclinic deformation radius. A typical value for γ_I is 10–20 km.

The external boundary layer is solved by setting

$$x = \gamma_E \eta$$

and obtaining

$$\frac{\gamma_I^2}{\gamma_E^2} u_{1\eta\eta\eta\eta}' - u_{1\eta\eta}' + u_1' = 0$$

The first term is small and the decaying solution for the barotropic scale is

$$u_1' = c_1\, e^\eta = c_1\, e^{x/\gamma_E}$$

For the internal boundary layer we scale

$$x = \gamma_I \zeta$$

and obtain

$$u'_{1\zeta\zeta\zeta\zeta} - u'_{1\zeta\zeta} + \frac{\gamma_I^2}{\gamma_E^2} u'_1 = 0$$

The last term may be dropped and the decaying solution for the internal boundary layer is

$$u'_1 = c_2 \, e^{\zeta} = c_2 \, e^{x/\gamma_I}$$

The total solution is

$$u'_1 = c_1 \, e^{x/\gamma_E} + c_2 \, e^{x/\gamma_I}$$

$$u'_2 = c_1 \, e^{x/\gamma_E} - c_2 \left(\frac{H_1}{H_2}\right) e^{x/\gamma_I}$$

Using the boundary conditions at $x = 0$, we obtain

$$c_1 = 0 \, !$$

$$c_2 = -\frac{H_2}{H_1 + H_2} U_E$$

The complete solution for this problem is

$$u_1(x, t) = -U_E \left[-\frac{H_2}{H_1 + H_2} + \frac{H_2}{H_1 + H_2} e^{x/\gamma_I} \right]$$

$$u_2(x, t) = U_E \left[-\frac{H_1}{H_1 + H_2} + \frac{H_1}{H_1 + H_2} e^{x/\gamma_I} \right]$$

$$v_1(x, t) = ft \, U_E \left(\frac{H_2}{H_1 + H_2}\right) e^{x/\gamma_I}$$

$$v_2(x, t) = -ft \, U_E \left(\frac{H_1}{H_1 + H_2}\right) e^{x/\gamma_I}$$

$$h_1 - H_1 = \frac{t \, H_1}{\gamma_I} U_E \left(\frac{H_2}{H_1 + H_2}\right) e^{x/\gamma_I}$$

$$h_2 - H_2 = -\frac{t \, H_2}{\gamma_I} U_E \left(\frac{H_1}{H_1 + H_2}\right) e^{x/\gamma_I}$$

The barotropic mode, γ_E, vanishes identically. [The approximation (15) is used in finding c_1 and c_2.]

The offshore flow, u_1, is brought to zero by the internal scale, γ_I. The onshore flow, u_2, is also brought to zero on this scale. The longshore flow, v_1, in the upper layer shows a coastal jet but the barotropic mode is zero. The most rewarding result is the existence of a poleward undercurrent, v_2, in the lower layer. The ratio of the longshore flow is

$$\frac{v_1}{v_2} = -\frac{H_2}{H_1}$$

The equatorward jet is therefore generally much stronger than the poleward undercurrent as found by observations.[14]

The upwelling, $h_2 - H_2 > 0$, occurs within the baroclinic radius of deformation, γ_I. This is typically 10 km for the real ocean.

It is hoped that this analytical solution will aid in understanding more complicated numerical studies. It is important to recognize that the upwelling occurs on the narrow baroclinic deformation scale, γ_I. As shown by Yoshida,[14] the maximum vertical velocity, W, is

$$W = \frac{H_1 U_E}{\gamma_I}$$

with typical values of 10^{-2} cm sec^{-1} or 10 m day^{-1}. The time scale, T, for the pycnocline to surface is

$$T = \gamma_I/U_E$$

As Garvine[13] and Hurlburt and Thompson[12] show, the north-south pressure gradient is essential to obtaining a poleward undercurrent in a coastal upwelling environment.

In simple terms, the prevailing equatorward winds, say off Peru, drive an equatorward surface flow and a poleward undercurrent. The latter has a more complicated physics than presented here (e.g., Smith,[16] Preller and O'Brien[17]). The upper layer is driven offshore due to the effects of the earth's rotation. In a thin 10–20 km region near the shore, the deeper water rises at a rate of 5–30 m day^{-1}. This cold water supplies the nutrients for the ecosystem. During El Niño, upwelling still occurs but brings to the surface warm, nutrient-poor water of equatorial origin.

A SIMPLE MODEL OF EQUATORIAL UPWELLING

In the Pacific Ocean there exist mean surface winds toward the west. These winds produce a region of cold water along the equator. Let us consider the equation set (1–6) of the last section. We consider for simplicity that no continents exist and the wind is everywhere westward across the ocean.

If we let $u = u_1 - u_2$, $v = v_1 - v_2$, $h = h_1$ and neglect east-west pressure gradients, Yoshida has shown that the equations

$$\frac{\partial u}{\partial t} = \beta yv + \frac{\tau_x^s}{\rho H_1}$$

$$\beta yu = -g\,\Delta\rho/\rho\,\frac{\partial h}{\partial y} \tag{16}$$

$$\left(\frac{1}{H_1} + \frac{1}{H_2}\right)\frac{\partial h}{\partial t} = -\frac{\partial v}{\partial y}$$

may be used to deduce some understanding of the physics. We have eliminated gravity waves by forcing the east-west flow to be geostrophic. When τ_x^s is a constant, impulsively applied, the north-south velocity, v, is a solution of the ODE

$$L^4\frac{\partial^2 v}{\partial y^2} - y^2v = aLy \tag{17}$$

The fundamental scale length, L, is known as the baroclinic equatorial radius of deformation. Time-dependent, baroclinic ocean motions always depend on the deformation radius. Here

$$L = (C/\beta)^{1/2}$$

$$C = \left[g\frac{\Delta\rho}{\rho}H_1H_2/(H_1 + H_2)\right]^{1/2}$$

$$a = \frac{\tau_x^s}{\beta LH\rho}$$

We expect a priori that the dominant baroclinic motions will be confined to within a distance L of the equator in the ocean interior. The solution of Eq. (16) is illustrated in Figure 6.

The measured values (or estimates) of these constants are

$$\beta = 2.25 \times 10^{-13} \text{ cm}^{-1}\text{ s}^{-1}$$
$$\frac{\Delta\rho}{\rho} = 0.003$$
$$H_1 = 10^4 \text{ cm}$$
$$H_2 > 4H_1$$

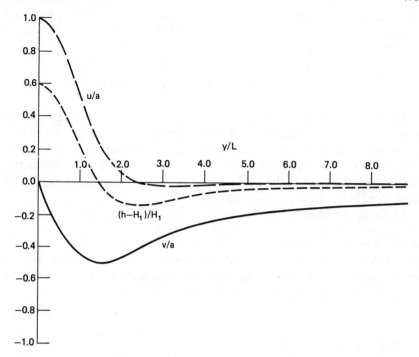

Figure 6 Solution to the linear model in nondimensional units. The eastward jet, u/a, and the pycnocline anomaly are plotted at time $(\beta L)^{-1}$. The jet flow decays to zero near $2L$; a weak westward flow is calculated from $2.3L$ and northward.

thus $L = 277$ km. The equatorial jet is symmetric about the equator. The width of the jet is *independent* of the shape and strength of the wind field. Equation (17) implies that far from the equator

$$v = \frac{-\tau_x^s}{\beta y \rho H}$$

which is the classical Ekman drift solution. Near the equator, v approaches zero. Thus from Eq. (16) the eastward jet is driven directly by the wind stress. If the wind acts for 20 days at 0.5 dynes cm^{-2}, we can expect the surface velocity to approach 100 cm s^{-1}. These estimates from the linear theory agree with the observations. After several days, this linear model is invalid if boundaries exist since substantial east-west pressure gradients develop.

The upwelling is confined to $\pm L$ of the equator and is maximum on the equator. In simple terms, the westward wind stress creates an Ekman divergence at the equator that pumps water upward. This model is too simple to be realistic. For understanding El Niño models we need to include the boundaries.

EQUATORIAL KELVIN AND ROSSBY WAVES

The presence of boundaries plays an important role in ocean circulation. At the equator, the effect of rotation changes sign and an important wave-guide, which traps waves exists. Long waves in the ocean will remain at the equator until dissipated and scattered by the turbulence and mean currents. In the previous section, we investigated equatorial upwelling in an ocean without boundaries.

We want to reconsider the ocean response to a patch of westward wind in the western Pacific (Figure 7). We introduce the concept of a reduced-gravity, β-plane ocean. In this example we solve for the transport (U—eastward, V—northward) and the depth, h, of a thin homogeneous ocean. We approximate the effect of the rotation of the earth and the sphericity by replacing the Coriolis parameters, f, by

$$f = y \frac{df}{dy} \equiv \beta y$$

where y is latitude.

The equations of this model are

$$\frac{\partial U}{\partial t} = \beta y V - g' \frac{\partial h}{\partial x} + \frac{\tau^x}{\rho} + A \nabla^2 U \tag{18}$$

$$\frac{\partial V}{\partial t} = -\beta y U - g' \frac{\partial h}{\partial y} + \frac{\tau^y}{\rho} + A \nabla^2 V \tag{19}$$

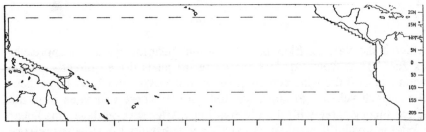

130E 140E 150E 160E 170E 180 170W 160W 150W 140W 130W 120W 110W 100W 90W 80W 70W

Figure 7 Comparison of model geometry in relation to the tropical Pacific Ocean. Zonal boundaries at 18°N and 12°S are open. Meridional boundaries are closed.

$$\frac{\partial h}{\partial t} = -U_x - V_y \tag{20}$$

where

$$U = uH \qquad \mathbf{V} = u\hat{i} + v\hat{j}$$

$$\tag{21}$$

$$V = vH \qquad g' = gH\frac{\rho_2 - \rho_1}{\rho_2}$$

We use the thickness, $H = 175$ m; the density contrast, $\rho_2 - \rho_1 = 2.0 \times 10^{-3}$ gm cm^{-3}; the wind stress, $\tau^x = 0.05$ Nm^{-2} or zero, $\tau^y = 0$; $\beta = 2 \times 10^{-11}$ m^{-1} s^{-1}. The horizontal eddy viscosity, A, is 10^2 m^2 s^{-1}. The eddy viscosity is small enough that its only physical effect is to bring the flow to zero at the east and west boundaries. The northern boundary at 18°N and the southern boundary at 12°S are open; that is, any waves incident on these boundaries will be transmitted.

At the initial time our ideal upper ocean is at rest. A westward wind stress is applied west of 170°E. We have prepared a series of pictures to show the unusual results (Figures 8–15).

In the first few weeks (not shown) we have well-behaved equatorial upwelling (dashed lines) west of 170°E. By the end of 30 days (Figure 8) the center of upwelling has *propagated* to 180°. There is about 30 m of upwelling in the center. By the third month, the upwelling patch has reached the coast of Ecuador and spread coastal upwelling along the coasts. The northern and southern boundaries are open to allow the coastal Kelvin waves to radiate out.

The velocity field associated with the Kelvin wave (Figure 9) has also propagated to South and North America. The wind continues to blow in the west but currents and upwelling have been generated all along the equator and the coastal boundary. If the wind had been toward the east in the patch, we would have found the identical pattern but, instead of 30 m of upwelling, we would find 30 m of downwelling!

By months 4–6 (Figure 10), we observe that the upwelled patch has started to propagate westward from Ecuador. This pattern is called an equatorial Rossby wave. There are actually a large number of waves but we will not make the subtle distinction here. The main Rossby wave propagates at one-third of the Kelvin wave speed; thus it will take 9 months to reach the western shore; while the Kelvin wave took 3 months. In Figure 11, the strong westward flow (120°W at 6 months) is associated with the main Rossby wave.

By the end of a year, the Rossby wave has reached the western

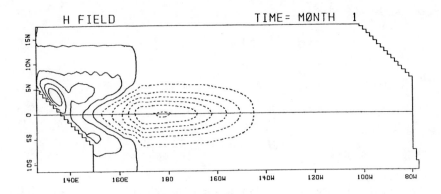

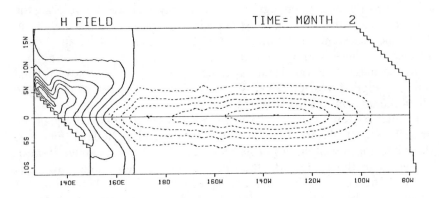

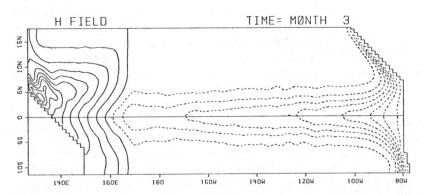

Figure 8 Interface displacements depicting the progression of an equatorially trapped Kelvin wave and resulting coastal Kelvin waves during the first three months of model integration. Contouring interval is 5 m; initial contours are ±2.5 m so as to suppress the zero contour. Dashed (negative) contours indicate upwelling.

178

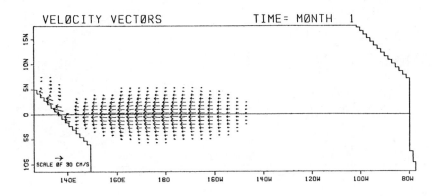

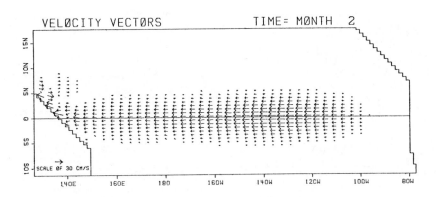

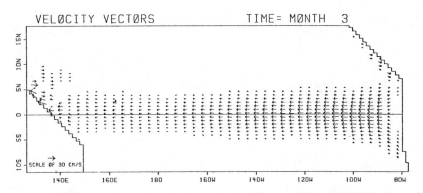

Figure 9 Velocity vectors illustrating the circulation associated with the height fields in Figure 8. Flow speeds less than 4 cm/s have been suppressed.

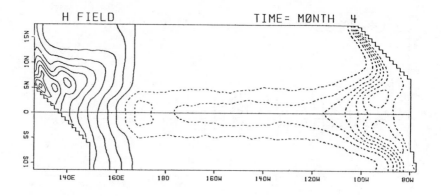

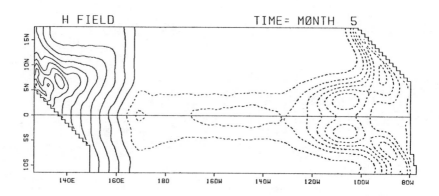

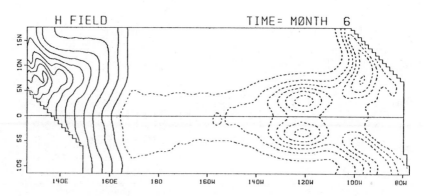

Figure 10 Westward progression of an equatorial Rossby wave as evidenced in the height fields.

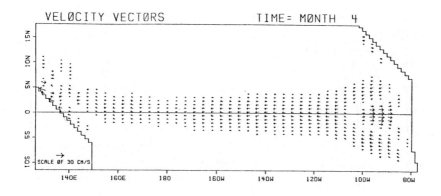

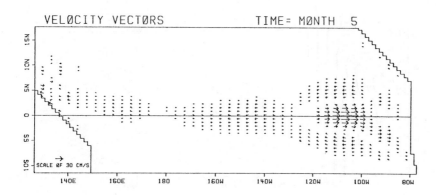

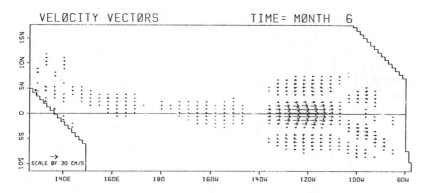

Figure 11 Velocity fields associated with a westward traveling Rossby wave.

181

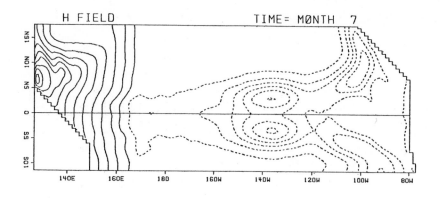

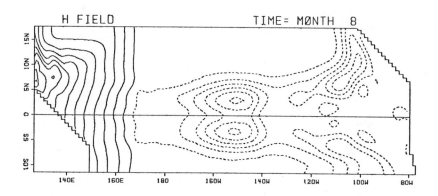

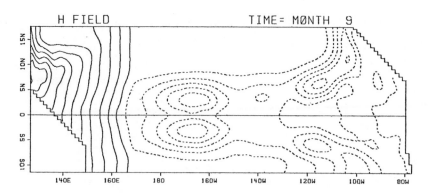

Figure 12 Westward progression of an equatorial Rossby wave as evidenced in the height fields.

182

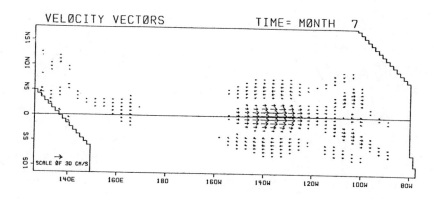

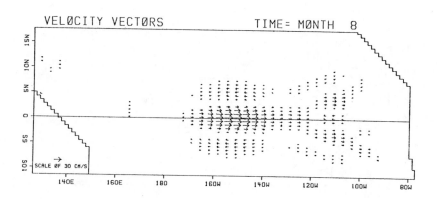

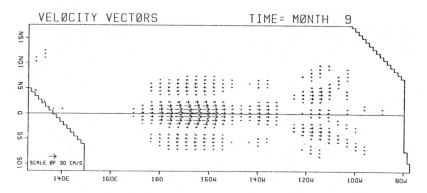

Figure 13 Velocity fields associated with a westward traveling Rossby wave.

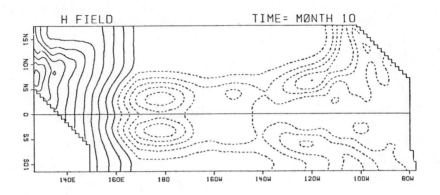

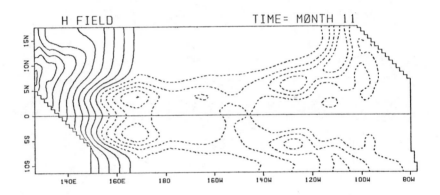

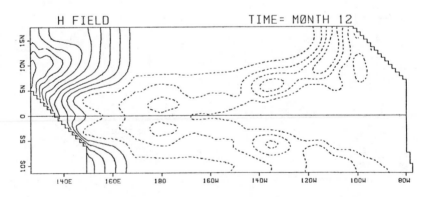

Figure 14 Interface displacements resulting during impingement of a Rossby wave on the western boundary.

184

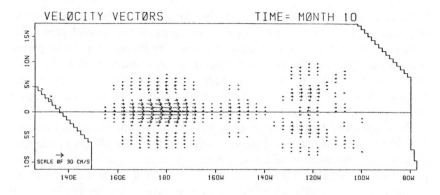

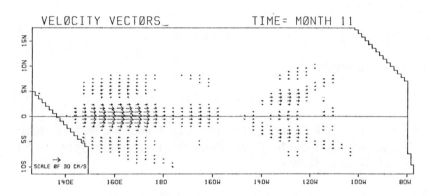

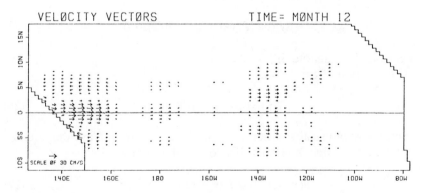

Figure 15 Depiction of the circulation during the impingement of a Rossby wave on the western boundary.

boundary where it will reflect in the form of another Kelvin wave (Figures 12–15). We are not including pictures of the second year. The solid lines in the west indicate downwelling. The thermocline is deepened toward the west in response to the wind. In this simple model, the east-west pressure gradient wants to equal the westward wind stress and reduce all motion to zero. This will occur except for the trapped equatorial waves.

How does the Kelvin wave arise as a solution to our model ocean? The free model response to Eqs. (18)–(20) includes the effects of intertia-gravity, Rossby, and mixed Rossby-gravity waves. Another interesting free response of Eqs. (18)–(20) is the flow on a β-plane with values of the meridional velocity identically equal to zero. With values of $v = 0$ and $A = 0$, cross differentiation of Eqs. (18) and (20) yields the wave equation for u

$$u_{tt} - g' u_{xx} = 0 \tag{22}$$

which has a solution

$$u(x,y,t) = G(y) F(x \pm ct) \tag{23}$$

where $c = g'^{\frac{1}{2}}$ and F is any function of $x \pm ct$. F need not be cosinusoidal.

It is possible to solve for G by eliminating in favor of u between Eqs. (19) and (20) and obtain

$$-\beta y u_t + g' u_{xy} = 0 \tag{24}$$

A substitution of Eq. (22) into Eq. (23) yields for G

$$-\beta y (\pm c[F]G) + g' [F][G] = 0 \tag{25}$$

where brackets denote differentiation.

The solution of Eq. (25) is

$$G(y) = a \exp[\pm 1/2(y/L_e)^2] \tag{26}$$

Since any physical solution must be bounded away from the equator only the minus sign is permissible so one may write

$$u(x, y, t) = a F(x - ct) e^{-\frac{1}{2}(y/L_e)^2} \tag{27}$$

where $L_e = (c/\beta)^{\frac{1}{2}}$ is the internal radius of deformation. A typical value of L_e is 300–350 km for the equatorial Pacific. The result is an eastward propagating Kelvin wave with phase speed c that is equatorially trapped with an e-folding distance of $\sqrt{2}\,L_e$.

We cannot do as simple an analysis for the Rossby waves. The reader is referred to Moore and Philander for a treatise on these free modes.[18]

It is important to recognize that the Kelvin waves are symmetric about the equator and thus can be excited by the time changes of the east-west winds. An increase in a westward wind excites an upwelled Kelvin wave; an increase in an eastward wind excites a downwelled Kelvin wave. The Kelvin wave only propagates toward the east. We might anticipate that the normal seasonal changes of the wind will excite a continual stream of Kelvin waves and a large change in the winds such as prior to an El Niño would excite a large anomalous signal. However the wind changes must be a decrease in the westward winds or an increase in eastward winds. The wind change must be over a broad area in order to excite a large patch of downwelled water.

SEASONAL VARIABILITY IN THE EASTERN EQUATORIAL PACIFIC

In this section, a numerical model is utilized to examine the seasonal variation of the thermal structure and the circulation of the equatorial Pacific Ocean. Attention is focused on the remote forcing due to equatorially trapped waves. In particular, theoretical evidence is presented which supports the hypothesis of Meyers that remote forcing may be responsible for the large semiannual component of the vertical motion of the thermocline in the eastern equatorial Pacific.[19] This section is entirely from Kindle.[20]

A recent paper by Meyers presents a very interesting analysis of the seasonal variation of the 14°C isotherm in the equatorial Pacific.[19] Bathythermographic data taken over many years between 2°S–2°N were used to determine the mean depth of the 14°C isotherm as well as the mean monthly variation. Meyers shows that (1) the 14°C isotherm is located in the lower portion of the thermocline; (2) variation of temperature within the depth range spanning the thermocline is due primarily to vertical motion of the thermocline rather than variation of the temperature gradients within it; and (3) vertical motion of the 14°C isotherm is essentially in phase with the vertical motion of the thermocline.

A scatter diagram of the observations of the depth of the 14°C isotherm averaged between 1°S to 1°N is shown in Figure 16. The east-west slope of the mean position of the isotherm is representative of the slope of the thermocline along the equator. The most striking aspect of this plot is

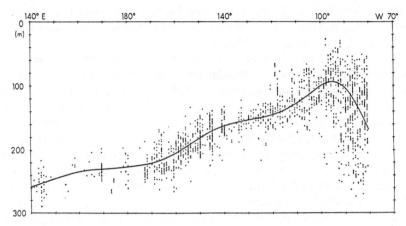

Figure 16 Scatter diagram of the observations of the depth of the 14°C isotherm between 1°S and 1°N in the Pacific Ocean (from Meyers[19]).

the large amplitude of the depth variations near the eastern boundary. The extreme values are due to the large El Niño events. However, even if one considers the region over which the distribution of data points is rather uniform, the largest amplitude is at the eastern boundary.

In order to study the seasonal variability of the thermocline, Meyers averaged the data over areas of 20° longitude (±2° about the equator) and calculated mean monthly values. Then for each region between 80°W and 140°E, he determined the mean depth as well as the annual and semiannual signals. A similar analysis was performed on the zonal component of the wind stress between 2°S and 2°N. The results of these calculations are shown in Figure 17. Note that both the annual and semiannual components of the 14°C isotherm depth variation have a maximum amplitude at the eastern boundary. This is particularly impressive in view of the fact that the semiannual component of the wind stress has virtually no amplitude in that region. This is what led Meyers to suggest that remote forcing may be responsible for the large amplitude of the semiannual component of the vertical displacement of the thermocline at the eastern boundary.

In what follows, the dynamical causes of the variation of the 14°C isotherm at the eastern boundary are examined in an effort to test Meyer's hypothesis. The mean seasonal variation of the isotherm at the boundary is shown in Figure 18. Note that the downwelling initiated in March is representative of a mini-El Niño.

The mean and seasonal components of the equatorial zonal wind stress as analyzed by Meyers are used to force a numerical model. The winds

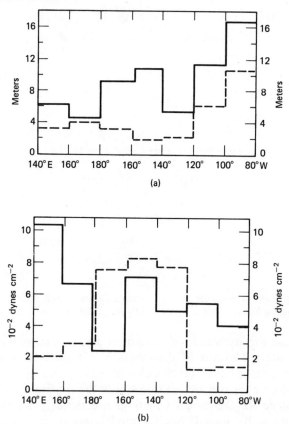

Figure 17 Comparison of amplitudes of the annual (solid line) and semiannual (dashed line) components of (a) vertical displacement of 14°C isotherm in the equatorial Pacific and (b) the zonal equatorial wind stress across the entire Pacific (from Meyers[19]). The data are averaged over 20° longitude.

are assumed to have no meridional variation. This is not a serious deficiency; the variation of the wind stress outside of the equatorial radius of deformation (~250 km) has little effect on the near equatorial response.[10] Meyers also found that the zonal winds in the near equatorial region were distributed symmetrically. Within each of the seven regions between 144°E and 80°W, the winds are assumed to be uniform.

The enormous size of the Pacific Ocean presents a formidable obstacle to the numerical modeling of its time-dependent circulation. Since we variation is consistent with observations in the central and western Pacific, there is only marginal agreement between the phase of the observed seasonal variability and the numerical simulation. This is not

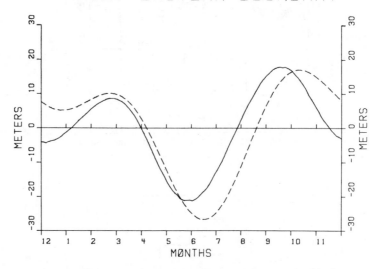

Figure 18 Comparison of nonlinear model solution (solid line) with observations (dashed line). The numerical solution is shown for the fourth year of integration.

will (1) examine the dynamics of equatorial and coastal regions on the scale of the internal radius of deformation; (2) perform a variety of numerical experiments over time scales of years; and (3) use a basin whose dimensions approximate the true size of the tropical Pacific Ocean, the logical choice for the model equations are those given by a single layer, reduced-gravity model. Such a model represents the baroclinic mode of a two-layer system in which the ocean is assumed to be hydrostatic, incompressible, and consists of two homogeneous layers of different density. The interface between the two layers represents the main pycnocline, which in the equatorial regions is very strong and shallow.

The resultant governing equations for the system are the nonlinear shallow water wave equations on an equatorial β-plane,

$$u_t + uv_x + vu_y - \beta yv = -g'h_x + A\nabla^2_h u + \frac{\tau^x}{\rho_1(H + h')}$$

$$v_t + uv_x + vv_y + \beta yu = -g'h_y + A\nabla^2_h v + \frac{\tau^y}{\rho_1(H + h')} \qquad (28)$$

$$h_t + (hu)_x + (hv)_y = 0$$

where u and v are the upper layer velocity components, g' is the reduced gravity given by $g(\rho_2 - \rho_1)/\rho_1$, h is the thickness of the upper layer, and h' is the pycnocline height anomaly (PHA), that is, the departure of the upper layer thickness from its initial uniform thickness, H.

The Cartesian coordinate system is righthanded with positive x directed eastward, and its origin is the western boundary of the model equator. The wind stress, whose components are (τ^x, τ^y), is included as a body force distributed uniformly throughout the upper layer. The effects of thermohaline mixing and thermodynamics are neglected.

The set of Eqs. (28) are solved in a rectangular basin as shown in Figure 19. The eastern and southern boundaries are at 80°W and 15°S, respectively.

Boundary conditions at the eastern and western walls are no slip. Symmetry is applied at the equator. The southern and northern boundaries are open, and a variant of the boundary conditions developed by Hurlburt is used there. The boundary condition yields an expression for the north-south pressure gradient at an open zonal boundary. All other meridional derivatives are set to zero except for the flux terms in the continuity equation.

The model parameters for the numerical simulations are those found in Table 1. The model is started from rest and run for four years in order to remove all transients initiated by the spin-up process. All solutions shown in this section are from year four.

The model PHA is now defined as the departure of the interface from its mean position. The comparison of the seasonal variation of the model PHA with the observed variation is shown in Figure 18. Considering the simplicity of the model, the close agreement is remarkable.

The numerical model simulation can be used to examine the seasonal variability of the circulation along the equator. The numerical results are applicable only within approximately ±5° of the equator, because we have assumed that the forcing has a uniform meridional distribution.

In Figure 20a, the seasonal variation of the upper layer thickness (ULT) along the entire equator is shown. The variability at the eastern boundary has been discussed above. The model data are also consistent with the observations of Taft and Jones,[21] who were unable to find a mean zonal pressure gradient east of 115.5°W during the Piquero Expedition (June 26–August 4, 1969). The numerical simulation predicts that the lack of a zonal thermocline slope east of 115°W is a seasonal phenomenon. The reader must realize that there are numerous Kelvin and Rossby waves propagating and interfering to give the pictures in Figure 20.

The ability of the model to simulate the seasonal variability of the thermocline depth is much better east of approximately 130°W than to the west of this region. Although the amplitude of the model thermocline

192

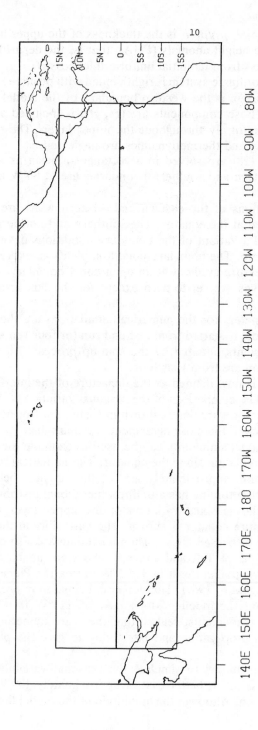

Figure 19 Comparison of model geometry in relation to tropical Pacific Ocean. The model basin extends 15,000 km zonally and ±1,500 km meridionally. The zonal boundaries are open; the meridional boundries are solid walls.

TABLE 1. Model Parameters

$\Delta\rho/\rho = 2.0 \times 10^{-3}$	N-S model
	extent $= \pm 1500$ km
$H = 200$ m	$A = 10^6$ cm^2 sec^{-1}
$\beta = 2.25 \times 10^{-11}$ m^{-1} s^{-1}	$g = 9.8$ m sec^{-1}
E-W model	$\Delta t = 5400$ sec
extent $= 1.3 \times 10^4$ km	

surprising in view of both the simplicity of the model and the much smaller amplitude of the seasonal cycle in the central and western Pacific.

The mean zonal slope of the thermocline, however, *is* accurately represented by the numerical model along the entire equatorial Pacific. The ability of the numerical model to simulate a realistic zonal slope of the equatorial thermocline will prove to be an important facet of the El Niño experiments in the following section.

The predicted seasonal variation of the zonal velocity component along the equator is shown in Figure 20b. This figure depicts only the departure from the mean flow, because the layered model does not possess a realistic steady-state solution for the current field. If the model is forced only by the mean winds, the steady state solution is a zero flow condition in which the mean wind stress balances the pressure gradient. The mean surface current at the equator is part of the westward South Equatorial Current.

The amplitude of the variability of the surface flow shows a distribution pattern similar to the pycnocline variability (Figure 20a), that is, the largest amplitude fluctuations are in the eastern third of the equator. The model predicts that the mean westward flow at the equator is weakened from April to June, and strengthened from July to October. This is consistent with the analysis of the EASTROPAC observation by Tsuchiya.[22] The observations taken during this year-long expedition in the eastern Pacific reveal that the westward surface current very near the equator was most intense in the fall season but weakest in the spring.

Vector plots for the velocity field in the eastern tropical Pacific are shown in Figure 21. These plots reveal that the equatorial flow periodically separates into two branches. This feature was also observed by Hurlburt et al. in their numerical simulation of El Niño.[3] They explained it as an effect due to the westward propagation of the pressure field as an internal Rossby wave. Thus, most of the variability of the upper Pacific Ocean at the equator can be simulated with a simple model.

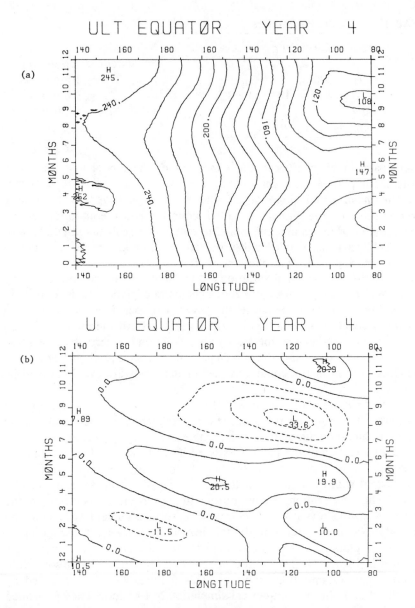

Figure 20 Seasonal variation along the equator of (a) pycnocline depth and (b) zonal velocity component as predicted by nonlinear numerical model. Components of wind stress for the calculations are from Meyers[19] as shown in Figure 17b. The contour interval for (a) is 10 m and for (b) is 10 cm sec^{-1}.

194

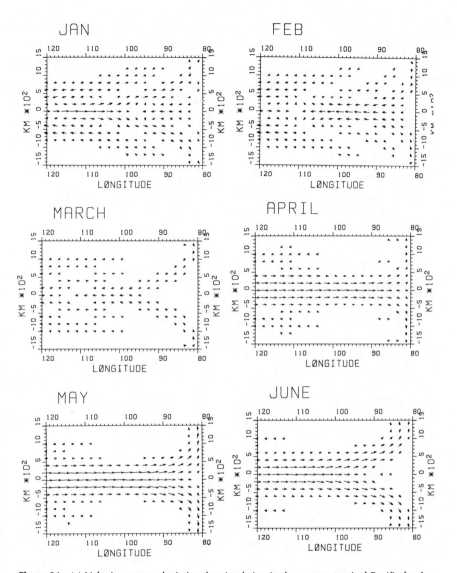

Figure 21 (a) Velocity vectors depicting the circulation in the eastern tropical Pacific for the first six months of the seasonal cycle. The numerical simulation is that which is represented in Figure 20.

195

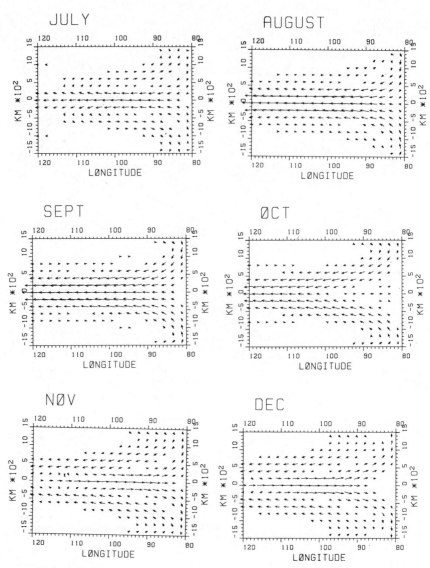

Figure 21 (b) Same as (a) except for second half of year. The numerical solution is shown at the mid-point of each month.

SOME SIMPLE MODELS OF EL NIÑO

El Niño is an anomalous oceanographic and meteorological event which is characterized by the sudden appearance of abnormally warm surface water on the scale of a thousand kilometers off the coast of Ecuador and Peru. These warm water anomalies, which can persist for up to a year or more, also extend thousand of kilometers along the equator. There are no criteria that clearly define El Niño, because (1) the onset of the event occurs during the normal seasonal warming period; and (2) the magnitudes of El Niño-like events vary from slightly greater than the seasonal perturbations to extreme fluctuations. Hence, most studies of this phenomenon focus on major El Niño events. The most recent occurrences of major El Niños were in 1957–1958, 1965, 1972–1973, and 1976.

To what extent El Niño is the cause of or the result of large-scale atmospheric anomalies is not well understood. Most studies of the oceanic manifestation of El Niño ignore complex air-sea feedback mechanisms and concentrate, instead, on the following question: Given a prescription of atmospheric forcing prior to and during El Niño, can one account for the oceanic observations of this event? Recent theoretical studies have been able to account only for the onset phase of the event, that is, the first few months.[3,10,23] In this section, the nonlinear numerical model is used to simulate El Niño from the onset phase to the termination of the event. In spite of the simplicity of the model, the numerical simulation is able to account for many of the observed features associated with El Niño.

There are a number of excellent descriptions of the observational evidence associated with El Niño. [8,9,24–31] It is not within the scope of this section to detail these works; instead, we will focus on the most salient features which appear to be common to all or most major El Niño events. The most prominent indicator of El Niño in the eastern Pacific is the sea surface temperature (SST) anomaly pattern. Both Wooster and Guillen[29] and Zuta et al.,[31] chronicle El Niño in terms of three phases of SST changes. The first phase is characterized by a very sudden appearance of warm water along the coasts of Ecuador and Peru. The warm water, which is of equatorial origin, advances rapidly along the coast. Although the onset of El Niño is coincident with the normal seasonal warming, the magnitude and extent of the warm equatorial water go well beyond the usual seasonal limits. The initial phase lasts for several months; its termination is characterized by the retreat of the warm water towards the equator. The intermediate phase is a quiescent period lasting for approximately 6 months. During this phase, warm SST anomalies persist but gradually diminish in magnitude. The final phase is similar to the initial period. Very warm SST anomalies suddenly reappear along the South American coast. The magnitude of SST anomalies during this phase are usually not as large as during the initial period. The final phase generally begins about 1 year after the initial onset of El Niño. The duration of the third phase is shorter than the initial period; frequently, the final stage terminates abruptly.

The occasional appearance of warm SST anomalies in consecutive years is one of the major mysteries of El Niño. Unlike the events in 1957–1958 and 1972–1973, the occurrences of El Niño in 1965 and 1976 evidenced only single peaks in temperature.[27] In fact, the SST in the eastern Pacific and along the South American coast were substantially colder than normal seasonal values during the first quarter of 1966.

Wyrtki describes El Niño in terms of the sea level response, which he also shows is representative of the vertical motion of the thermocline along the equator.[9,27] Wyrtki finds that sea level rapidly peaks at the start of El Niño along the coast of Peru. The western Pacific, however, evidences a very different response of sea level during the event. In this region, a very slow decline of sea level occurs throughout the entire El Niño year. This sequence of events occurred in both the 1972 and 1976 El Niño years. However, substantial differences were encountered between those two events during the year which followed each of these occurrences. In 1973 a second peak in sea level was observed, which was similar in nature to the peak in the preceding year. No successive sea level peak was observed in 1977. In both 1973 and 1977, the return to a normal sea level distribution occurred very rapidly.

Bjerknes was the first to suggest that the occurrence of El Niño is related to changes in the Pacific trade wind system.[25] It was Wyrtki, however, who proposed the specific theory that El Niño is caused by the sudden relaxation of the equatorial trade winds in the central and western Pacific Ocean.[32] Analysis of the equatorial winds by Wyrtki and Meyers revealed that (1) major El Niño events are often preceded by a stronger than normal wind which lasts for a year or more; (2) subsequent to relaxation, the winds in the central and western Pacific remain weak for approximately a year; and (3) the winds in the eastern equatorial Pacific and along the coast of Peru are close to seasonal values during El Niño events.[33] Hence, Wyrtki hypothesized that the relaxation of the wind in the central and western equatorial Pacific would allow warm water, which had accumulated in that part of the ocean prior to El Niño, to flow eastward.[25] The eastward flow would be accomplished either by an internal Kelvin wave or by an increase in the transport of the North Equatorial Countercurrent. Eventually, the warm water would accumulate in the eastern Pacific and subsequently affect the coastal regions off Ecuador and Peru.

Wyrtki's hypothesis received theoretical support from Hurlburt et al.[2] and McCreary.[10,23] Both these modeling efforts showed that a Kelvin wave, excited by the relaxation of the wind stress, could account for the rapid, pronounced warming during the onset phase of El Niño. The numerical solutions of Hurlburt et al. reveal that nonlinear effects may help to explain the extremely rapid response observed during El Niño. The linear, analytic model of McCreary shows that a realistic distribution of the wind stress about the equator does not only produce the necessary Kelvin wave, but also can account for possible cross equatorial flow at the eastern boundary during El Niño.

Wyrtki's hypothesis, as well as the above modeling efforts, pertain only to the onset of El Niño. Generally, it is believed that more complicated models are required to account for the longer time-scale aspects of El Niño, such as the slow decline in sea level at the western boundary, the reoccurrence of temperature peaks in successive years and the rapid end to the El Niño event. The very simple numerical model from the last section is utilized to examine a model El Niño event from its inception to its termination.

It was shown that the numerical model results could account for a number of observed features of the seasonal cycle in the equatorial Pacific. Among other aspects, the model realistically represents the zonal slope of the main thermocline and the seasonal variation of its depth at the eastern boundary. The ability of the model to represent these features makes it a very attractive tool with which to examine fundamental dynamics related to El Niño. The strategy of the El Niño experiments in this chapter is to initialize the model with the seasonal solution at the end of year 2, and, subsequently, examine the anomalous model solutions resulting from the application of El Niño-type winds.

There are several detailed accounts of the variability of the zonal equatorial winds related to El Niño.[4,26,32,33] In an attempt to compromise between the complexity of directly applying real wind data and the over-simplicity of using highly idealized winds, we have made the following assumptions about the zonal equatorial winds associated with El Niño.

1 The zonal equatorial wind stress at interannual time scales has a much smaller amplitude in the eastern Pacific than in the western regions.

2 Even during El Niño periods, the seasonal cycle continues un-changed throughout the tropical Pacific.

3 The zonal distribution of the wind stress amplitude at interannual

time scales is similar to the distribution of the mean zonal wind stress.

4 During El Niño events, the magnitude of the wind relaxation is comparable to the value of the mean wind stress in that region.

There may be no single El Niño event in which all these assumptions apply. However, it is felt that these assumptions embody the fundamental behavior of the forcing during a major El Niño event. In addition, only the symmetrical response about the equator is examined. Although asymmetrical affects may be important to El Niño dynamics, it is the response associated with the Kelvin waves which is dominant, that is, the symmetrical solution.

The initial state for the standard El Niño case is the seasonal solution at the end of year 2. Subsequent numerical integration over 3 years represent the intensification of the winds prior to El Niño, the El Niño year, and the return to normal conditions, respectively. Only the mean component of the winds west of 140°W is modified during the numerical integrations. The seasonal components of the trade winds are maintained at their long-term values as well as the mean winds west of 140°W. During the third year of integration, the mean wind components west of 140°W are increased linearly over a 12-month period, at which time they are a factor of 1.5 greater than their long-term values. The sudden relaxation of the equatorial winds occurs at the start of year 4. The mean winds west of 140°W are decreased over a 2-month period to one half their normal mean values and are maintained at these values throughout the remainder of the year. At the beginning of year 5, the relaxed winds are increased to their long-term magnitudes over a 2-month period. After the first 2 months of the last year, the forcing is identical to the wind stress used in the seasonal simulation, and no further modifications are made.

The details of the standard run are described below. Two additional numerical simulations are performed; the differences between these cases and the standard run are shown in Table 2.

The intensification of the mean winds west of 140°W causes a westward zonal acceleration relative to the seasonal response (Figures 20b and 22a). The jet continues to accelerate rapidly until the arrival of an equatorially trapped Kelvin wave excited by the time variation of the wind stress divergence. The Kelvin wave, which propagates eastward, causes the zonal pressure gradient in the vicinity of the wind to increase, thereby helping to balance the increased wind stress force (Figure 22b). In the region to the east of the increased forcing, the Kelvin wave produces upwelling relative to the seasonal solution (Figure 23). As the

TABLE 2.   Description of Winds for El Niño Cases

	Description of Mean Winds West of 140°W During El Niño Experiments		
Case	Year 3	Year 4	Year 5
1	Winds increase over a 12 month period to 1.5 times their normal value (NV)	Relax over first 2 months to 0.5 their NV	Strengthen over first 2 months to NV
2	Same as Case 1	Relax over first month to 0.5 NV	Strengthen over 2 month period to NV beginning at month 3.5
3	Same as Case 1	1.5 NV for first 2 months. Subsequent relaxation is similar to Case 2	Same as Case 1

wind continues to increase, so do the westward flow and upwelling at the eastern boundary. At the end of year 3, the model thermocline is nearly 40 m shallower than its normal seasonal value at the eastern boundary (Figure 23). The east-west slope of the thermocline across the entire Pacific has intensified (Figure 22b). These conditions are representative of the anomalously cold periods observed prior to El Niño events in the eastern equatorial Pacific and along the coast of South America.

The mean winds west of 140°W decrease to one-half their normal value at the start of year four. The maximum westward wind stress in the model central Pacific Ocean changes from approximately 0.75 dynes cm^{-2} to 0.25 dynes cm^{-2}. As stated above, the magnitude of the relaxation is dependent upon the value of the mean wind stress in a given longitudinal band. The response to this decrease in the wind in the central and western Pacific is what Wyrtki has hypothesized to be the initiating mechanism of El Niño. Detailed discussions of the dynamics associated with the onset of the event can be found in Hurlburt et al.[3] and McCreary.[10,23] Only the major features of the onset of El Niño are described below.

The relaxation of the wind stress west of 140°W destroys the approximate balance between the wind stress and the zonal pressure gradient. Subsequently, the unbalanced pressure gradient produces an eastward

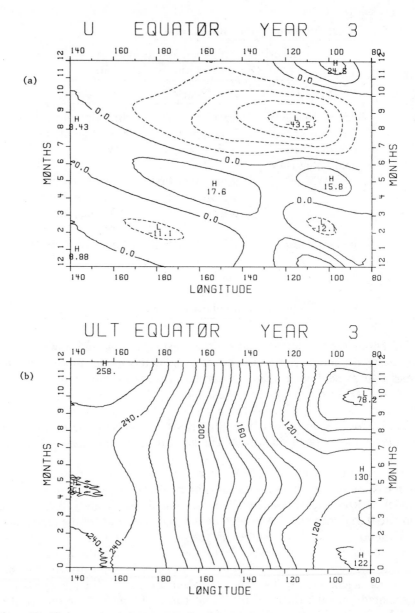

Figure 22 The x-t sections along equator for (a) zonal velocity component and (b) the upper layer thickness during the intensification phase of Case 1. During this year the mean winds west of 140°W are linearly increased to 1.5 times their long-term value. The contour interval is 10 cm sec⁻¹ for (a) and 10 m for (b).

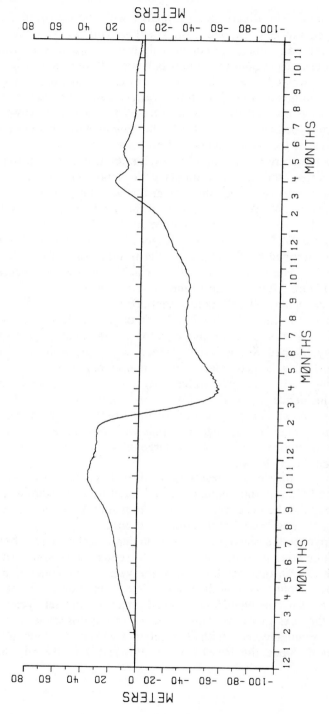

Figure 23 Time series of the departure of the ULT from the long-term seasonal solution during years 3, 4, and 5 of Case 1. The solution is evaluated at the eastern boundary of the equator.

acceleration. The divergence associated with the eastward driven flow excites equatorially trapped Rossby and Kelvin waves. In the vicinity of the decreased forcing, the equatorially trapped waves tend to bring the zonal pressure gradient into a new balance with the reduced wind stress. Hence, downwelling is initiated in the eastern portion of the anomalously forced region, and upwelling in the western half. The leading edge of the Kelvin wave excited by the wind relaxation induces downwelling as it propagates eastward from 140°W. The downwelling is stopped only by the arrival of the trailing edge of the Kelvin wave.

Since there are no waves in the system that propagate faster than the Kelvin wave, the eastern boundary does not feel the effects of the reduced wind stress until the arrival of the leading edge of the Kelvin wave (Figure 24). The combination of the incident Kelvin wave and the reflected Rossby waves create intense downwelling at the eastern boundary (Figure 25). The pycnocline deepens by 100 m within 2 months and is finally stopped by the arrival of the trailing edge of the Kelvin wave (Figure 24). The slight downwelling which continues between months 4 and 5 (Figure 25) is a result of the seasonal signal.

The deepening of the model thermocline at the eastern boundary represents the intense warming event observed during El Niño. The downwelling response at the equator is propagated poleward by a coastally trapped Kelvin wave. Hence, along the eastern boundary, a poleward flow is generated which, in the real ocean, is capable of advecting warm equatorial water along the coasts of Ecuador and Peru.

The incidence of the Kelvin wave onto the eastern boundary creates Rossby waves. The propagation of the Rossby waves away from the boundary broadens the region of poleward flow near the equator; this feature is representative of the large offshore scale of the warming observed during El Niño.

Thus far, most of the described features have been aspects that were modeled by the earlier simulations.[3,10,23] Let us now examine the model event beyond the onset time scale to see if such a model can continue to account for observed features during El Niño.

The pycnocline reaches its maximum depth at the eastern boundary in early June and, subsequently, begins to return to its long-term position (Figure 25). Throughout the rest of the year, the pycnocline appears to be strongly influenced by the seasonal cycle. However, Figure 23 shows that even if we disregard the seasonal effect, the model pycnocline rises during the next 4 months. This is a result of the Rossby wave which was excited simultaneously with the Kelvin wave at the time of the wind relaxation. Since the Rossby wave propagates westward, the leading edge initiates upwelling, while the trailing edge halts the motion. After reflection from the western boundary, this upwelling response can propagate rapidly across the basin as a Kelvin wave. Although the effects of this wave are not nearly as dramatic as the downwelling Kelvin wave, this upwelling event causes the pycnocline to shallow by about 20 meters.

Between June and October of year 4, the pycnocline at the eastern

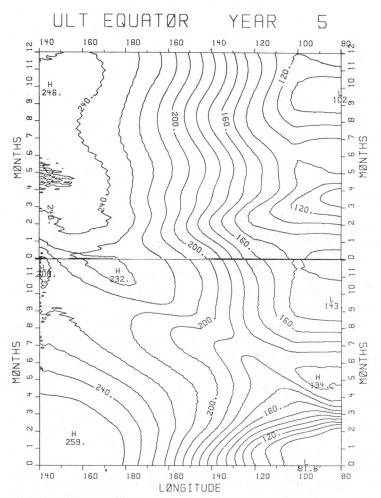

Figure 24 The x-t section of upper layer thickness along equator during years 4 and 5 of Case 1. The internal Kelvin wave excited by the wind relaxation west of 140°W reaches the eastern boundary near month 2.

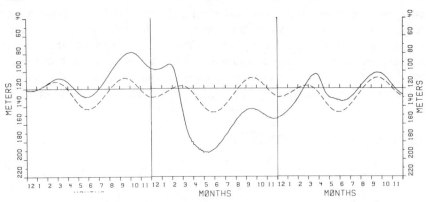

Figure 25 Time series of upper layer thickness at the eastern boundary during years 3, 4, and 5 of Case 1. The dashed line is the long-term seasonal value given by the numerical solution (Figure 18). The difference between the initial and final values in year 3 of the seasonal solution is caused by remnants of the initial spin-up transients.

boundary rises by about 50 meters (Figures 24, 25). The poleward flow along the eastern boundary is replaced by weak equatorward flow. Although the pycnocline remains well below its normal seasonal value, a certain semblance of the seasonal condition is reestablished. The numerical solution during the second half of year 4 is consistent with the quiescent period reported by Wooster and Guillen[29] and Zuta et al.[31] During this period it was observed that, although warm water anomalies persisted, the normal seasonal cycle had a significant effect. If, however, the seasonal increase of the Southeast Trades is anomalously weak during the Southern Hemisphere summer, the thermocline might remain far below its normal seasonal depth.

Finally, we examine the return to normal conditions during year 5 of the model El Niño. The strengthening of the trade winds west of 140°W induces a response that is qualitatively similar to the relaxation phase. An equatorial Kelvin wave is excited and propagates eastward. This Kelvin wave produces an upwelling response which, as before, is most intense at the eastern boundary (Figures 24, 25). The duration of the upwelling response is shortened by the incidence of the Kelvin wave emanating from the western boundary (Figures 23, 24). This Kelvin wave was generated by the reflection of the large amplitude Rossby wave near the end of year 4. The combined effects of the nearly simultaneous arrival of the downwelling Kelvin wave and the recently induced upwelling response tend to cancel at the eastern boundary. The remainder of the year at the eastern boundary is characterized by a gradual return to normal conditions (Figure 23).

The model El Niño event does not produce the second peak in sea level which is frequently observed during major El Niño events. Therefore, it is interesting to inquire whether a difference in the timing between the relaxation and restrengthening phases might produce such a response. Hence, Case 2 is designed to examine the solution at the eastern boundary for an event in which the relaxed winds remain at anomalously low values for a longer period than in Case 1.

The intensification of the winds prior to the relaxation phase is identical for both Cases 1 and 2. At the start of year 4, the winds are decreased in the exact manner as Case 1, except that the relaxation takes place over a 1-month period instead of 2 months. The wind remains at low values until month 3.5 of year 5, at which time they are increased as in the standard El Niño run. Hence, the difference in the duration of anomalously low winds between Case 2 and 1 is 4.5 months. The response at the eastern boundary is shown in Figure 26. The downwelling Kelvin wave reaches the eastern boundary prior to the upwelling event, and the double peak in pycnocline depth is observed. The second warming event is rapidly terminated by the upwelling Kelvin wave, which produces shallower thermocline depths at the eastern boundary during the remainder of the year.

Case 3 is an El Niño event in which the time between the relaxation phase and the reintensification period is shorter than in Case 1. Cases 3 and 1 are identical except that the decrease of the trade winds is delayed by 2 months, and the relaxation of the winds take place over a 1-month period. The corresponding El Niño events are quickly terminated by the upwelling response induced by the restrengthening of the wind (Figure 27). The pycnocline rises to levels that are much shallower than the normal seasonal values. This anomalously cold period is ended by the incidence of the downwelling Kelvin wave; conditions are very quickly returned to normal values.

It is recognized that the model equations for this section are the simplest possible representation of the equatorial Pacific circulation which can examine equatorial wave dynamics. Future numerical simulations of El Niño should include such features as the effects of the mean current field, a more realistic boundary configuration and forcing distribution, bottom topography, thermodynamics, and thermohaline mixing. However, the essential aspects of equatorial wave dynamics will still be presented in these more sophisticated modeling efforts. Hence, the dynamical scenarios of El Niño are offered as an attempt to understand the fundamental physics of the El Niño event. It is hoped that they will

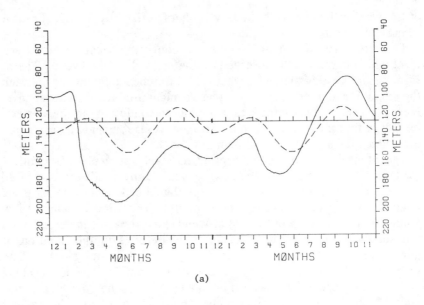

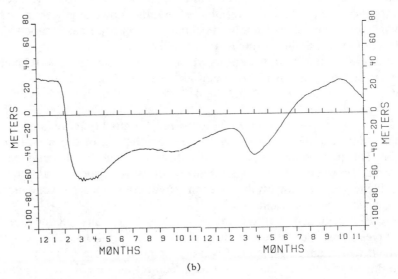

Figure 26 (a) Time series of upper layer thickness and (b) departure of ULT from long-term seasonal value during years 4 and 5 of Case 2. The dashed line in (a) is the average seasonal solution. The relaxation of the wind is similar to Case 1 except that the time scale for the relaxation is one month. The mean wind was increased to its normal value at month 3.5 in year 5.

208

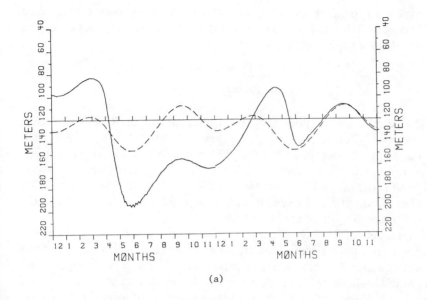

(a)

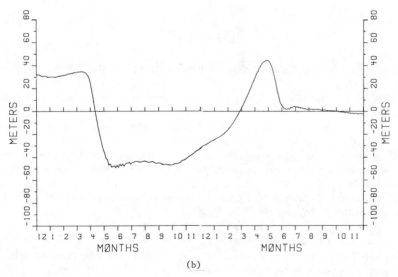

(b)

Figure 27 Same as Figure 26 except solution is for Case 3. The relaxation of the wind is the same as in Case 2 but delayed 2 months. The return to normal winds at the beginning of year 5 is the same as in Case 1.

209

be useful not only as an aid in the interpretation of more complicated models of El Niño but also as a starting point in the examination of the observations of this complex event.

ACKNOWLEDGMENTS

This work was supported by the Office of Naval Research, Ocean Science and Technology Branch, under contract N000-14-75-C0201. Partial support was provided by the International Decade of Ocean Exploration (IDOE) through the Coastal Upwelling Ecosystems Analysis (CUEA) program under Grant No. OCE78-00611 and NSF Grant ATM-7920485. Computations were performed on the CDC Cyber 74 at Florida State University, Tallahassee, Florida.

The models of El Niño in the last section are taken from a dissertation by one of us, John Kindle. The free Kelvin wave case will be described in Antonio Busalacchi's thesis in early 1980. Figure 5 was prepared by Mr. Stan Goldenberg for his thesis which is in preparation. J. J. O'Brien very humbly and gratefully thanks his wonderful students for their hard work and imagination.

REFERENCES

1 J. J. O'Brien, "El Niño—an example of ocean-atmosphere interactions," *Oceanus*, 21, 40–46 (1978).

2 W. H. Quinn, "Monitoring and predicting El Niño invasions," *Journal of Applied Meteorology*, 13, 825–904 (1974).

3 H. E. Hurlburt, J. C. Kindle, and J. J. O'Brien, "A numerical simulation of the onset of El Niño," *Journal of Physical Oceanography*, 6, 621–631 (1976).

4 B. Hickey, "The relationship between fluctuations in sea level, wind stress and sea surface temperature in the Equatorial Pacific," *Journal of Physical Oceanography*, 5, 460–475 (1975).

5 K. Wyrtki, E. Stroup, W. Patzert, R. Williams, and W. Quinn, "Predicting and observing El Niño," *Science*, 191, 343–346 (1976).

6 B. C. Weare, A. R. Novato, and R. G. Newell, "Empirical orthogonal analysis of Pacific sea surface temperature," *Journal of Physical Oceanography*, 6, 671–678 (1976).

7 T. P. Barnett, "The principal time and space scales of the Pacific trade wind fields," *Journal of Atmospheric Science*, 34, 221–236 (1977).

8 T. P. Barnett, "An attempt to verify some theories of El Niño," *Journal of Physical Oceanography*, 7, 633–647 (1977).

9 K. Wyrtki, "Sea level during the 1972 El Niño," *Journal of Physical Oceanography*, 7, 779–787 (1977).

10   J. P. McCreary, "Eastern tropical ocean responses to changing wind systems: With application to El Niño," *Journal of Physical Oceanography*, **6**, 632–645 (1976).

11   J. J. O'Brien and H. E. Hurlburt, "A numerical model of coastal upwelling," *Journal of Physical Oceanography*, **2**, 14–26 (1972).

12   H. E. Hurlburt and J. D. Thompson, "Coastal upwelling on a β-plane," *Journal of Physical Oceanography*, **3**, 16–32 (1973).

13   R. W. Garvine, "A simple model of coastal upwelling dynamics," *Journal of Physical Oceanography*, **5**, 460–475 (1975).

14   C. N. K. Mooers, C. A. Collins, and R. L. Smith, "The dynamic structure of the frontal zone in the coastal upwelling region off Oregon," *Journal of Physical Oceanography*, **6**, 3–21 (1976).

15   K. Yoshida, "Circulation in the Eastern Tropical Pacific Ocean with special references to upwelling and undercurrents," *Japan Journal of Geophysics*, **4**(2), 1–75 (1967).

16   R. L. Smith, "Poleward propagating perturbations in currents and sea levels along the Peru coast," *Journal of Geophysical Research*, **83**, 6083–6092 (1978).

17   R. Preller and J. J. O'Brien, "The influence of bottom topography on upwelling off Peru," *Journal of Physical Oceanography* (1980, in press).

18   D. W. Moore and S. G. H. Philander, "Modeling of the Tropical Oceanic Circulation," in E. D. Goldberg, I. N. McCave, J. J. O'Brien, and J. H. Steele, Eds., *The Sea, Vol. VI*, Wiley, New York, 1977, pp. 319–362.

19   G. Meyers, "Annual variation of the slope of the 14°C isotherm along the equator in the Pacific Ocean," *Journal of Physical Oceanography*, **9**, 885–891 (1979).

20   J. C. Kindle, "Equatorial Pacific Ocean variability—seasonal and El Niño time scales," Ph.D. Dissertation, Florida State University, Tallahassee, 1979.

21   B. Taft and J. Jones, "Measurements of the Equatorial Undercurrent in the Eastern Pacific," *Progress in Oceanography*, Vol. 6, Pergamon Press, Elmsford, 1974, pp. 47–110.

22   M. Tsuchiya, "Variations of the surface geostrophic flow in the eastern intertropical Pacific Ocean," *Fishing Bulletin*, **72**, 1075–1086 (1974).

23   J. P. McCreary, "Eastern ocean response to changing wind systems." Ph.D. Dissertation, University of California, San Diego, 1977.

24   J. Bjerknes, "El Niño study based on analysis of ocean surface temperatures, 1935–1957," *Bulletin of Inter-American Tropical Tuna Commission*, **5**, 217–303 (1961).

25   J. Bjerknes, "Survey of El Niño 1957–58 in its relations to tropical Pacific meteorology," *Bulletin of Inter-American Tropical Tuna Commission*, **12**, 25–86 (1966).

26   K. Wyrtki, "El Niño—the dynamic response of the equatorial Pacific Ocean to atmospheric forcing," *Journal of Physical Oceanography*, **5**, 572–584 (1975).

27   K. Wyrtki, "The response of sea surface topography to the 1976 El Niño," submitted to *Journal of Physical Oceanography* (1979).

28   C. Caviedes, "El Niño 1972: Its climatic, ecological, human and economic implications," *Geographical Review*, **65**, 493–509 (1975).

29   W. S. Wooster and O. Guillen, "Characteristics of El Niño in 1972," *Journal of Marine Research*, **32**, 387–404 (1974).

30   J. Namias, "Some statistical and synoptic characteristics associated with El Niño," *Journal of Physical Oceanography*, **6**, 130–138 (1976).

31 S. Zuta, D. B. Enfield, J. Voldivia, P. Lagos, and C. Blandin, "Physical aspects of the 1972–73 El Niño phenomenon," in the Guayaquil Workshop *Report on the Phenomenon Known as El Niño*, U.N. Food and Agriculture Organization, Rome, 1975.

32 K. Wyrtki, "Fluctuations of the dynamic topography in the Pacific Ocean," *Journal of Physical Oceanography*, **5**, 450–459 (1975).

33 K. Wyrtki and G. Meyers, "The trade wind field over the Pacific Ocean," *Journal of Applied Meteorology*, **15**, 698–704 (1976).

CHAPTER 8

El Niño
Pacific Eastern Boundary Response
to Interannual Forcing

DAVID B. ENFIELD

Abstract

Observational results from the Pacific eastern boundary region are presented and discussed in relation to previous observations and theory concerning El Niño. The historical development of the wave dynamical theory of El Niño is reviewed and a numerical simulation of an equatorial Kelvin wave impingement on the eastern boundary is described. Observational evidence of wavelike propagation of sea level fluctuations poleward along the eastern boundary at both high and low frequencies is reviewed. El Niño-related interannual fluctuations of sea level are coherent from northern Chile to the Canadian border and have a phase structure consistent with poleward long wave propagation. Hydrographic data from the region off Peru and Ecuador confirms the occurrence of thermal structure depression and its poleward spreading during El Niño, consistent with theory and previous observational results. The anomalies of monthly averaged wind speeds along the Peru coast confirm that the winds remain strong and upwelling favorable during El Niño, decreasing slightly at Talara, and increasing at Lima. Evidence is given for the existence of continued upwelling during El Niño. The interaction of isotherm depression, winds, upwelling, and sea surface temperature anomalies is discussed.

INTRODUCTION

The most recent phase of research on the oceanic aspects of the El Niño phenomenon essentially began with the work of Wyrtki in the mid 1970s.[1] Long before, it had been recognized that large-scale interannual

213

fluctuations occurred in the atmospheric pressure distributions over the South Pacific, one aspect of which was the marked decrease of surface pressure at Santiago, Chile and Easter Island that occurred during El Niño disturbances. Easter Island lies near the climatic center of the South Pacific anticyclone, hence the pressure decrease there implied a general weakening of the South Pacific trade wind system, which extends north of the equator across most of the Pacific. Also, prior to the 1960s the dramatic effects of El Niño were generally thought to be limited to a region along the coast of South America, especially off Peru. It was natural then to assume that the phenomenon resulted from weakened southeast trade winds off South America.

Data taken during the strong El Niño of 1957–1958 suggested that a much larger area was affected, including the mid-equatorial Pacific and its current system. It was Bjerknes who suggested a large scale ocean relaxation in which equatorial trade wind weakening played a key role.[2] But it was still thought that the ocean-atmosphere relaxation occurred in the eastern tropical Pacific as well, with the cessation of coastal upwelling being cited as a key characteristic of El Niño.

Then Wyrtki noted that trade wind weakening occurred mainly west of 140°W, meaning that a Bjerknes-type ocean relaxation could not take place all the way to the South American coast.[1] Wyrtki therefore proposed that a mid-equatorial relaxation *propagated* eastward in the form of a downwelling Kelvin wave that impinges on the eastern boundary, depressing the thermal structure there and raising sea level. Since Wyrtki's work, a number of theoretical and numerical studies have demonstrated the dynamical feasibility of the proposed mechanism, hence the Wyrtki hypothesis has come to be known as the wave dynamical theory of El Niño.

Unfortunately, although Wyrtki himself presented evidence in support of his hypothesis, observational treatments of the eastern boundary response during El Niño have lagged the theoretical and numerical work. More analysis must be done on wind and hydrographic data from the eastern boundary region to confirm the wave dynamical mechanism and demonstrate how it results in the anomalies encountered there. The purpose of this chapter is to present such an analysis, synthesizing a more coherent concept of the eastern boundary response and its relation to the wave dynamical theory from the earlier and more recent results of others as well as new findings by this author.

In the following section the development of the wave dynamical concept of El Niño is reviewed and a description is given of an idealized wave impingement on the eastern boundary. The sections on Eastern Boundary Sea Level, Ocean Thermal Structure Anomalies, Interannual Variability of Winds, and Peru Coastal Upwelling During El Niño Events discuss observational results concerning eastern Pacific sea level, hydrography, and winds, respectively. The Summary synthesizes these results in a discussion of coastal upwelling during El Niño and how it is related to the anomalous features observed.

BACKGROUND

The observational results presented in the sections on Eastern Boundary Sea Level, Ocean Thermal Structure Anomalies, Interannual Variability of Winds, and Peru Coastal Upwelling During El Niño Events relate to Wyrtki's hypothesis that El Niño is primarily an oceanic response to anomalous trade wind weakening in the central equatorial Pacific and that the response takes place in a wave dynamical fashion.[1] The purpose of this section is to set the stage for the results to be presented by reviewing the developments that led to this hypothesis as well as the subsequent theoretical and numerical work by which the proposed mechanism has been better understood. Although the El Niño phenomenon is indeed a large-scale process, involving most of the Pacific Ocean, the focus here is on the region where that process got its name and where the anomalies and their impact are strongest: the eastern tropical Pacific.

Observational Results and Conceptual Models

Bjerknes noted that the prevailing easterly winds in the equatorial Pacific force a westward drift of surface water and result in a zonal sea surface setup such that warm surface water accumulates toward the west resulting in a west-to-east sea surface downslope.[2] At the equator, where Coriolis effects are nonexistent, the sea surface downslope is associated with an eastward flow (the Equatorial Undercurrent, or Cromwell Current) immediately below the westward wind drift; to either side of the equator this eastward flow also has a slight equatorward component, geostrophically balanced by the zonal sea surface slope. This geostrophic convergence balances the Ekman divergence in the wind drift layer above. Bjerknes argued that in the event of a cessation of the easterly winds, the westward wind drift, the surface Ekman divergence and the equatorial upwelling must also cease. This should allow both the subsurface eastward flow and the geostrophic convergence to extend their effects into the surface layer. The mass imbalance resulting from the net convergence should then cause the ocean surface to bulge upward and water to downwell, depressing the thermocline.

Bjerknes gave evidence from Pacific island wind measurements (Can-

ton Island) and from basin-wide fluctuations of atmospheric pressure that
the Pacific trade wind system undergoes a large scale weakening during
El Niño occurrences, specifically in the case of the 1957–1958 El Niño.[2]
The above mechanism of successive atmosphere and ocean relaxation
was therefore proposed to explain the isotherm depression observed in
the mid-equatorial Pacific in 1957.[3] Although Bjerknes did not show
direct wind measurements for the eastern Pacific and the Pacific coast of
South America, it is clear that he believed both the trade wind weakening
and an oceanic relaxation mechanism to apply to that region as well:

> Weakness of the trade winds has more sudden and spectacular ocean
> effects in terms of temperature induced by the cessation of upwelling. The
> areas directly affected are a narrow coastal strip off Chile and Peru plus an
> equatorial, not quite so narrow, strip extending westward to the dateline.[2]

Thus, Bjerknes visualized the El Niño as being an ocean-atmosphere
process occurring in much the same way (by trade wind weakening and
the cessation of upwelling) over two ocean areas characterized by
anomalously warm sea surface temperature (SST): the equatorial Pacific
and the Pacific coast of South America. The assumption that warm SST
anomalies along the South American coast are caused by the cessation
or weakening of coastal upwelling (and upwelling favorable winds) is
often encountered in the literature.[4–6]

Zuta et al. found little evidence from the monthly averaged winds at
Peruvian airports (1971–1973) that the upwelling favorable winds weakened along the Peru coast during the strong El Niño of 1972–1973.[7] Only
the Talara winds (4½°S) showed any signs of variation—they were
somewhat weaker and/or from a different octant during 1972—but nevertheless remained upwelling favorable. Thus, although occasional southward invasions of tropical surface waters as far as 7–9°S could perhaps
be explained by brief periods (less than a month) of weak winds off
Ecuador and northern Peru, the orthodox explanation of high SSTs—
over the coast and for more than a year—was not supported by wind
data for the 1972–1973 event. It appeared that the classical relaxation
mechanism, epitomized in Bjerknes' explanation,[2] was either not valid in
general or did not apply to events in the eastern tropical Pacific.

Wyrtki questioned the existing orthodoxy concerning El Niño.[1] He
examined 25 years of ship-reported winds averaged monthly over 30°
longitude by 10° latitude zonal bands in the equatorial Pacific, and over
a 10° square off the coast of Peru (10–20°S; 70–80°W). The zonal
component of the SE trades from 120–150°W, and to a greater extent
from 150–180°W, became stronger during the characteristically cold

years preceding El Niño, and decreased markedly at the onset of El Niño events. Wyrtki concluded—as had Bjerknes—that during the periods of strong SE trades there was a sea level setup in the western equatorial Pacific and that during the subsequent wind relaxation there would be a release of the zonal potential energy available in the west-to-east sea level downslope.

However, Wyrtki's analysis showed no weakening of the zonal wind component in the eastern equatorial Pacific (0–10°S; 90–120°W) nor of the upwelling favorable winds of Peru. Thus it was not possible for a basinwide ocean relaxation to take place simultaneously as suggested by Bjerknes.[2] Rather, Wyrtki argued that the energy release would take place intially in the mid-equatorial Pacific (the region of wind weakening) and would propagate rapidly eastward in the form of a baroclinic disturbance, much like an equatorially trapped internal Kelvin wave. The basic scenario for oceanic relaxation proposed by Bjerknes would apply at each point of passing of such a disturbance, that is, the sea surface would bulge up at the equator, the thermocline would be depressed, and eastward flow would accelerate and become more extensive, laterally and vertically.[2] Upon arrival of the disturbance off South America there would be a deepening of isotherms and a massive influx of warmer water. To support his arguments, Wyrtki found evidence from the reports of the Instituto del Mar Peru (IMARPE) that the 15° isotherm was significantly deeper along the Peru coast in 1972 than in the relatively cold (anti-El Niño) years of 1964 and 1967.

Theoretical and Numerical Models

Wyrtki's proposed mechanism has provided the observational paradigm for most of the subsequent theoretical and numerical models, which have in turn contributed to our conceptualization of the El Niño process. Lighthill had already proposed that equatorially trapped waves play an important role in the response of the Indian Ocean to monsoon forcing.[8] Godfrey analytically and numerically investigated the spindown of a two-layer subtropical gyre.[9] During the deceleration process, a warm pool initially in the western portion of the gyre spread eastward along the equator and then poleward along the eastern boundary (i.e., in the antigyral sense) in the form of downwelling waves similar to boundary trapped internal Kelvin waves. However, because of the highly idealized nature and limited basin dimensions of the model, it is difficult to make a direct application of it to El Niño. Another theoretical model of ocean spindown was examined by White and McCreary.[10] Salient features of their model included the deceleration of the anticyclonic gyre and the offshore movement of the eastern boundary current system, but boundary trapped internal Kelvin waves were not produced because the pycnocline was fixed along the boundaries. Godfrey found internal Rossby waves propagating westward from the initial disturbance and White and McCreary found them propagating westward from the eastern boundary. A good discussion of these models is found in Hurlburt et al.[11]

Hurlburt et al. numerically simulated an El Niño response by spinning up a two-layer, nonlinear, primitive equation model with a uniform easterly wind stress, and subsequently turning off the wind forcing over the entire ocean.[11] Because the wind forcing and relaxation are basinwide, after the stress is turned off Kelvin waves are excited at both ends of the equator. At the eastern end, a Kelvin wave front immediately propagates poleward along the eastern boundary from the equator. As the front passes a given point on the eastern boundary, the alongshore flow there reverses from equatorward to poleward and the pycnocline deepens. Subsequently there is a separation of the poleward flow from the coast due to internal Rossby wave propagation westward from the eastern boundary; this separation occurs initially and most rapidly close to the equator, then successively and more slowly at higher latitudes due to the phase speed dependence on latitude of the Rossby waves (see also Cane and Sarachik[12] and the disucssion in the next section). At the moment the wind is turned off, the internal Kelvin wave excited at the western boundary propagates eastward along the equator, arrives at the eastern boundary a few months after the onset of relaxation, and propagates poleward along the eastern boundary.

The importance of the Hurlburt et al. model is that it resolves the internal radius of deformation near the equator and the boundaries and that it takes into account the effect of a varying pycnocline depth (due to the waves themselves) on propagation speeds. Along the eastern boundary, the latter effect causes a sharp internal wave front to develop. In effect, since the phase speed increases in proportion to the upper layer thickness, the trailing portion of the disturbance (where the pycnocline is deeper) tends to "catch up" to the leading portion, resulting in a sharpened downwelling front. Therefore, the model reproduces the suddenness of the El Niño onset observed by others, for example, Wooster and Guillen.[5]

A limitation of the model is the fact that the wind effects are felt basinwide and are not confined to the western and mid-equatorial ocean. This results in a bias in the timing of Kelvin wave effects at the eastern boundary (O'Brien, personal communication).

McCreary used a linear reduced-gravity primitive equation model to investigate the dynamic response of the eastern Pacific Ocean to different symmetries of the wind field relaxation over the interior ocean.[10] Variations of meridional winds produced no El Niño response, nor did the zonal winds outside of a relatively narrow ($\pm 5°$ of latitude) equatorial band. An asymmetric change of the equatorial zonal wind field produced the most realistic El Niño effects, including isotherm deepening and crossequatorial (southward) flow in the eastern Pacific and poleward flow along the eastern boundary.

The Wave Dynamical Process at the Eastern Boundary

It is appropriate here to recapitulate and gain a better understanding of the process that takes place as the eastward propagating equatorial disturbance arrives at the eastern boundary. As originally shown by Moore[13] (see also Moore and Philander[14]), theoretical results indicate that equatorially trapped internal waves at a single frequency incident on the eastern boundary may be partially transmitted poleward along the coast in the form of coastal trapped internal Kelvin waves. As already noted, this feature was reproduced by the numerical models of Godfrey[9] and Hurlburt et al.[11] for the linear and nonlinear cases, respectively. Cane and Sarachik show theoretically how at very low frequencies (such as characterize the El Niño) the energy radiation off the eastern boundary in the form of Rossby waves becomes more important.[12] This distinction regarding the low frequency wave phenomenon should be kept in mind in the discussion that follows.

It will be useful to imagine a single large scale disturbance as it arrives at the eastern boundary, albeit comprised of superimposed component waves. Figure 1 is redrawn from a portion of Figures 8 and 10, discussed by O'Brien (Chapter 7), and illustrates the following discussion. It shows the pycnocline anomaly field of an equatorial internal Kelvin wave disturbance as it arrives at the eastern boundary of an ocean initially at rest. It is from a numerical simulation using a linear model consisting of a two-layer ocean with basin dimensions similar to the Pacific and with open boundaries at 18°N and 12°S. In the actual simulation the Kelvin wave disturbance was an upwelling pulse generated by abruptly turning on a uniform wind field in the western Pacific, but for the purposes of this discussion it is depicted as a downwelling pulse and is shown only for the portion of the basin where direct wind forcing is absent and the disturbance is freely propagating.

The incident disturbance (first panel) is confined to an equatorial band 5–6° of latitude on either side of the equator, as determined by the equatorial radius of deformation. The leading edge, or front, at the eastern extremity of the disturbance is not a sharp one as in Hurlburt et

David B. Enfield

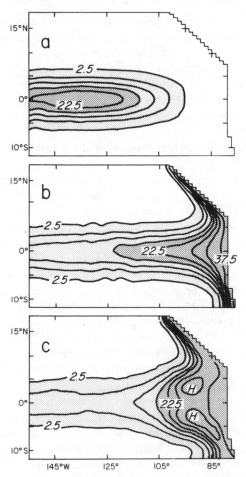

Figure 1 Numerical simulation of a freely propagating, linear, equatorially trapped internal Kelvin wave that impinges on the eastern Pacific boundary, depicted for the downwelling case. Depression of the two layer interface is given in meters. After the simulation discussed in Chapter 7.

al.,[11] because the simulation is linear. Within less than a month (after the first panel) the disturbance arrives at the eastern boundary and the pycnocline deepens as the disturbance spreads poleward into both hemispheres. The poleward propagation speed of the leading edge (downwelling front) is equal to the Kelvin phase speed as determined by the layer thickness and is independent of latitude.[12] It is also pointed out by Cane and Sarachik that the coastal internal radius of deformation, and therefore also the offshore width of the disturbance in the vicinity of the leading edge, decreases with increasing latitude.[12]

The second panel (one month after the first panel) shows the situation after the leading edge has passed beyond the open boundaries to the north and south. Behind each advancing downwelling front there is an elongated zone where the pycnocline slopes sharply upward offshore, from the nearshore depressed area toward the undisturbed ocean interior. This quasi-alongshore front is contiguous with the advancing downwelling front, and begins to advance offshore via internal Rossby wave propagation westward, as shown by Godfrey,[9] Hurlburt et al.[11] and Cane and Sarachik.[12]

By the time of the third panel (two months after the first panel), the downwelling fronts have propagated poleward far beyond the open boundaries and the thickened upper layer has expanded offshore in the coastal region. A westward propagating bulge develops near the equator—the effect of the Rossby waves, which propagate more rapidly at low latitudes. At higher latitudes the rate of offshore migration of the alongshore pycnocline front is progressively less, because Rossby waves with progressively smaller propagation speeds prodominate with increasing distance from the equator. This effect is treated in theoretical detail by Cane and Sarachik.[12]

Summary

Theoretical and numerical studies in the last 5 years tend to support Wyrtki's argument that El Niño is primarily a wave dynamical response to trade wind forcing over the interior of the equatorial Pacific Ocean.[1] Rapid thermocline depression is to be expected in the eastern equatorial Pacific at the onset of El Niño, as the equatorial disturbance arrives at the eastern boundary. The incident energy will be partitioned into two phases: a Kelvin-like downwelling front that propagates rapidly poleward along the eastern boundary, and a Rossby wave reflection westward off the eastern boundary, perceived as a gradual offshore widening of the thickened upper layer along the coast.

EASTERN BOUNDARY SEA LEVEL

One approach to testing the wave dynamical theory of El Niño is to look for observational evidence of poleward propagation along the eastern boundary. Using reasonable values for an upper layer thickness and density contrast between layers, the propagation speed of a coastally trapped internal Kelvin wave in a two-layer ocean is about 100–200 km/ day. It should be possible to distinguish the propagation of such a disturbance from that due to advective processes alone, which involve speeds an order of magnitude smaller. Since the passage of a downwelling (upwelling) coastal baroclinic wave is accompanied by a sea surface rise (fall), the sea level oscillations should be detectable at coastal tide stations, and the propagation speed determined from the delay time of arrivals between stations. Moreover, sea level is a particularly appropriate variable to examine because its fluctuations tend to go as the integral of the changes in the upper layer alongshore flow between the coast and one radius of deformation offshore. Sea level is therefore relatively insensitive to local processes occurring in the vicinity of the tide gauge, unlike, for example, shore temperature.

Smith analyzed 6-hourly time series of coastal sea level and alongshore shelf currents (100 m isobath) between 10°S and 15°S along the Peru coast for a 45-day period during 1976.[15] The fluctuations were not well correlated with the local wind, but sea level and the alongshore currents were coherent between 12°S and 15°S in the subtidal frequency band (0.05–0.25 cpd). Phase spectra indicate a poleward propagation of 200 km/day in this band, which also agrees with the maximum lagged cross correlations between stations. The propagating fluctuations were associated with variability in the vertical temperature structure over the shelf and slope in a manner consistent with the thermal wind relation. Smith concluded that the observations available were consistent with baroclinic Kelvin waves that originate equatorward of 10°S and propagate at least to 15°S.[15] Christensen and Rodriguez found sea level oscillations off Baja California (29–32°N) to be largest at periodicities of 15–30 days and also uncorrelated with the local weather, but their study did not consider the possibility of alongshore propagation.[16] There is evidence that similar, nonlocally forced sea level oscillations propagate poleward along the Pacific coast of mainland Mexico, but don't emerge from the Gulf of California to continue up the Pacific coast of Baja California (Christensen, personal communication).

While these studies suggest that the low latitude Pacific eastern boundary may be an efficient waveguide for baroclinic disturbances, they involve time series of less than a year and are therefore incapable of resolving the very low frequency (interannual) oscillations characteristic of the El Niño time scale. Roden found alongshore coherence at frequencies of ≤ 6 cpy between long records (50 years or more) of monthly averaged sea level at Pacific North American stations as far apart as San Francisco (38°N) and San Diego (33°N) or Sitka (57°N) and Neah Bay (48°N).[17] But he did not consider propagation effects, the relationship of sea level variability to El Niño cycles, nor alongshore coherence involving stations south of San Diego.

Enfield and Allen analyzed 25 year time series (1950–1974) of monthly averaged sea level anomalies along the Pacific eastern boundary from Yakutat, Alaska (59°N) to Valparaiso, Chile (33°S).[18] The visual coherence of the high sea level events associated with El Niño (e.g., 1957–1958, 1965, 1969, and 1972–1973) are evident in Figure 2. Enfield and Allen found that the sea level fluctuations were lag correlated at above the 99% significance level between stations from Prince Rupert (54°N) to Matarani (17°S).[18] The variability at each station from Crescent City (42°N) to Antofagasta (23°S) was lag correlated at above the 99% significance level with the Southern Oscillation Index, as defined by Quinn,[19] indicating a statistical correspondence to El Niño-related fluctuations over a large latitude range. The areal correlation between sea level stations at separations of 3,000 km or more appeared to be associated mainly with this interannual time scale (i.e., the El Niño-related variability). The sea level fluctuations at North American stations were found to lag those at near-equatorial stations by 0–2 months, indicating poleward propagation and leading to speed estimates of 180 ±100 km/day. This is consistent with baroclinic wave propagation and is much faster than can be expected from purely advective processes.

Enfield and Allen[18] also did a spectral analysis of the sea level anomalies of Figure 2 that extends the results of their correlation analysis into the frequency domain. Their results are summarized below, illustrating the main features of the eastern boundary sea level response. Figure 3 shows the distribution of coherence between anomalous sea level along the eastern boundary and the anomalies of the Southern Oscillation Index (SOI). The SOI was calculated as the difference in surface atmospheric pressure between Easter Island and Darwin, Australia. Quinn[19] showed that the SOI is representative of Pacific trade wind forcing and is anomalously low (high) during the El Niño (anti-El Niño) years characterized by high (low) SST off Peru and Ecuador. For most of the eastern boundary, the coherence is consistently greater at interannual frequencies (2–5 years) than at higher frequencies. The interannual coherence exceeds the 95% significance level from northern Chile to the Canadian border, except in the Panama Bight.* Thus the effects of interannual trade wind forcing and El Niño-related variability are reflected in eastern Pacific sea level considerably poleward of the

---

*Enfield and Allen attribute the lower coherence in the Panama Bight to the anomalous behavior of the trade wind regime there, which tends to counteract the sea level oscillations associated with El Niño occurrences.[18]

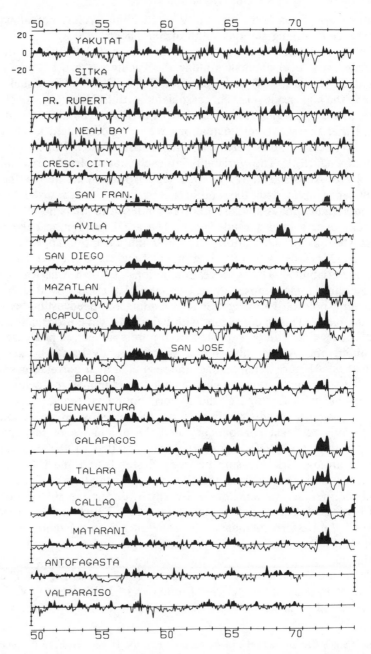

Figure 2 Time series of the monthly anomalies of sea level (cm) at 19 stations along the Pacific eastern boundary from Yakutat, Alaska (59°N) to Valparaiso, Chile (33°S). Positive anomalies are shaded black. From Enfield and Allen.[18]

224

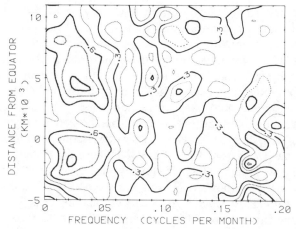

Figure 3 Distribution of coherence between the Southern Oscillation Index[19] and the anomalies of sea level at 19 stations from Yakutat, Alaska (59°N) to Valparaiso, Chile (33°S). Positive ordinate values give the alongshore distance north of the equator, and shading indicates significance at or above the 95% level. From Enfield and Allen.[18]

Peru-Ecuador region where the oceanic manifestations of El Niño are sterotyped.

Figure 4 gives the distribution of areal coherence of anomalous sea level as a function of the alongshore distance between stations. The plot is based on 78 coherence spectra from both hemispheres, obtained by

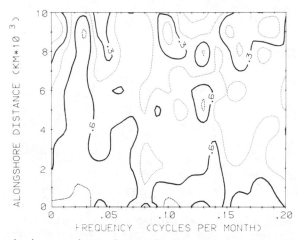

Figure 4 Areal coherence of anomalous sea level as a function of the alongshore distance between stations. Shading indicates coherence at or above the 95% level. From Enfield and Allen.[18]

crossing the record for each station equatorward of 40° latitude with those of all stations poleward of that station. Sea level at stations separated by less than 2000 km tends to be coherent at or above the 95% significance level over most of the frequency range. At separations of 2000–4000 km significant coherence occurs only at subannual frequencies. For separations in excess of 4000 km significant coherence is limited to a narrower interannual band corresponding to periodicities of 2.5–4 years. It is remarkable that this coherent band extends to separations of nearly 10,000 km, and that it also occupies the frequency range for which sea level is most coherent with the SOI (Figure 3). Enfield and Allen[18] attributed the wide-band (high frequency) sea level coherence at short separations to the coupling effects of local synoptic scale forcing, such as weather patterns or localized ocean circulation effects.[18] The large scale coherence at interannual frequencies could not be accounted for by atmospheric coupling mechanisms such as the teleconnective processes proposed by Bjerknes.[20]

In Figure 5 the phase lags corresponding to all significant (> 95%) interannual (2–5 years) coherence estimates from Figure 4 are grouped into 2000 km intervals of station separation. Sea level records are

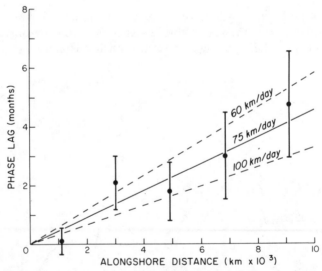

Figure 5 Distribution of the mean (±95% confidence interval) phase lag between sea level records corresponding to significant (≥95%) coherence estimates in the interannual band (2–5 years) of Figure 4, shown as a function of station separation and grouped by 2000 km intervals of alongshore distance. The zero-intercept least squares propagation line (solid) and its ±95% confidence interval (dashed) are also shown. From Enfield and Allen.[18]

consistently phased in the sense of poleward propagation and a zero-intercept regression yields a slope corresponding to a phase speed of 75 km/day. Such propagation speeds are inconsistent with purely advective processes but can be explained by wave phenomena of the sort discussed in the section on the wave dynamical process at the eastern boundary.

A noteworthy feature of Figure 2 is that the El Niño-related sea level anomalies appear to be only slightly smaller in amplitude along the coast of California and Mexico than along the Peru coast. This is qualitatively possible, even though energy may be "leaked offshore" in the form of Rossby waves, because the poleward propagating fraction is confined to a narrower coastal band with increasing latitude (due to the poleward decrease of the internal radius of deformation). Thus, if the total energy in the poleward propagating fraction is reduced by bottom friction and offshore radiation of Rossby waves at a rate no larger than that of the reduction of the internal radius of deformation ($\propto f^{-1}$), then the energy density (and therefore the sea level amplitude) of the disturbance will not decrease. This appears to hold approximately in the northern hemisphere. The inverse of this relationship was pointed out by Hurlburt et al. to explain why their simulation did not produce an *increase* in amplitude with increasing latitude.[11]

There is some indication from Figure 2 that the sea level fluctuation amplitudes decrease faster with latitude south of the equator than to the north (compare, for example, Antofagasta and Valparaiso with Talara). Enfield and Allen found the areal coherence of sea level to be weaker between the Chilean stations and near-equatorial stations than is the case for comparable stations in the northern hemisphere.[18] Hence there may be a hemisperic asymmetry in the effectiveness of poleward propagation. For example, one can speculate that frictional dissipation and/or the reflection of incident equatorial wave energy in the form of westward propagating Rossby waves may be more important south of the equator, leaving less energy available for poleward propagation along the South American coast.

The results reviewed in this section strongly suggest that the Pacific eastern boundary sea level response to interannual forcing is wave dynamical in character, consistent with the observational and theoretical considerations discussed in the Background section. But two qualifications must be made. First, the possibility of other mechanisms for the large-scale areal coherence of sea level has not yet been adequately tested. Such mechanisms tend to fall in the class of atmospheric teleconnections (e.g., Bjerknes[20]) or direct atmospheric forcing of sea level fluctuations along the eastern boundary. Preliminary results by Enfield and Allen suggest that these mechanisms may not be important.[18]

A second consideration is that wave propagation may not be the only type of eastern boundary sea level response at interannual frequencies. For example, observations taken during El Niños usually reveal the anomalous presence of low salinity surface water off the Peru coast, south of 5°S (see Wooster and Guillen[5] and Zuta et al.[7]). The only source for these salinities is the tropical water north of the Equatorial Front, clearly indicating an advective process. This aspect of El Niño events seems to occur only episodically during the summer-fall season (southern hemisphere) and can perhaps be explained as due to the anomalous poleward advection in the wake of a poleward propagating Kelvin wave front.[11] Current meter measurements off central Peru (12–15°S) during the El Niño of 1976 indicated strong poleward velocities to depths of over 100 m over the shelf and slope.[21] Assuming that these currents are anomalous,* they may also be explained in terms of a wavelike response and may drive the surface flows. But the possibility exists that localized surface advection occurs episodically, independent of wave dynamical effects. It seems unlikely that large-scale circulation anomalies could bring about significant coherence in coastal sea level over the large distances and short time scales suggested by Figure 5.

In summary, there are strong indications from sea level records that the Pacific eastern boundary response to interannual forcing is of a large scale and involves a wave dynamical mechanism. The interannual coherence of sea level appears to extend farther into the northern hemisphere than into the southern hemisphere, suggesting an asymmetric response. An additional asymmetry arises from surface advective effects that occur south of the equator, although such advection is not necessarily inconsistent with the wave dynamical mechanism.

OCEAN THERMAL STRUCTURE ANOMALIES

Under the wave dynamical theory of El Niño, as the easterly winds in the central and western Pacific relax, a downwelling equatorially trapped Kelvin wave forms in the mid-Pacific and quickly propagates eastward. The nonlinear numerical model of Hurlburt et al.[11] suggests that as the disturbance arrives in the eastern Pacific, the downwelling of the pycnocline there should be sudden. The observation of Wyrtki that sea level rapidly peaks off Peru at the onset of El Niño is consistent with this.[22] The numerical results also indicate a poleward spreading of the down-

---

*Current meter measurements have never been made along the Peru coast during a "cold" (anti-El Niño) year, therefore the abnormality of the measured flow remains open to question.

welling event along the eastern boundary. The downwelling process should be most clearly manifested by anomalous isotherm depression.

Historical documentation of El Niño-related thermal structure anomalies in the eastern equatorial Pacific is scarce. The 15° isotherm topographies of Wyrtki[1] and the temperature sections of Zuta et al.[7] and Enfield[23] provide evidence of isotherm depression off Ecuador and Peru during the 1972 El Niño. For the most part, the existing analyses are inadequate in their time and space coverage to provide a clear picture of the process.

One problem is that the seasonal oscillation of eastern Pacific thermal structure is imperfectly known, although it has been treated by Wyrtki based on earlier, relatively scarce data.[24] More recently, Meyers has documented the seasonal behavior of the 14° isotherm along the equator in the Pacific, using bathythermograph data.[25] The 14°C isotherm is located near the base of the tropical thermocline, just above the 13°C thermostad described by Stroup.[26] Between Ecuador and the Galapagos Islands, the 14°C isotherm has been observed as shallow as 50 m and as deep as 260 m (Meyers' Figure 3), whereas the long term mean depth is about 150 m and the average seasonal oscillation is from 130–170 m (Meyers' Figures 3 and 4). Hence the nonseasonal departures from the mean of the 14°C isotherm depth east of the Galapagos appear to be much greater than its seasonal displacement. Assuming a two layer ocean system with the lower layer at rest, this is also consistent with the observation by Enfield and Allen that most of the variability in monthly averaged sea level at the Galapagos and Peru tide stations is anomalous, rather than seasonal.[18] Seasonal averages in this geographic region therefore tend to be less meaningful than elsewhere, since the ocean is so often in one extreme state or the other, rather than near the numerical mean. This poses a problem when one attempts to compare an isolated temperature section with a mean profile for the same season, since the latter is itself comprised of "anomalous" data and is difficult to interpret in terms of a true climatic average. An alternate approach is to examine a time series of temperature-depth data covering the period before, during and following a particular event, or, lacking such a series one can also compare temperature distributions for the same season from two contrasting (but not necessarily consecutive) years.

Isotherm Depression East of the Galapagos Islands

A remarkable time series of temperature sections exists for the 1971–1973 period spanning the 1972 El Niño event. Starting in 1971 the Instituto Oceanografico de la Armada (INOCAR) conducted quarterly hydrographic cruises for serveral years off the mainland of Ecuador. The section most consistently completed was along the 82–83°W meridional band, from 1.5°N–3°S (see Figure 6). The uninterrupted sequence of quarterly temperature sections from November-December 1971 through August-September 1973 is shown in Figure 7. The data consisted of hydrographic (bottle) stations alternated with bathythermograph drops. The fact that these two data sources agree precludes the possibility of serious data quality problems (e.g., erroneous bottle depths, thermometry errors, etc.). The series covers the cool pre-El Niño period (1971), the February 1972 through March 1973 El Niño event, and the post-El Niño period; these phases were also described by Zuta et al.[7].

In late 1971 (first panel) the Tropical Surface Water north of the Equatorial Front occupied the upper 30–40 m. Immediately below this a sharp thermocline extended from 35–50 m, comprised of the 15–20°C isotherms (shaded). Below the thermocline the stratification was minimal

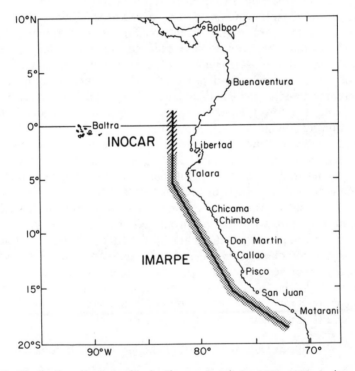

Figure 6 Distribution of hydrographic data from cruises during 1971–1973, used to construct the temperature section of Figures 7 and 9. Diagonal-line shading indicates data taken by the *Instituto Oceanográfico de la Armada del Ecuador* (INOCAR); stippled shading indicates data taken by or in cooperation with the Instituto del Mar del Peru (IMARPE).

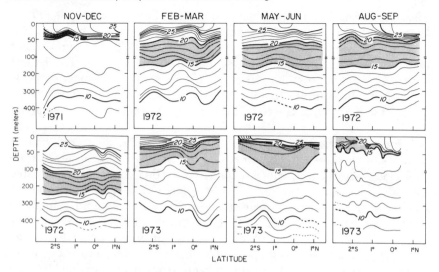

Figure 7 Temperature sections constructed from data taken by the *Instituto Oceanográfico de la Armada del Ecuador* (INOCAR) during 1971–1973 along the 82–83°W meridional band, from 1½°N to 3°S. The 15–20°C temperature band is shaded.

from 50–250 m, including the 14°C and 13°C isotherms. This thermostad seems to be the result of influx of water from the Equatorial Undercurrent, which typically has this structure (see Stroup[26] and Jones[27]). Below 250 m was a second stratified zone.

The first signs of an El Niño onset were found by the cruise of February-March 1973 (second panel). The surface water was considerably warmer (>27°C) and the Equatorial Front, if it existed, was well to the south. The most notable changes took place at greater depths: the sharp thermocline/thermostad structure had disappeared, all of the isotherms from 13–20°C had deepened, and the 15–20°C temperature zone had expanded to a thickness of about 100 m. Similar conditions persisted through the two following cruises, except that the Equatorial Front reappeared in August-September 1972.

The isotherm depression intensified in November-December 1972. The water was uniformly stratified below 50 m. The 15–20°C zone was over 100 m thick and had deepened by an additional 50 m. Then, during the 1973 cruises, conditions gradually returned to what appeared—in August-September 1973—very similar to the pre-El Niño conditions of 1971.

Except for February-March, all the panels of Figure 7 show interannual differences, with the El Niño period having a relatively depressed thermal structure. That such differences are not as clearly evident for February-

March is because this period was anomalous in both years, as shown by Zuta et al.[7] The greatest interannual difference occurred between November-December, 1971 and the same quarter of 1972. The temperature increase over that 12-month period is shown in Figure 8. The largest increase occurred from 30–120 m, with a maximum warming of 8–9°C near 60 m. In contrast to this, the largest interannual differences in surface temperature observed off Peru during the same period were about 5°C.[7]

The cause of the 1972 warming was advective. The quarterly changes in heat storage during the 2-year period covered by Figure 7 ranged from 250–1350 W/m². These values cannot be accounted for by reasonable estimates of surface heat flux. Clearly, a large-scale mass redistribution had taken place. The geographic source of the water required to increase the volume between 15 and 20°C is not clear. Salinities from the 1972 cruises off Ecuador are of doubtful quality due to equipment malfunctions and very few oxygen data were taken that year. Hence, a water mass tracing of the salinity and oxygen characteristics is not possible.

The Poleward Spread of Isotherm Depression

There is also evidence that the subsurface warming spread poleward along the Peru coast from the equator. Several cruises along the Peru coast coincided with Ecuadorian cruises in 1972–1973, enabling Zuta et al.[7] to construct composite alongshore sections along the line shown in Figure 6. These sections are redrawn in Figure 9.

In November-December 1971 (first panel, Figure 9) the thermal structure off northern Peru was basically similar to that off Ecuador, with a near-surface thermocline and a thermostad below it. Farther south both

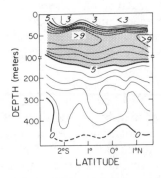

Figure 8   Distribution of the temperature increase along 82–83°W from November–December 1971 to November–December 1972, shown as a function of depth and latitude. Increases of 5°C or more are shaded.

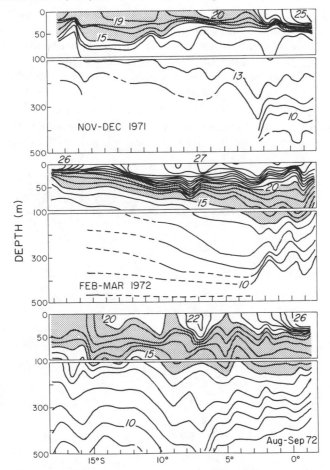

Figure 9 Temperature sections constructed from data taken by the *Instituto Oceanográfico de la Armada* (INOCAR) and by or in cooperation with the *Instituto del Mar del Peru* (IMARPE), along the shaded zone shown in Figure 6. The 15–20°C temperature zone is shaded.

these features weakened, with the stratification being more uniform near 15°S. The Equatorial Front was found near 0–3°S.

By February-March 1972 the isotherm depression that occurred off Ecuador (Figure 7) had also spread as far as 10°S along the Peru coast. The Equatorial Front was not evident and considerable surface warming had occurred off Peru. A remnant of the 13–14°C thermostad was still found off northern Peru, but at least 100 m deeper; the 15–20°C water had also deepened and its volume was increased north of 7°S.

By August-September 1972 the isotherm depression had spread poleward along the entire coast of Peru. Water between 13 and 17°C had deepened by 50–100 m and its volume was considerably increased over what it was in November-December, 1971. In contrast to the subsurface changes, the surface water off Peru was cooler in August-September by some 5°C over the February-March values, and the Equatorial Front had reformed off Ecuador and northern Peru.

The temperature changes between the cruises of Figure 9 are shown in Figure 10. From November-December 1971 to February-March 1972 the warming off northern Peru was most intense in the surface layer

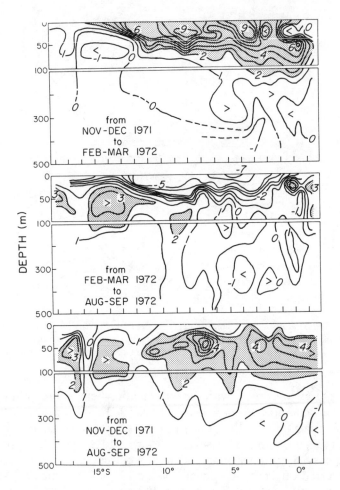

Figure 10 Distribution of the temperature increase in time (positive) along the sections shown in Figure 9. Increases of 2°C or more are shaded.

(0–30 m) as far south as Callao (12°S). The surface warming was associated with a tongue of low salinity (<34.5 ‰) water that had spread poleward as far as 10°S from the area of warm tropical water off northern Ecuador.[7]

The warming below 30–40 m off northern Peru was not associated with low salinities and appeared to be exclusively the result of isotherm depression. The depth of the subsurface warming decreased poleward and was not evident beyond 10°S. From February-March to August-September (second panel) the surface temperatures had nearly returned to their November-December 1971 values. Some additional subsurface warming had occurred north of 10°S, but most of the warming was south of there and below 30 m. The overall change from November-December 1971 (third panel) was maximum near 50 m and was 2°C or more in a layer from about 30 to 100 m.

The subsurface conditions off Peru in August-September 1972, were almost certainly anomalous, even though it is not possible to compare them to a long-term mean or to the same season in 1971. The ocean climate is normally coolest in August-September,[28] yet this period in 1972 was much warmer than November-December 1971, which is the warming season off South America. Also, it is noteworthy that the isotherm depression was a persistent process, whereas in the surface layer the anomalies were more episodic and restricted to the summer-fall season when the upwelling favorable winds off northern Peru are normally weaker.

The Offshore Extent of Isotherm Depression

The 15°C isotherm, which lies near the base of the thermocline, appears to be one of the most affected by isotherm depression (see Figures 7 and 9). Throughout the 1970s the Instituto del Mar del Peru (IMARPE) has constructed topographies of the 15°C isotherm from the oceanographic cruises of its research vessels and from synoptic fisheries explorations involving many fishing vessels, which leave their respective ports simultaneously and drop bathythermographs along offshore tracks.

The 15°C topographies of 1972[7] indicate the effects of isotherm deepening along the coast, consistent with the sections of Figures 7–10. Wyrtki[1] compared these topographies to those of February-March 1967 (EASTROPAC) and September 1964 (IMARPE), which were both cool (anti-El Niño) years. During the cool years the 15°C isotherm was found near 80–100 m beyond 200–300 km offshore and sloped upward toward the coast to a depth of 20–40 m (1967), or intersected the sea surface over the shelf (1964). In 1972 the 15°C isotherm initially deepened off

Ecuador and nearshore off northern Peru. During the rest of 1972 and within 200 km of the coast, the 15°C isotherm was found at depths of 100–200 m off Peru and 60–120 m further offshore.

Figure 11 shows that during the El Niño of 1976 there occurred a similar depression of the 15°C isotherm relative to 1971, a cold year. The months compared are July and August, respectively, so that seasonal differences between the two topographies are minimal. In August 1971 the 15°C isotherm depth was 60–80 m offshore and shoaled to 20 m over the shelf. In July 1976 the isotherm was found near 60–100 m offshore of Peru and deepened to 120–160 m near the shelf and slope, where it intersected the bottom. In the far north the 15°C depth was over 200 m off the Gulf of Guayaquil. The distribution of 15°C difference between the two periods is also shown. The depth difference is only about 20–40 m at 200 km from the coast, but reaches 100–160 m near the shelf break. A gradient of isotherm depression between the 2 years stretches alongshore in a band some 100–200 km from the coast, just offshore of the continental margin. The effects of isotherm depression appear to extend somewhat farther offshore in the north.

Timing

According to the wave dynamical theory of El Niño, isotherm deepening should proceed quickly. The numerical simulations of El Niño by Kindle[29] indicate that the process should be essentially completed off Peru by the second month following the initial impingement of the equatorially trapped Kelvin wave on the eastern boundary. There is some indication from the time-depth sections of Zuta et al.[7] that this was true in 1972. The first indication of isotherm deepening was in late February and early March. Isotherms had deepened off norther Peru considerably more by the time additional data was taken in May. In July they were somewhat deeper, but most of the depression had already taken place. Off southern Peru the isotherms were deepest in April (15°S) and late May (17°S), the latter being the first time data had been taken since February. It appears that the process, from the onset in the north to the maximum depression in the south, took place in less than 2–3 months.

Figure 9 suggests that a second wave of isotherm depression had occurred by November-December 1972. This is also depicted in the time-depth sections of Zuta et al.[7] for points along the Peru coast (not shown). In the north (6°S) the deepening took place between late October and early December. Off Chimbote (9°S) the isotherms were deepest in December. If the process extended to southern Peru, it was apparently

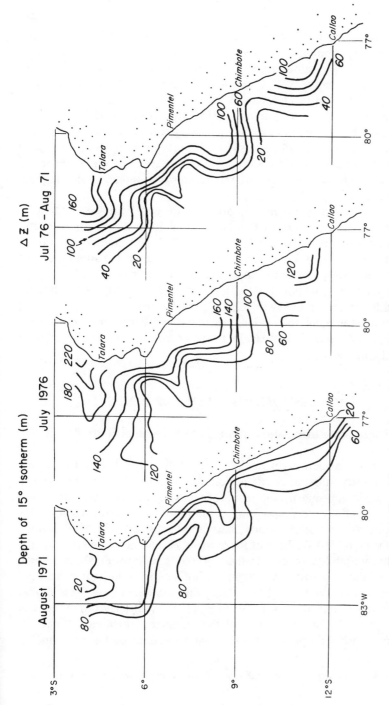

Figure 11 Two leftmost maps: topographies of the 15°C isotherm depth (meters) in August 1971 and July 1976 (from the *Instituto del Mar del Peru*). Right map: distribution of the 15°C isotherm depth difference between August 1971 and July 1976 (positive values indicating an increase with time).

237

missed because no data were taken between early November and the end of January. Anomalies of monthly mean sea level at Callao (12°S) went from a relative minimum of +6.9 cm in October to a maximum of +18.6 cm in December. The sea level anomalies at Matarani (17°S) varied from +7.2 cm in October to +18.9 cm in December, decreasing thereafter to +6.5 cm in January 1973. The available data therefore suggests that the second disturbance probably went to completion in less than 2 months. The second depression event (typical of strong El Niños) was numerically simulated by Kindle[29] and appears to be due to a second Kelvin wave which originates at the western boundary following the initial relaxation.

The available hydrographic data off Ecuador and Peru seems compatible with the wave dynamical theory of El Niño in nearly all respects. Isotherm depression is evident to depths of 200–300 m. Vertical excursions of the order of 100 m agree well with the numerical simulations of Kindle.[29] The process of isotherm depression spreads poleward along the eastern boundary, consistent with Hurlburt et al.,[11] and goes to completion quickly, in agreement with Kindle's work. Finally, the confinement of isotherm deepening to a nearshore band off Peru also agrees with theory and the second depression event seems to be consistent with numerical simulations.[29]

INTERANNUAL VARIABILITY OF WINDS

The Large-Scale Circulation in the Eastern Tropical South Pacific

As noted in the section on Background, Wyrtki questioned the widely held (but poorly substantiated) notion that the El Niño in the eastern tropical Pacific is primarily the result of weakened trade winds off South America and of the corresponding cessation of coastal upwelling.[1] He found from ship reported winds that the strongest trade wind forcing occurred west of 140°W. East of there to 90°W, and near the equator, the interannual variability is less. In fact, Wyrtki's time series show a slight *increase* in the easterly component during several relatively warm periods: 1957–1958, 1969, and 1972. Using the same data base for a similar zone, Hickey found a slight *weakening* of the southerly component of the SE trades.[30]* This suggests that the anomalous behavior of the SE trades during El Niño periods (in the eastern equatorial Pacific) consists

*In this region the surface winds normally blow from the south and east during the entire year.

of a slight counterclockwise turning of the mean wind vector rather than a significant change in magnitude. These tendencies must be treated with caution, however, since the data base from ship reports is sparse for the eastern equatorial Pacific.

Krueger and Winston examined the global anomalous 700 mb and 200 mb circulations during 1971 and 1972.[31] In this case the data base consisted of the National Meteorological Center's objective tropical analysis, which includes data from atmospheric soundings, aircraft observations and satellite-derived cloud motions. The tradewind circulation at 700 mb was characterized by the following features during the 1972 El Niño: (a) the North and South Pacific anticyclones were displaced poleward of their "normal" positions; (b) these movements left a broad area near the equator with reduced trade winds; (c) the trades just south of the equator over the eastern Pacific did not appear weaker than normal; (d) a relative weakness in the flow occurred at 5–10°S in the eastern Pacific; and (e) the SE trades south of 10°S, off Peru and Chile were above the mean.

The first two features noted by Krueger and Winston were oceanwide and associated with the general trade wind weakening in the central and western Pacific,[31] also noted by Wyrtki.[1] The last three features were characteristic of the eastern Pacific: an equatorial band of unaltered trades and an area of above normal trades (south of 10°S), separated by a band of weakened winds (5–10°S). The latter coincides geographically with the normal convergence zone noted by Wyrtki and Meyers just south of the equator (about 3–7°S).[32] Hence the weakening noted by Krueger and Winston along 5–10°S may have been related to strengthened convergence activity in that area. Their observation that in 1972 the equatorial winds just north of this convergent zone were not weaker than normal is consistent with the interpretation given above to the anomalous zonal and meridional components analyzed separately by Wyrtki[1] and Hickey.[30] According to Krueger and Winston the stronger than normal trades south of 10°S were the result of the southward displacement of the South Pacific anticyclone.[31]

Coastal Winds

Whereas the above features may apply to the atmospheric circulation offshore of South America, they are not necessarily characteristic of the winds in the coastal zone where land-sea interactions become important. Lettau has presented a thermotidal forcing model for the coastal winds along the tropical west coast of South America, by which he explains the extremely arid climate of that region.[33] He proposes the existence of an alongshore wind jet above the coastline that is forced by the thermally induced pressure gradient across the land-sea boundary, and is quasi-decoupled from the general circulation offshore. Hence it is desirable to inquire separately into the interannual variability of the coastal winds.

Wyrtki analyzed a 24 year time series of monthly averaged easterly and southerly wind stress components compiled from ship records for the area between 10°S, 20°S, 80°W, and the coast of South America.[1] Such averages are representative of the coastal zone because the great majority of ship reports from that area come from within 100 km of the coast. Both components of stress tend to be somewhat stronger than normal during El Niño years, in agreement with Krueger and Winston.[31] Wyrtki also found a similar wind stress behavior in the 10° square to the northwest (0–10°S, 80–90°W).[1] He therefore concluded that this wind pattern was characteristic of the entire region and that the idea of weaker than normal winds off Peru during El Niño must be abandoned.

The pattern noted by Wyrtki for the coastal winds resembles the large-scale circulation anomalies of 1972 in most respects.[31] However, in Wyrtki's averages there was apparently no manifestation of the offshore weakening observed by Krueger and Winston near 5–10°S. Since such weakening, at a given time, might be associated with a relatively narrow convergence zone, it is conceivable that Wyrtki's 10° square data did not resolve it.

Wind Data from Coastal Airports

To gain a better understanding of the coastal winds, this author obtained the montly averaged wind speeds and prevailing (modal) wind directions for Peru coastal airports, from the Corporation Peruana de Aviacion Civil (CORPAC). Most of the time series have nonstationarities and apparent shifts in the means that do not correlate between airports and are probably the result of instrument drift, site relocations, etc., for which no information has been obtained. Therefore, analysis of much of this data is at least temporarily impossible. Even so, anomalies of these time series do not show any identifiable characteristics that can be related to pattesenormally warm or cold years. Two of the time series appeared particularly good and deserved further analysis: Talara (4.5°S) from 1957 through 1977, and Lima (12°S) from 1956 through 1977. Time series of the 3-month running means of monthly anomalies are shown in Figure 12 for the Southern Oscillation Index (SOI) and for the scalar surface wind speed and sea surface temperature (SST) at Talara and Lima-Callao, respectively.

Because the monthly wind data was not vector averaged, vector

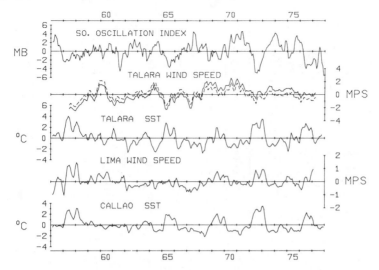

Figure 12   Time series of the 3-month running means of monthly anomalies for the Southern Oscillation Index,[19] and the scalar surface wind speed (CORPAC) and shore temperature (IMARPE) at Talara and Lima-Callao, respectively. Dashed curve indicates the anomalies of the adjusted Talara wind speeds.

speeds and alongshore components cannot be shown. However, the behavior of modal directions as a function of time is relatively invariant at both airports, never varying by more than one octant from the most frequent monthly direction. Moreover, the interannual variability of the monthly modal direction shows no pattern that can be related to the occurrence of El Niño events or warm years. A more rigorous test of directional invariance at Lima is made possible by the existence of a 20-year time series of rawinsonde measurements taken at that site (1958–1977). Analysis of the monthly averaged wind data at 950 mb (500 m above the ground) yields very high correlation coefficients between both the vector speed (0.98) and alongshore component (0.96), respectively, and the scalar speed. Therefore, for the purpose of discussing Figure 12, it will be assumed that the average scalar wind speeds are representative of the alongshore wind.

The statistics of the wind speeds and SST for which anomalies are depicted in Figure 12 are given in Table 1. The mean wind at Talara is nearly three times as strong as at Lima. The minimum wind at Talara occurs in the late southern hemisphere summer and the maximum in the late winter. This approximately corresponds to the known seasonal behavior of the large scale SE trades off South America. At Lima the

TABLE 1. Basic Statistics of the Monthly Mean Wind Speeds and Sea Surface Temperature (SST) at Talara and Lima-Callao

	Talara Wind (ms^{-1})	Talara SST (°C)	Lima Wind (ms^{-1})	Callao SST (°C)
Overall mean	8.4	19.1	3.3	16.5
Seasonal min.	5.9	17.6	2.7	15.5
(Month)	(Mar.)	(Aug.–Sept.)	(June)	(Oct.–Nov.)
Seasonal max.	9.5	21.7	3.7	18.3
(Month)	(Sept.)	(Feb.)	(Dec.–Jan.)	(Mar.)
Standard deviation of Residuals	1.2	1.5	0.5	1.3
Primary direction mode	South	—	South	—

minimum wind occurs in the early winter and the maximum in early summer, considerably out of phase with the large scale SE trades. At Talara the standard deviation of the anomalies is 14% of the long-term mean, whereas at Lima it is 16%. At both stations the wind blows mainly from the southern octant. Thus, although the standard deviation of the wind anomalies at Talara is over twice the value at Lima, the variabilities at the two stations are similar in proportion to the long term means. Together with the invariance of the modal directions, this implies that the winds at Talara weaken slightly during El Niño, but still remain strong and upwelling favorable on time scales of a month or more.

The anomalies of wind speed are clearly quite different between Talara and Lima (Figure 12). The maximum cross correlations between the variables are shown in Table 2a. The wind speed anamalies at Talara behave in a manner roughly similar to the SOI but opposite to sea surface temperatures: when the SOI is low and SST is high (El Niño periods), the Talara winds tend to be anomalously weak. At Lima the opposite is true: during such periods the Lima winds are anomalously strong. Thus, the winds at both stations have an anomalous behavior associated with El Niño occurrences, but in an opposite manner.

The time series of wind speed anomaly at Talara (Figure 12) suggests that there may have occurred an artificial increase in the mean, starting in 1968. To test if such a shift would seriously alter the statistical characteristics of the data, a bias of 1.3 ms^{-1} was subtracted from the data after 1967 to bring the series mean into agreement with the pre-1968 mean. The anomalies computed from the adjusted data are shown in Figure 12 (dashed curve). The overall mean of the adjusted data is 7.8 mps, the standard deviation of the anomalies is 1.0 ms^{-1} and the seasonal extremes occur as before (see Table 1). The maximum lagged cross correlations of the adjusted Talara winds with other variables are given in Table 2b. It appears that statistical inferences about the Talara winds are essentially unaffected by the possibility of a bias in the data.

In view of the possibility that trade wind weakening may occur in the latitude zone of Talara[31] and that this weaking may be related to unusual convergence activity, one can speculate that the lower monthly wind averages at Talara during El Niño periods reflect a greater intermittency at shorter time scales. Significant wind relaxations of a week or less in this area could explain the transequatorial invasions of tropical water that tend to occur around March, when the Talara winds are at their seasonal minimum. Yet such relaxations might not significantly alter the modal direction or greatly reduce the average speed for the month. Long time series of wind data taken at higher sampling rates (e.g., daily) are needed to explore this possibility.

TABLE 2a. Below-diagonal: the maximum lagged cross correlations between the variables for which anomalies are shown in Figure 12; above-diagonal: The corresponding below-diagonal values normalized by the large lag standard error (as described by Sciremammano[34])[a]

	SOI	Talara Wind	Talara SST	Lima Wind	Callao SST
SOI	—	2.7	-4.0	-3.3	-3.9
Talara Wind	0.28	—	-3.2	-0.9	-2.5
Talara SST	-0.45	-0.42	—	3.9	5.8
Lima Wind	-0.33	-0.12	0.46	—	4.2
Callao SST	-0.49	-0.36	0.82	0.54	—

[a]Normalized values of 1.3, 1.7, 2.0, 2.6, 3.3 indicate significance at the 80, 90, 95, 99, and 99.9% levels, respectively.

TABLE 2b. Maximum lagged correlation of the adjusted Talara wind with the variables shown (normalized values in parentheses)

	SOI	Talara Wind	Talara SST	Lima Wind	Callao SST
Adjusted Talara Wind	0.28 (2.5)	0.84 (6.2)	-0.51 (-4.1)	-0.19 (-1.7)	-0.42 (-3.0)

At Lima, only the El Niño of 1965 is not associated with an increase of wind speed in Figure 12. This appears to be due to erroneous data, however, because the atmospheric sounding data indicate that the wind at 950 mb (500 m) increased considerably during that year (not shown). The wind anomalies are so similar to the SST anomalies that even the double peaked structure of the SST anomalies is reproduced.

It appears unlikely that the Lima wind anomalies, which mimic the SST anomalies along the Peru coast, are merely a coastal manifestation of the large-scale trade circulation of the eastern tropical South Pacific. The larger scale wind anomalies of Wyrtki[1] do not show this characteristic structure (although this may be a reflection of inadequate sampling by ships). Other indices of the trade circulation, such as the Southern Oscillation Index, do not behave in this way either. Kindle has been able to numerically simulate a double-peaked effect in the ocean thermal structure along the eastern boundary[29] that explains the corresponding feature observed in the Peru SST anomalies during the larger El Niño events. This numerical simulation is forced only by the average seasonal easterlies in the central and western Pacific, combined with a very simple reduction of them that is completely lacking in any double-peaked structure. Kindle's double-peaked effect is a uniquely oceanic phenomenon based on the timing of the mid-Pacific wind relaxation and subsequent recovery, and on successive reflections of the excited Kelvin and Rossby waves off the eastern and western ocean boundaries.[29] Hence, a likely explanation for the wind anomalies at Lima is that they are caused in some way by the sea surface temperature anomalies in the coastal zone.

A possible mechanism for oceanic forcing of the Lima wind involves the vertical wind structure in the marine boundary layer. The vertical wind shear can be considerable when the lower atmosphere is normally very stable, as it is at Lima. If this stability is reduced due to surface heating and moisture flux from a warm ocean surface, it is conceivable that horizontal momentum will be mixed downward from higher levels with greater efficiency, thus resulting in greater wind speeds near the surface. This mechanism might also explain why a similar forcing is not observed at Talara. Talara is only about 250 km south of Guayaquil, where the lower atmosphere is normally much less stable than at Lima. The winter stratus conditions at Lima, associated with the inversion there, are absent at Talara. It therefore seems at least qualitatively consistent to say that the normal efficiency of downward momentum flux at Talara is relatively great and that surface heating effects would not have the same impact.

A second mechanism for oceanic forcing of the coastal wind is through the surface atmospheric pressure gradient, which is normally directed from sea to land along the Peru coast and is associated with the geostrophic component of the alongshore wind (Lettau[33]). If the pressure gradient is somehow increased during El Niño—for example, by a greater surface heating over land than over sea—the alongshore wind can also be expected to increase. Anomalies of SST during El Niño might act to partially dissipate the coastal stratus clouds that normally cover much of the Peru coast from June to October, resulting in greater air temperature anomalies over land than over sea. This mechanism is also consistent with the contrasting behavior of the winds at Talara (annual cycle and anomalies) since low-level cloudiness is virtually nonexistent at that station and, therefore, would not play a role in wind variability.

In summary, it is clear from multiple data sources on eastern Pacific winds that the classical orthodoxy regarding the cessation of coastal winds as the cause of El Niño-related ocean anomalies is untenable. The evidence shows that on time scales of a month or longer the winds remain upwelling favorable along the entire Peru coast during El Niño. A slight weakening does occur at Talara, however, consistent with other indications of weaker winds in a relatively narrow zonal band from 5–10°S. It is possible that this weakening is due to increased convergence activity there and that it consists of an increased intermittency in the upwelling favorable winds on time scales of less than a month. The wind at Lima tends to increase during El Niño, consistent with the large-scale circulation south of 10°S. However, this similarity of the near coastal winds to the large-scale circulation may be fortuitous, since the close resemblance of the Lima wind anomalies to the local sea surface temperature anomalies suggests that the atmosphere is being locally forced by the ocean. Possible mechanisms for this forcing can be proposed, but remain speculative.

PERU COASTAL UPWELLING DURING EL NIÑO EVENTS

As we have seen, the existing information on winds in the eastern tropical South Pacific indicates that during El Niño occurrences the Peru coastal winds remain strong and upwelling favorable on time scales of a month or longer. One would suppose from this that the physical process of upwelling continues as well, albeit in such a fashion that drastic anomalies in biological productivity may result. Indications to this effect came as a result of the JOINT-2 upwelling experiment, conducted along the Peru coast near 15°S in 1976–1977 by the Coastal Upwelling Ecosystems Analysis Program (CUEA). Since the JOINT-2 experiment took place during an El Niño period, it is quite relevant to the question. Additional information related to upwelling during El Niño can be gained by looking at the historical hydrographic data gathered by IMARPE.

The Joint-2 Upwelling Experiment

Important results of the JOINT-2 experiment and their implications are discussed by Barber and Smith in Chapter 9. It is appropriate here to cite some of the JOINT-2 results. The physical environment of the upwelling zone from 12–15°S included the following characteristics in 1976–1977:

1 The mean alongshore flow over the outer shelf and slope was poleward from just below a shallow wind drift layer (about 15–20 m) to 100–200 m.[21]

2 Fluctuations of days to weeks in the alongshore flow and sea level were observed to propagate poleward in a manner consistent with coastally trapped baroclinic waves. The variability in the alongshore flow did not correlate well with local winds.[15]

3 The wind near 15°S was upwelling favorable and had a mean close to the long term (climatic) average.[35] The Callao (12°S) winds also remained upwelling favorable.[15]

4 The mean cross-shelf flow was offshore and in the surface layer and onshore at greater depths. Fluctuations in the cross-shelf flow, at midshelf and nearshore, correlated with local winds in a manner consistent with coastal upwelling.[21]

5 Fluctuations in ocean temperature, at midshelf and nearshore, correlated with local winds in a manner consistent with coastal upwelling.[35]

The above results indicate that in 1976–1977 the winds were upwelling favorable and that upwelling (in the physical sense) continued to occur (3–5). Superimposed on the upwelling process were the poleward flow (1) and its propagating fluctuations (2). The juxtaposition of upwelling effects and strong variability unrelated to upwelling has not been observed in other coastal upwelling regions, such as Oregon and Northwest Africa. It is not yet known whether this juxtaposition is a normal condition along the Peru coast or not. The possibility that the mean alongshore flow is equatorward or significantly less poleward during cold (anti-El Niño) years is supported by three facts: (1) numerical simulations predict that a reversal of the mean flow occurs at the onset of El Niño conditions at the eastern boundary; (2) coastal sea level is considerably lower during non-El Niño years than during El Niños; and (3) the thermal structure in a coastally confined zone is probably considerably depressed during El Niño years, relative to other years. However, it is not clear that the propagating fluctuations in the alongshore flow at time scales of less than a month would not also exist in non-El Niño years.

Other Indications of Coastal Upwelling During El Niño

Figure 13 shows the sea surface temperature distribution and its anomaly for the months of April and September 1972 (from IMARPE[36]). The fact that strong offshore thermal gradients continued to exist in spite of rather uniform positive anomaly fields suggests that upwelling was taking place, but that the subsurface temperatures of the upwelling source waters were anomalously high. This is consistent with the anomalous isotherm depression noted in the section on Ocean Thermal Structure Anomalies (Figures 9 and 10). During 1976–1977 the offshore depth of upwelling source waters was about 75 m,[35] consistent with the estimate by Wyrtki[37] that Peru upwellings are limited to the upper 100 m. Figure 10 suggests that in 1972 the source water temperatures (i.e., near 75 m) were possibly 2–4°C above normal (1) off northern Peru at the onset of El Niño conditions and (2) along the entire coast in August-

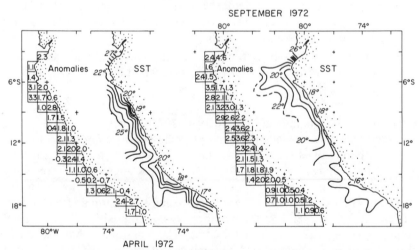

Figure 13 Sea surface temperature (SST) distributions and their anomalies (by one degree squares) during April and September 1972. From the *Instituto del Mar del Peru*.[36]

September. The poleward decrease of SST anomalies in April and September (Figure 13) is also in agreement with similar decreases at 75 m offshore in February-March and August-September (Figure 10). Similar distributions of SST and its anomalies occurred during the rest of 1972.[36]

Figure 14 shows the distribution by lag and station of the lagged correlations of shore temperature anomalies (1950–1974) between Talara and 10 other ports from the Galapagos Islands (Baltra) to Valparaiso. Time series of shore temperature anomalies like the ones shown in Figure 12 are coherent from the equator to northern Chile and occur within 0–1 months of each other. If one assumes that upwelling continues unabated during El Niño occurrences, a reasonable explanation for such observations is that the source waters for coastal upwelling become anomalously warm on a relatively short time scale. Although anomalous poleward surface advection does occur during El Niño, it tends to be episodic and more limited in latitudinal extent (north of 10°S). Anomalous surface advection from offshore may also occur but it seems unlikely that it would produce a nearly simultaneous and highly coherent response over some 3000 km of coastline. Moreover, in either case the anomalous

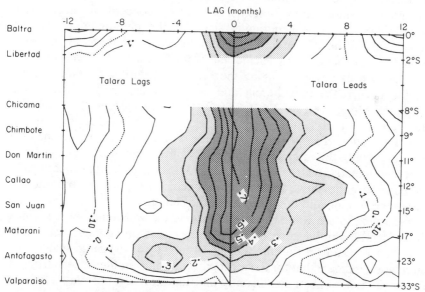

Figure 14 Distribution by lag and station of the lagged correlations of shore temperature anomalies (1950–1974) between Talara and ten other ports from the Galapagos Islands (Baltra) to Valparaiso. Light and dark shading indicate significance above the 95 and 99% confidence levels, respectively.

surface water from advective sources would be unlikely to affect shore temperatures in view of the continual renewal and offshore transport of upwelled water close to shore. It seems similarly improbable that anomalous heat fluxes at the ocean surface can explain the observed shore temperature anomalies in the presence of persistent coastal upwelling, although such fluxes may be important in affecting SST offshore. However, at specific times and places, away from areas of active coastal upwelling, the distributions of SST anomaly may also be affected by surface advection and air-sea heat fluxes.

SUMMARY

The work of Wyrtki[1] marks the beginning of a revolution in oceanographers' understanding of the physical characteristics of El Niño. It still can be said that the data base is far short of ideal and that much work remains to be done. But theoretical and numerical work, as well as data from a number of independent sources, are converging on a unified concept of the phenomenon. As usually happens when important progress is made in science, a number of otherwise enigmatic features of El Niño can now be understood in terms of a single conceptual base. These include (1) the large geographic scale of the phenomenon; (2) the coherence and near simultaneity of interannual oscillations of sea level from Peru to the western Pacific and from California to Chile; (3) the continuation of upwelling along the Peru coast during El Niño events; (4) the occurrence of coherent and near-simultaneous nearshore SST anomalies along the Peru coast, in spite of continued upwelling; and (5) the depression of isotherms in the eastern tropical Pacific and the associated subsurface thermal anomalies. While the climatic mechanism associated with the interannual variability of the equatorial Pacific winds is not yet understood, the wave dynamical theory of El Niño successfully explains how the above features (1–5) occur in response to the interannual wind forcing and how they are interrelated.

According to the wave dynamical theory, the weakening of the winds in the central and western Pacific is sufficient to precipitate an El Niño occurrence. Data on the interannual variability of the large-scale trade circulation indicate that for the most part the winds in the eastern tropical Pacific do not weaken significantly and therefore cannot play a role in exciting the El Niño response. The response includes, initially, an equatorially trapped internal Kelvin wave that propagates eastward from the mid-Pacific. Upon arrival at the eastern terminus of the equatorial waveguide, density structures in the eastern equatorial Pacific are depressed and a coastally trapped baroclinic disturbance begins to propagate poleward into each hemisphere. The leading edge of each disturbance takes the form of a downwelling front that advances with the phase speed of a Kelvin wave (in the case of an idealized ocean). Behind the advancing front, the alongshore flow at the boundary reverses from equatorward to poleward; isopycnal deepening there is confined to a coastal band which gradually widens as an alongshore isopycnal depression front slowly migrates offshore via Rossby waves. The lowest order Rossby waves, near the equator, propagate westward most rapidly. As Rossby waves impinge on the western boundary, successive Kelvin waves are reflected eastward. It is the arrival of the second Kelvin wave that is believed to result in the characteristic second anomaly peaks during the larger El Niño.

Upon examination of existing data, the wave dynamical theory holds up well. Quasi-simultaneous and coherent sea level fluctuations at interannual frequencies are associated with El Niño cycles and can be traced into both hemispheres. A lagged correlation and spectral study of these fluctuations reveals that they advance poleward at speeds much larger than is possible with advective processes alone. Hydrographic data off Ecuador and Peru confirm that thermal structures were significantly depressed during the 1972–1973 and 1976–1977 El Niños. The isotherm depression process appears to be associated with maximum temperature anomalies of 2–8°C near 50–60 m depth. The existing data for the 1972–1973 El Niño indicate that the onset of isotherm depression spread poleward along the entire Peru coast within 2–3 months, and that a second depression event went to completion in less than 2 months. The association of sea level anomalies with isotherm depression events and their mutual propagation poleward at large speeds gives considerable support to the idea of a wave mechanism.

Data from a number of independent sources indicate that the winds along the Peru coast remain favorable for coastal upwelling during El Niño occurrences. Continued upwelling was confirmed by the CUEA JOINT-2 experiment off Peru during the 1976–1977 El Niño. At Talara there is some weakening of the wind during El Niño, on time scales of a month or more, possibly due to increased convergence activity in a zonal band near 5–10°S. Although the average Talara winds remain strong and upwelling favorable, intermittency may occur over short intervals that would permit the southward invasions of tropical and equatorial surface waters that are noted episodically off nothern Peru. At Lima the monthly averaged wind is observed to increase during El Niño, closely mimicking the positive shore temperature anomalies.

It is proposed here that a coupling exists between the coastal wind and the nearshore ocean temperature, over much of the Peru coast. As isotherm depression sets in and spreads southward along the Peru coast during an El Niño, surface temperature distributions confirm that coastal upwelling continues under the upwelling favorable wind. But the southward spread of SST anomalies suggests that anomalously warm water is upwelled nearshore from the abnormally warm subsurface source waters offshore. It appears probable that the SST anomalies trigger an excess heat flux into the lower atmosphere in such a way that near surface wind speeds increase. This process guarantees the continuation of (physical) upwelling and hence the persistence of nearshore SST anomalies, over the alongshore extent of, and for the duration of anomalous isotherm depression.

ACKNOWLEDGMENTS

The preparation and original research of this chapter were supported by National Science Foundation grant OCE 78-03382 as part of the Coastal Upwelling Ecosystems Analysis (CUEA) Program, and also by National Science Foundation Grant ATM 78-20419 of the Climate Dynamics Program. The author has had many thought-provoking discussions with Drs. W. H. Quinn and R. T. Barber, and wishes to thank Dr. R. L. Smith for his continuing support and encouragement of this work.

REFERENCES

1   K. Wyrtki, "El Niño—the dynamic response of the equatorial Pacific Ocean to atmospheric forcing," *Journal of Physical Oceanography*, **5**, 572 (1975).

2   J. Bjerknes, "Survey of El Niño 1957–58 in its relation to tropical Pacific meteorology," *Bulletin of Inter-American Tropical Tuna Commission*, **12**, 3 (1966).

3   T. S. Austin, "II. Summary 1955–1957 ocean temperatures, central Pacific," *California Cooperative Oceanic Fisheries Investigations Report*, **7**, 52 (1960).

4   W. S. Wooster, "El Niño," *California Cooperative Oceanic Fisheries Investigations Report*, **7**, 43 (1960).

5   W. S. Wooster and O. Guillen, "Characteristics of El Niño in 1972," *Journal of Marine Research*, **32**, 387 (1974).

6   C. N. Caviedes, "Secas and El Niño: Two simultaneous climatical hazards in South America," *Proceedings of the Association of American Geographers*, **5**, 44 (1973).

7   S. Zuta, D. Enfield, J. Valdivia, P. Lagos, and C. Blandin, "Aspectos fisicos del fenomeno 'El Niño' 1972–1973," in U. N. FAO, Ed., *Actas de la Reunion de Trabajo sobre el Fenomeno Conocido como "El Niño"*, FAO Inf. Pesca, 185, Rome, 1976, p. 3.

8   M. J. Lighthill, "Dynamic response of the Indian Ocean to onset of the southwest monsoon," *Philosophical Transactions of the Royal Society of London*, **265**, 45 (1969).

9 J. Godfrey, "On ocean spindown. I: a linear experiment," *Journal of Physical Oceanography*, 5, 399 (1975).

10 J. McCreary, "Eastern tropical ocean response to changing wind systems: with application to El Niño," *Journal of Physical Oceanography*, 6, 632 (1976).

11 H. E. Hurlburt, J. C. Kindle, and J. J. O'Brien, "A numerical simulation of the onset of El Niño," *Journal of Physical Oceanography*, 6, 621 (1976).

12 M. A. Cane and E. S. Sarachik, "Forced baroclinic ocean motions: II. The linear equatorial bounded case," *Journal of Marine Research*, 35, 395 (1977).

13 D. W. Moore, "Planetary-gravity waves in an equatorial ocean," Ph.D. Thesis, Harvard University, 1968.

14 D. W. Moore and S. G. H. Philander, "Modeling of the tropical ocean circulation," in Goldberg et al., Eds. *The Sea* (Chapter 8, Vol. 6), Interscience, New York, 1977.

15 R. L. Smith, "Poleward propagating perturbations in currents and sea levels along the Peru coast," *Journal of Geophysical Research,* 83, 6083 (1978).

16 N. Christensen and N. Rodriguez, "A study of sea level variations and currents off Baja California," *Journal of Physical Oceanography*, 9, 631 (1979).

17 G. I. Roden, "Low-frequency sea level oscillations along the Pacific coast of North America," *Journal Geophysical Research*, 71, 4755 (1966).

18 D. B. Enfield and J. S. Allen, "On the structure and dynamics of monthly sea level anomalies along the Pacific coast of North and South America," *Journal of Physical Oceanography*, 10, 557 (1979).

19 W. H. Quinn, "Monitoring and predicting El Niño invasions," *Journal of Applied Meteorology*, 13, 825 (1974).

20 J. Bjerknes, "Monitoring and predicting El Niño invasions," *Tellus*, 18, 820 (1966).

21 K. H. Brink, J. S. Allen, and R. L. Smith, "A Study of low-frequency fluctuations near the Peru coast," *Journal of Physical Oceanography*, 8, 1025 (1978).

22 K. Wyrtki, "Sea level during the 1972 El Niño," *Journal of Physical Oceanography*, 7, 779 (1977).

23 D. B. Enfield, "Oceanography of the region north of the Equatorial Front: physical aspects," in *Proceedings of the Workshop on the El Niño Phenomenon*, Guayaquil, Ecuador, December 4–12, 1974. FAO Inf. Pesca, 185, 1976, p. 229.

24 K. Wyrtki, "The thermal structure of the eastern Pacific Ocean," *Deutschen Hydrographischen Zeitschrift*, Erg. Reihe A. (8°), No. 6, 84 pp. (1964).

25 G. Meyers, "Annual variations of the slope of the 14°C isotherm along the equator in the Pacific Ocean," *Journal of Physical Oceanography*, 9, 885 (1979).

26 E. D. Stroup, "The thermostad of the 13°C water in the equatorial Pacific Ocean," Ph.D. Dissertation, University of Hawaii, 1969, 202 pp.

27 H. J. Jones, "Vertical mixing in the Equatorial Undercurrent," *Journal of Physical Oceanography*, 3, 286 (1973).

28 S. Zuta and W. Urquizo, "Temperatura promedio de la superficie del mar frente a la costa Peruana, periodo 1928–1969," *Boletin del Instituto del Mar del Peru*, 2, 462 (1972).

29 J. C. Kindle, "Equatorial Pacific Ocean variability—seasonal and El Niño time scales," Ph.D. Dissertation, Florida State University, Tallahassee, 1979, 139 pp.

30 B. Hickey, "The relationship between fluctuations in sea level, wind stress and sea surface temperature in the equatorial Pacific," *Journal of Physical Oceanography*, 5, 460 (1975).

31 A. F. Krueger and J. S. Winston, "Large-scale circulation anomalies over the tropics during 1971–72," *Monthly Weather Review*, **103**, 465 (1975).

32 K. Wyrtki and G. Meyers, "The trade wind field over the Pacific Ocean," *Journal of Applied Meteorology*, **15**, 698 (1976).

33 H. H. Lettau, "Explaining the world's driest climate," in H. H. Lettau and K. Lettau, Eds., *Exploring the World's Driest Climate*, IES Rep. 101, University of Wisconsin, 1978, 263 pp.

34 F. Sciremammano, "A suggestion for the presentation of correlations and their significance levels," *Journal of Physical Oceanography*, **9**, 1273 (1979).

35 F. A. Medina, "A study of wind, oceanic temperature and hydrography in the upwelling area off Peru," M. S. Thesis, Oregon State Univerity, Corvallis, 1979, 62 pp.

36 Instituto del Mar del Peru, "Cartas promedio de temperatura superficial del mar," *Serie de Informes Especiales*, Nos. 102, 109, 114, 118, 123, 126, 138, 139, 140, 147. Callao, Peru (1972–1973).

37 K. Wyrtki, "The horizontal and vertical field of motion in the Peru Current," *Bulletin of the Scripps Institution of Oceanography*, **8**, 313 (1963).

CHAPTER 9

Biological Productivity and El Niño

OSCAR GUILLÉN G.
AND
RUTH CALIENES Z.

Abstract

The El Niño phenomena of 1965, 1972, and 1976 had limiting effects on the primary productivity, and the 1972 El Niño also had a major effect on the fisheries, due, in part, to its long duration.

In 1965 it appeared with greatest intensity during March and April and was caused by the invasion of the equatorial warm waters (24–27°C; salinity 33.0–34.8 ‰) associated with low concentrations of nutrients, of chlorophyll (0.18–1.0μg/l), and of productivity (< 0.05 gC/m²/day).

The 1972 El Niño off the coast of Peru appeared in two phases. The first phase took place in February-March and was characterized by the invasion of the equatorial waters approximately to latitude 10°S (salinity < 34.5 ‰) and low concentrations of nutrients (2.0–4.0 μg-at/l of silicates; 0.50–0.75 μg-at/l of phosphates; and 0.75–1.00 μg-at/l of nitrates). The phytoplankton contained less chlorophyll than during the 1965 El Niño and the total average production in the euphotic zone was 0.39 gC/m²/day, three times less than the average for the region. The composition of the phytoplankton also changed markedly throughout the year, especially in the Pimentel-Chimbote region (7–9°S), where diatoms and dinoflagellates characteristic of warm waters were identified. The second phase of the 1972 El Niño appeared in December and extended to latitude 10°S.

In January and February of 1976, a warming of the surface temperatures of the ocean was observed north of 15°S, with the equatorial surface waters (salinity < 34.8 ‰) advancing to 9°S and retreating in March. The maximum thermal deviation occurred in June-August, associated in this case with the subtropical surface waters characterized by low concentration of nutrients and productivity. The distribution of chlorophyll a was very different from normal patterns, especially in July, with values of < 1.0 μg/l at almost the entire coast, showing tendencies of increasing in August between 7–9°S and 15°S, with some coastal concentrations (3.0–10.0μg/l) associated with species similar to those that appeared in 1972, and especially with the dinoflagellate *Gymnodinum splendens*.

The results are compared to the average "normal" environmental conditions of 1966, showing also the negative effects of the extremely cold years.

INTRODUCTION

The El Niño phenomenon appears irregularly off the west coast of South America with large repercussions in the trophic levels of the ecosystem, affecting the yield of the fisheries, especially of the anchoveta. Among the largest El Niño events one can mention those of 1891, 1925–1926, 1940–1941, 1957–1958, 1965, 1972–1973, and 1976. Events of less intensity occurred in 1953 and 1969. These events have been described by many authors.[1–18] The biological aspects of the phenomenon have been studied by Vildoso,[19] Valdivia,[20] and more recently by Cowles et al.[21]

Although this anomalous condition has been studied for a long time, only in the last few years has there been an increase in the effort of coordination among scientists to understand better the causes of these complex fluctuations that occur in the marine environment of the Pacific coast of South America.

This work describes variations of the environmental conditions and of the primary production observed during the El Niño phenomena of 1965, 1972, and 1976, as well as the effects of the extremely cold years.

MATERIALS AND METHODS

For the years under study the information has been taken from the data obtained by the cruises of the BAP UNANUE (7202, 7211-12, 7612), the R/V PROFESSOR MESYATSEV (7208-09), the SNP-1 (7604, 7607, 7609-11), and the R/V EASTWARD-LEG 7 (CUEA) in the region of 4–12°S. The oxygen samples were analyzed according to the Winkler method with the modifications of Carpenter.[22] The nutrient analyses were done using techniques developed by Strickland and Parsons.[23] The chlorophyll samples were filtered through Whatman GF/C filters and analyzed according to the spectrophotometric techniques of Lorenzen.[24]

Primary production was measured using the classic C_{14} method developed by Strickland and Parsons.[23] The samples were collected at depths corresponding to 100, 50, 25, 10, and 1% incident surface radiation and injected with a solution of 4 μCu. The incubation time was 24 hours and Sartorius membrane filters of 27 mm and 0.2 μm porosity were used.

THE 1965 EL NIÑO PHENOMENON

The 1965 El Niño phenomenon appeared with greatest intensity during March and April. At the beginning of this period the advance of a tongue

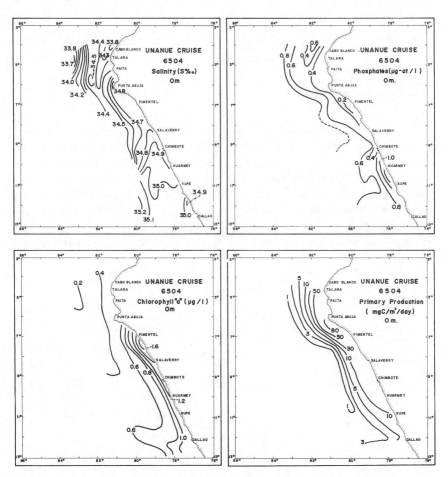

Figure 1 Horizontal distribution of salinity (o/oo), phosphates (μg-at/1), chlorophyll a (μg/1) and primary production (mgC/m³/day) at the ocean surface. Data obtained by the cruise UNANUE 6504.

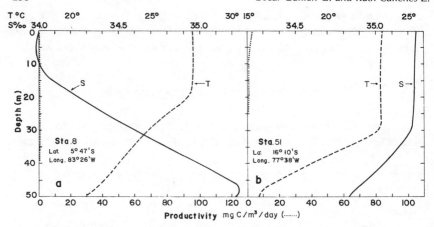

Figure 2 Productivity, temperature and salinity profiles in (a) surface equatorial waters and (b) subtropical surface waters, during the 1965 El Niño (UNANUE 6504).

of surface equatorial waters approximately 30 m deep was observed, with its southern tip off Callao. The tongue was characterized by temperatures of 24–27°C and salinity less than 34.8 ‰. The cold coastal waters (< 20°C and salinity of 34.8–35.1 ‰) were confined to a very near-shore band as far south as Punta Falsa. In April (Figure 1), the anomalous conditions continued but a retreat of the surface equatorial waters toward the north (10°S) was observed. The phosphate values at the surface were lower than usual in the area studied, and the lowest

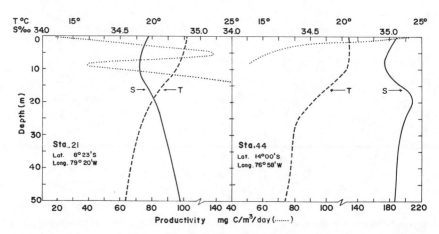

Figure 3 Productivity, temperature and salinity profiles off Salaverry and Pisco during the 1965 El Niño (UNANUE 6504).

concentrations of chlorophyll a and productivity, 0.2 μg/1 and 1.0 mgC/ m^3/day, respectively, were found in the area occupied by the equatorial surface waters.

The highest concentrations of phytoplankton at the surface were found near the coast (1.6 μg/l of chlorophyll a and > 80 mgC/m^3/day). The total production was high in Salaverry, 1.56 gC/m^2/day. On the other hand, the equatorial waters had values of 0.05 gC/m^2/day.[25] Figures 2 and 3 show a station occupied by equatorial surface waters (station 8) and another station (51) occupied by subtropical surface waters, both with very low productivity (0.01 gC/m^2/day). Stations 21 and 44 represent regions of greater productivity, located in Salaverry and Pisco, but with values below the average.

THE 1972 EL NIÑO PHENOMENON

In this year the El Niño was very intense; two invasions (of low salinity water) from the equatorial region toward the southwest, one in February-March and one in December, were observed.[17,26] Zuta et al. have shown that during the first phase, the tongue of warm water extended approximately to latitude 10°S (< 34.5 ‰ and > 23°C).[13] Figure 4 shows as characteristic the low-nutrient contents (2–4 μg-at/l of silicates, 0.50–0.75 μg-at/l of phosphates and 0.75–1.00 μg-at/l of nitrates). The subtropical surface water (salinity > 35.1 ‰) contained the lowest concentrations of nutrients. The average productivity at the surface for the region between 6–9°S of latitude was 29 mgC/m^3/day, the highest values corresponding to the region closest to the coast. The vertical distribution (Figure 5) was irregular with the highest concentrations at depths of less than 3 m with values of 125 and 118 mgC/m^3/day off Pimentel and Chimbote (stations 80 and 110). The total average production in the euphotic layer was 0.39 gC/m^2/day, three times smaller than the average for the region.

The phytoplankton biomass measured in terms of chlorophyll averaged 20.0 μg/l of chlorophyll a in the region 9–10°S (Figure 6a) with the largest concentration off Chimbote (> 3.0 μg/1). In the phytoplankton composition the diatom *Thalassiosira parthenensis* (Pimentel-Chimbote) was found for the first time and the following dinoflagellates characteristic of warm waters were identified: *Ceratium azoricum, C. gravidum, C. hexacanthum f.s. contortum, C. paradoxides, C. lunula, C. carriense, C. gibberun f.s. disper,* and *C. trichoceros.*[19,27–29]

The vertical distribution of chlorophyll offshore (Figure 6b) was very low and rather homogeneous with depth. Normally, the distribution near

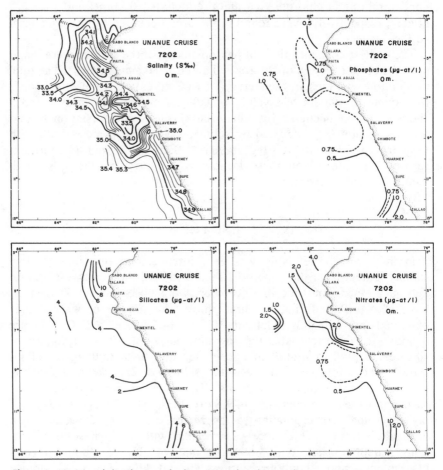

Figure 4 Horizontal distribution of salinity (‰), phosphates, silicates, and nitrates (μg-at/l) at the ocean surface. Data obtained during the cruise UNANUE 7202.

the coast is irregular and shows more than one maximum in the water column, very different from what was observed in this case (Figure 6c). Similarly, the chlorophyll *a* content of the euphotic zone in 1972 was less than the averages for the same regions given by Guillen et al.[30,31] and Guillen,[32] with a maximum value of 47.6 mg/m² off Chimbote and an average of 7.40 mg/m².

In the second phase of the 1972 El Niño (Figure 7), positive anomalies as large as 6° above the average were found; these waters advanced as

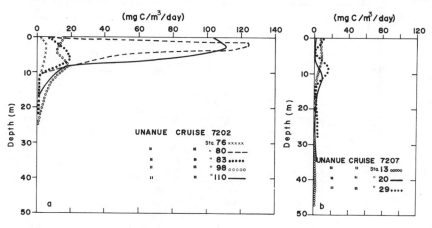

Figure 5 Vertical distribution of productivity during (a) summer and (b) winter of 1972.

far as latitude 10°S, with salinities of 33.8 to 34.8 per mil and temperatures of 22 to 25°C.[14] Two tongues of low salinity waters extended toward the south, one at longitude 82° (salinity < 34.4 ‰) and the other to the west of longitude 83° (> 34.5 ‰). In the first one, the distribution of nutrients had values of less than 1.0 μg-at/l of phosphates and less than 8.0 μg-at/l of silicates. The chlorophyll concentration of 0.0–0.3 μg/1 was less than the spring concentration, possibly due to a lack of vertical transference to the upper layer.

The warming began to disappear in March 1973. A cooling along the entire coast south of 4°S was observed in April.

THE 1976 EL NIÑO PHENOMENON

In January and February a warming of the surface temperatures of the ocean was observed north of 15°S. This trend lessened in the following month but a new anomaly occurred in the fall and beginning of winter, reaching its maximum thermal deviation of the year between June and August, associated with the subtropical surface waters that dominated a large part of the Peruvian coast. The effects of this anomaly can be perceived in the values of nutrients and chlorophyll a measured in July (Figure 8).

The chlorophyll a distribution for July was below average (0.0–2.8 μg/1) with concentrations of < 0.5 μg/1 prevailing over the greater part of

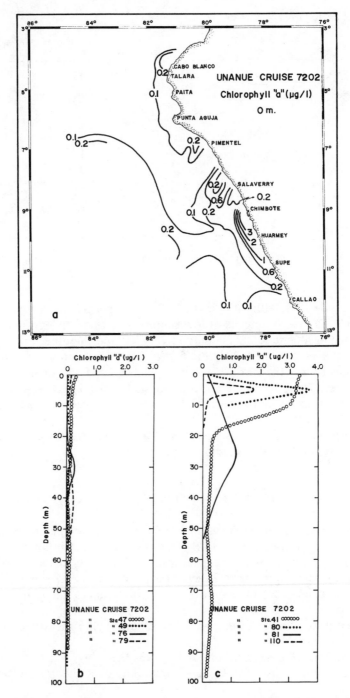

Figure 6 Horizontal distribution at 0 m. and vertical distribution of chlorophyll a. Data obtained during the cruise UNANUE 7202.

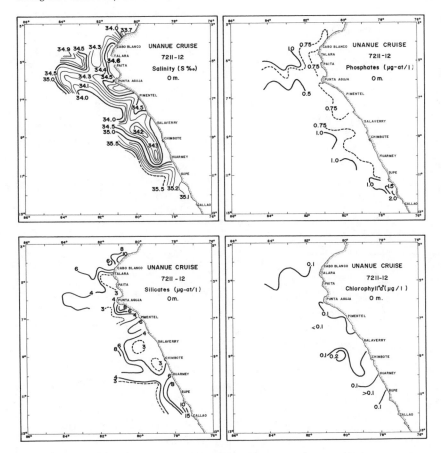

Figure 7 Horizontal distribution of salinity (o/oo), phosphates, silicates (μg-at/l) and chlorophyll a (μg/l) at the ocean surface. Data obtained by the cruise UNANUE 7211-12.

the region. The largest concentration of phytoplankton was found off Supe-Callao (> 2.0 μg/1), where the largest values for nutrients, especially nitrates (> 15 μg-at/l) and phosphates (> 2.0μg-at/l) were found. In the area of Pimentel there were also values (of phytoplankton biomass) associated with low concentrations of nitrate (9.00 μg-at/l), silicate (< 5.0 μg-at/l) and phosphate (> 1.00 μg/at/1). The phytoplankton of the region was composed of *Thalassiosira parthenesis, Arterionella japonica, Skeletonema costatum, Chaetoceros socialis, Gymnodinum splendens,* and oceanic species not common to this region such as *Eucampia*

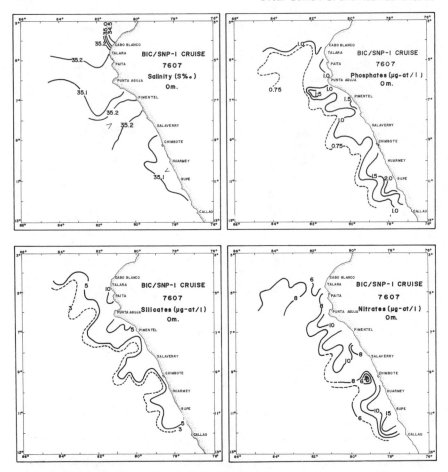

Figure 8   Horizontal distribution of salinity (‰), phosphates, silicates and nitrates (μg-at/l) at the ocean surface. Data obtained during the cruise UNANUE 7607.

cornuta, Thalassiothrix delicatula. Also warm water dinoflagellates were observed, such as *Ceratium gibberium f.s. disper, C. azoricum, C. hexacanthum f. s. contortum, C. gravidum, C. lunula, C. extensun, C. insisum,* and *C. longirrostrus.*[29]

Generally, in the region between 4 and 12°S nutrient values were lower than the monthly average: phosphates 0.64–2.93 μg-at/l, silicates 0.51–10.51 μg-at/l, and nitrates 3.82–15.3 μg-at/l, with the lower values prevailing over most of the region due to the influence of the subtropical surface waters which are characterized by the low concentrations of nutrients and low productivity.

The conditions of the Peruvian coast during the second half of the year could be summarized through the measurements of nutrient and chlorophyll content which were characterized by a distribution very different from normal patterns, especially in July ($<$ 1.0 $\mu g/1$). In August and September a sharp change was noted with an increasing trend (in chlorophyll and nutrients) in the Pimentel-Salaverry and San Juan up-welling regions. During the following months (October-December) there were high concentrations found but in very confined areas right next to the coast, still with a distribution lower than average, but less significantly so than in the earlier period of July to September.

DISCUSSION

The distribution of salinity, temperature, nutrients, chlorophyll a, and primary production (Figures 1, 4, 7, and 8) off the coast of Peru are influenced by the circulation patterns of the region. It is not known as yet with certainty whether the warm waters come as a result of advection or originate locally. This aspect has been discussed by Wooster who suggested two regions of origin of the warm surface waters, one nearshore north of the Equatorial Front (low salinity) and the other offshore west of central Peru (high salinity).[5] Subsequently Wooster and Guillen considered the possibility that the low salinity surface waters may also be coming from near the Galapagos Islands.[17] Bjerknes hypothesized that the weakening of the southeast tradewinds favors a transequatorial circulation that displaces southward the equatorial front, setting up the conditions for El Niño.[6] On the other hand, Namias has shown that the variations in transport in the countercurrent could be related to an index of northeast tradewind strength, weak values of the index being associated with strong countercurrent transport.[16] Also, Wyrtki proposed a theory that the El Niño phenomenon occurs as a dynamic response of the Pacific Ocean to atmospheric forces.[33]

The nutrient distribution in the El Niño years under consideration (Figures 1, 4, 7, and 8) suggests that the flow proceeding from the equatorial region enters from the northwest with characteristics somewhat different from the southward coastal flow. Productivity studies along the Peru coast indicate that the period of greatest production is summer-autumn (February to April). Consequently, the effects of any extreme anomaly will be greater during this season.

Effects of Cold Years on Primary Production

Temperature anomalies greater than 2°C, as shown by the SCOR Working Group on El Niño, are indicative of the presence of El Niño. Using a similar criterion, it could be noted that negative anomalies of less than 2°C have also negative effects on the primary production ("La Niña"?). The biological effects of El Niño have been studied by a number of authors, whereas the effects of extremely cold years are considered for the first time. In these years, when the phytoplankton growth is more intense than normal during the summer and fall, the waters, even though rich with nutrients, contain low quantities of phytoplankton due to the general instability of the waters,[34] to the rapid sinking of organisms, or probably to the more intense grazing by the zooplankton.

From the distribution of thermal anomalies (Figure 9) one can note on the one hand the El Niño phenomena taking place in 1965, 1969, 1972, and 1976 (> 2°C), and on the other hand the cold years of negative anomalies (< 2°C), 1964, 1968, 1970–1971, and 1974–1975. Intermediate between these two groups of years, 1966 can be considered as a year of average environmental conditions and the highest concentration of phytoplankton (> 4 μg/1 of chlorophyll).

Comparing the phenomena of 1965, 1972, and 1976, it is evident that the most intense one occurred in 1972. This El Niño had a disastrous effect on the anchoveta fishery. It also showed some similarities with the 1965 El Niño, including the invasion of tropical surface waters during the summer.[17] The chlorophyll a distribution during these two years was less than the corresponding monthly average (Figure 10) and was related to the flow of waters from the equatorial region. Cowles et al., referring to the 1975 weak El Niño, found that the primary production in the equatorial region was less than 0.2 gC/m²/day, which represents one-fifth of the normal values.[21] The 1975 occurrence did not affect the coastal region of Peru (Figure 11), where the coastal phytoplankton growth was comparatively normal, with production values of > 2.5 gC/m²/day, but over a limited distribution area. The biological effect of the 1975 phenomenon was short, as by April-May the equatorial region started to reach normal primary production values.

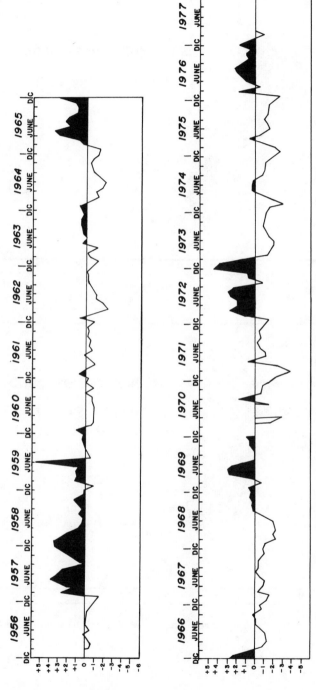

Figure 9 Annual temperature variation (°C) of the ocean surface at Puerto de Chicama (07°43'S).

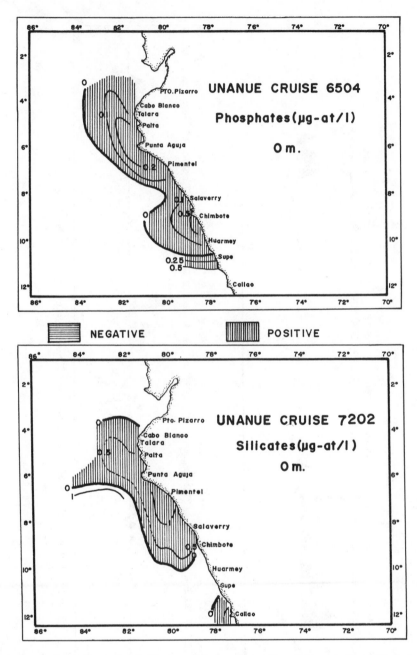

Figure 10 Anomalies in the distribution of nutrients (μg-at/l) and chlorophyll a (μg/l) during

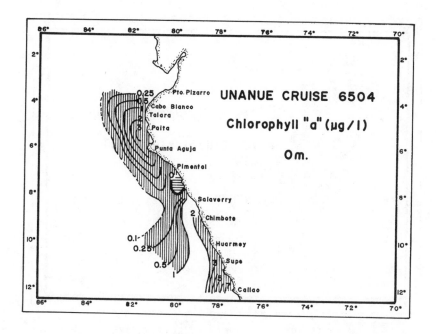

UNANUE CRUISE 6504

Chlorophyll "a" (μg/l)

0m.

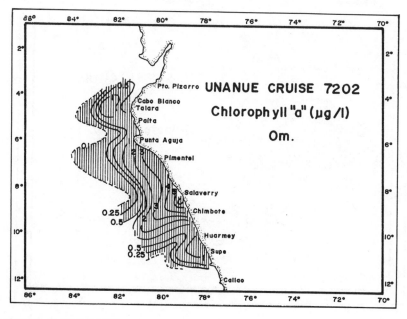

UNANUE CRUISE 7202

Chlorophyll "a" (μg/l)

0m.

the 1965 and 1972 El Niño.

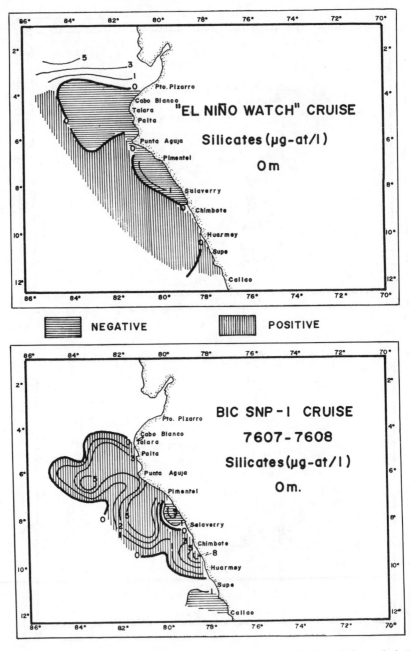

Figure 11 Anomalies in the distribution of nutrients (µg-at/l) and chlorophyll a (µ/l) during

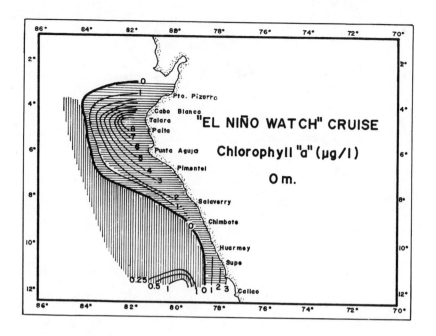

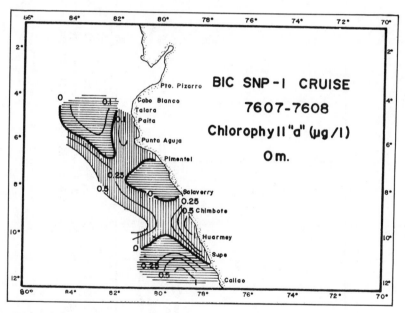

the 1975 and 1976 El Niño.

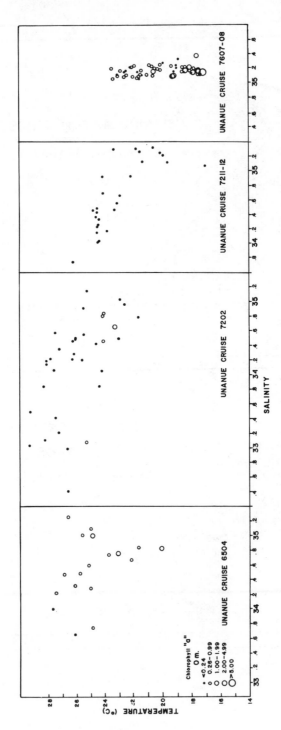

Figure 12 Diagram of the relation of temperature and salinity to chlorophyll a at 0 m, during the El Niño phenomena of 1965, 1972 (first and second phase), and 1976.

272

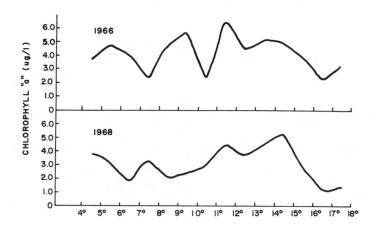

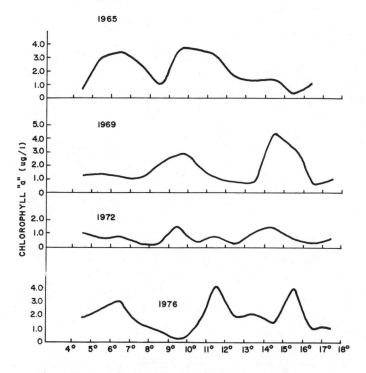

Figure 13 Latitudinal variation in chlorophyll a at the ocean surface.

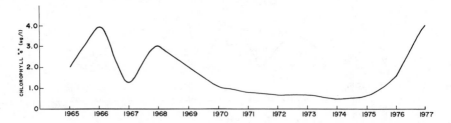

Figure 14 Annual variation of chlorophyll *a* in the 60-mile band along the coast.

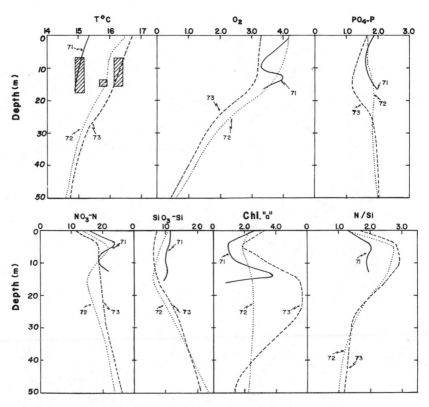

Figure 15 Nutrient, chlorophyll a, and nitrate/silicate profiles during autumn 1968 (UNANUE 6805).

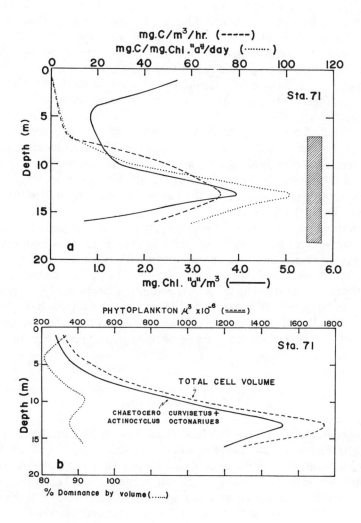

Figure 16 Vertical distribution during autumn 1968 (UNANUE 6805) of (a) chlorophyll a, productivity, and index of production; and (b) total volume of cells, percentage of dominance per volume, and dominant species.

Chlorophyll Variation During El Niño

The effect of the water characteristics on the chlorophyll during the abnormal years can be seen in Figure 12. The latitudinal variation of chlorophyll *a* off the coast of Peru during the years 1965, 1969, 1972, and 1976 (Figure 13), shows a steady low harvest of phytoplankton relative to the 1964–1978 average and much lower than the average of 1966. In the annual variation of phytoplankton production in the 60-mile band off the coast (Figure 14) the influence of both the cold years and the warm years can be seen.

When considering the year 1966 as a period of "model" conditions, it can also be noted that the dominant species of phytoplankton in that year were similar to those of 1975 in the coastal region of Peru,[21] whereas during the 1972 and 1976 phenomena other dominant species were found. Unfortunately, sufficient information regarding the phytoplankton composition during 1965 is lacking.

Even though 1968 was a cold year (summer-autumn), a good production was observed resulting from the good growth during the autumn in a very limited region and a better photosynthetic efficiency during spring. For example, during a cruise in autumn (6805) in the upwelling area of Chimbote, it was found that the distribution of acoustically determined density of anchoveta (Figures 15 and 16) coincided with maximum values of chlorophyll and depletion of nitrates and silicates, with an average production of 0.93 gC/m²/day, corresponding to an annual production of 330 gC/m².

Fisheries and El Niño

The amounts of yearly catches of anchoveta and of other species shown in Figure 17 and Table 1 allow us to follow in general terms the annual fluctuations relative to cold and warm years. The anchoveta catch declines during an El Niño phenomenon, while the yield of other species such as hake increases notably during the following year. On the other hand, Lasker has shown that the survival of anchoveta larvae *(Engraulis mordax)* is dependent on the availability of phytoplankton cells of the appropriate size above a critical concentration.[35] It is reasonable to assume a similar survival mechanism for the Peruvian anchoveta *(Engraulis ringens)*. If survival of the year's population depends on having a high concentration of the right kind of phytoplankton, it can be assumed that the effects of both factors, quality and quantity of food, could be noted in the anchoveta fishery during 1972 and 1976.

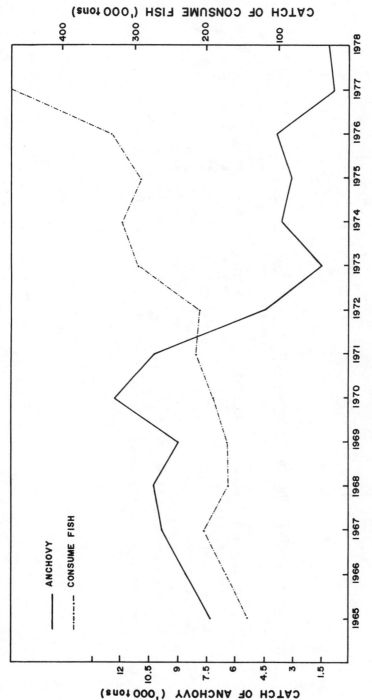

Figure 17 Annual catches of anchoveta and fish for consumption along the Peru coast.

277

TABLE 1. Catches of the Principal Species in Peru[a]

	1971	1972	1973	1974	1975	1976
Engraulis ringens J.	10,276.8	4,447.4	1,513.0	3,583.5	3,078.8	3,863.0
Sardinops sagax sagax	6.1	6.3	132.2	72.6	62.9	174.7
Trachurus symmetricus murphy	9.2	18.8	42.8	129.2	37.9	54.2
Scomber japanicus	10.1	8.7	65.0	63.3	23.6	40.2
Merluccius gayi peruanus	27.4	12.5	132.8	109.3	84.4	92.8
Mustelus whitneyi and M. mento	9.3	8.4	19.1	13.5	11.6	8.4

[a]FAO (1977), Anuarios Estadísticos Pesqueros 1971–1976 Ministerio de Pesqueria, Peru. Vol. 42.

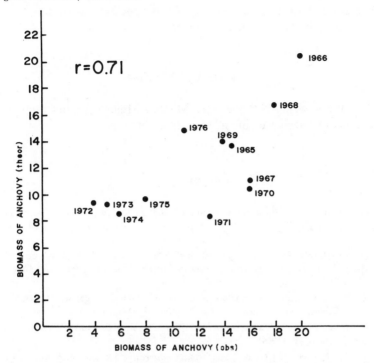

Figure 18 Correlation of the anchoveta biomass calculated on the basis of chlorophyll a data and the anchoveta biomass based on stock evaluations.

Relating the biomass of anchoveta to the availability of food (in terms of chlorophyll), a correlation of $r = 0.71$ was found. Figure 18 shows the anchoveta biomass calculated on the basis of chlorophyll data versus that based on stock evaluations, suggesting that chlorophyll a could be considered a good variability index for the anchoveta biomass. However, it is not possible to estimate the fishery potential on the basis of primary production measures, as can be concluded from the example of 1970–1971 when the anchoveta were abundant but primary production was low. The variation in the latter has its effect on the anchoveta, but the nature and intensity of this effect depends on the extent to which the anchoveta is dependent at any stage on the phytoplankton. According to Mathisen et al., the food availability would be limiting to the larvae but not to the adult population, although it can influence its distribution.[36]

A more detailed knowledge of the phytoplankton cycle, as well as experimental studies of the feeding of this species in its different stages (especially in the larval stages), is needed.

ACKNOWLEDGMENT

The authors wish to thank Mr. Marcos Melero for his help in the preparation of the figures included in this chapter.

REFERENCES

1 G. Schott, "Der Peru-Strom und seine nördlichen Nachbargebiete in normaler und abnormaler Ausbilding," *Annalen Hydrographie und Marine Meteorologie,* **59**, 161–253 (1931).

2 E. H. Schweigger, "Las irregularidades de la Corriente de Humboldt en los años 1925 a 1941, una tentativa explicación," *Boletin Sociedad Administradora del Guano* **18**, 27–42 (1942).

3 E. H. Schweigger, "Anomalías térmicas en el Oceano Pacífico Oriental y su pronóstico," *Sociedad Geográfica de Lima Boletin,* **78** (314), 3–50 (1961).

4 E. H. Schweigger, "El litoral peruano," Universidad Nacional "Federico Villarreal," Lima, Peru 1–414 (1964).

5 W. S. Wooster, "El Niño," *California Cooperative Oceanic Fisheries Investigations Rept.* **7**, 43–45 (1960).

6 J. Bjerknes, "El Niño. Study based on analysis of temperatures 1935–1937," *Bulletin of the Inter-American Tropical Tuna Communications,* **5** (3), 219–303 (1961).

7 J. Bjerknes, "Survey of El Niño 1957–58 in its relation to tropical Pacific meteorology," *Inter-American Tropical Tuna Commission Bulletin* **12** (2), 1–42 (1967).

8 O. Guillén, "Anomalies in the waters off the Peruvian Coast during March and April 1965," *Studies of Tropical Oceanography Miami,* **5**, 452–465 (1967).

9 M. R. Stevenson, O. Guillén, and J. Santoro, *Marine Atlas of the Pacific Coastal Waters of South América,* University of California Press, Berkeley, 23 pages, 99 charts (1970).

10 S. Zuta and O. Guillén, "Oceanografía de las Aguas Costeras del Perú," *Boletin Instituto del Mar del Perú* **2** (5), 157–324 (1970).

11. S. Zuta, "El fenómeno El Niño," *Revista de Estudios del Pacífico,* Chile (5), 27–42 (1972).

12 S. Zuta, "El fenómeno El Niño 1972–73," *Revista de la Asociación de Oficiales Generales ADOGEN* Peru (35), 29–44 (1973).

13 S. Zuta, Urquizo and V. Liendo, "Informe preliminar del Cr. BAP 'Unanue' 7202-03 (15 febrero–13 marzo de 1972)," *Serie Informes Especiales Instituto del Mar del Perú* (IMP-142) (1973a).

14 S. Zuta and W. Urquizo, "Condiciones oceanográficas anormales frente al Perú en la primavera de 1972," *Serie Informes Especiales Instituto del Mar del Perú* (IMP-160) (1973b).

15 K. Wyrtki, "Teleconnections in the equatorial Pacific," *Science,* **180**, 66–80 (1973).

16 J. Namias, "Response of the equatorial countercurrent to the subtropical atmosphere," *Science,* **181**, 1244–1245 (1973).

17 W. S. Wooster and O. Guillén, "Características de "El Niño" en 1972," *Boletin Instituto del Mar del Peru,* **3** (2), 44–72 (1974).

18 O. Guillén and M. Farfán, "Anomalías oceanográficas durante el año 1972 (por publicarse 1974).

19 A. Ch. Vildoso, "Aspectos biológicos del fenómeno El Niño 1972–73, Parte I: Distribución de la fauna. Actas de la Reunión de Trabajo sobre el fenómeno conocido como El Niño," Guayaquil, Ecuador, FAO Informes de Pesca No. 185 (1976).

20 J. Valdivia, "Aspectos biológicos del fenómeno El Niño 1972–73, Parte II: La población de anchoveta. Actas de la Reunión de Trabajo sobre el fenómeno conocido como El Niño," Guayaquil, Ecuador, FAO Informes de Pesca No. 185 (1976).

21 T. J. Cowles, R. T. Barber and O. Guillén, "Biological consequences of the 1975 El Niño," *Science,* **195**, 285–287 (1977).

22 J. H. Carpenter, "The Chesapeake Bay Institute technique for the Winkler dissolved oxygen method," *Limonology and Oceanography* **10** (1), 141–143 (1965).

23 J. D. Strickland and T. R. Parsons, "A practical handbook of seawater analysis," *Bulletin of the Fisheries Research Board of Canada* **167** (1972).

24 C. J. Lorenzen, "Determination of chlorophyll and phaeo-pigments: Spectrophotometric equations," *Limnology and Oceanography* **12** (2), 343–346 (1967).

25 O. Guillén, "The El Niño phenomenon in 1965 and its relations with the productivity in coastal Peruvian waters," *Fertility of the Sea,* **1**, 187–196 (1971).

26 O. Guillén, "El fenómeno El Niño 1972 y su relación con la productividad. III Simposio Latinoamericano sobre Oceanografía Biológica, Caracas, Venezuela (1975).

27 Instituto del Mar, Operación Eureka XXIII, 3-6 Agosto. *Serie Informes Especiales Instituto del Mar del Perú* (IMP-112) (1972a).

28 Instituto del Mar, Operación Eureka XXIV, 5-8 Septiembre. *Serie Informes Especiales Instituto del Mar del Perú* (IMP-116) (1972b).

29 Instituto del Mar, Simposio interno sobre anchoveta IMARPE, Junio 1977 (por publicarse) (1977).

30 O. Guillén, R. Calienes, and R. de Rondán, Contribución al estudio del ambiente de la anchoveta (Engraulis ringes J.). *Boletin Instituto del Mar del Perú* **2** (2), 49–76 (1969).

31 O. Guillén and R. de Rondán, Distribution of chlorophyll *a* in the Peru coastal current. Oceanography of the South Pacific 1972, Comp. R. Fraser, New Zealand Commission for UNESCO, Wellington, 387–395 (1973).

32 O. Guillén, El sistema de la corriente Peruana, I Parte: Aspectos físicos. Actas de la Reunión de Trabajo sobre el fenómeno conocido como "El Niño." Guayaquil, Ecuador, FAO Informes de Pesca No. 185, (1976).

33 K. Wyrtki, "El Niño—the dynamic response of the equatorial Pacific Ocean to atmospheric forcing," *Journal of Physical Oceanography,* **5** (4), 572–584 (1975).

34   O. I. Koblentz-Mishke, "On phytoplankton production in the western north Pacific Ocean in spring, 1955," *Doklady Acad. Sci. USSR.,* **116**, 1029–1031 (1957).

35   R. Lasker, Paper presented at the Joint Oceanographic Assembly, Edinburgh, Scotland, September 1976. *Fisheries Research Bulletin* **73** 453 (1975).

36   O. Mathisen, R. Thorner, R. Trumble, and M. Blackburn, Food consumption of pelagic fish in an upwelling area. Reprinted from *Upwelling Ecosystems.* Springer-Verlag, Heidelberg, in German, 1978, pp. 111–123.

PART III
SOCIETAL IMPLICATIONS OF EL NIÑO

CHAPTER 10

Postwar Production, Consumption, and Prices of Fish Meal

JOHN VONDRUSKA

Abstract

Postwar use of fish meal, production and consumption trends, and price behavior are analyzed, including El Niño effects. Fish meal is a high-quality, high-protein animal feed ingredient. World production and exports grew rapidly in the postwar period until about 1970 and have stabilized since then, because of fishery resource constraints in major producing-exporting countries.

INTRODUCTION

Fish meal is a high-value, high-protein animal feed ingredient that is manufactured from raw material consisting of small, whole, oily fish; other whole fish; and fish offal which is a byproduct of processing various fish for direct human consumption or other uses. In the manufacturing process, fish meal, fish oil, and fish solubles are all outputs. In 1978, U.S. output of these three products was valued as follows: fish meal, $122 million; fish solubles, $23 millon; fish oil, $61 million; totaling $205 million. The United States also imports fish meal and fish solubles, and is traditionally a net importer (imports exceed exports), but in 1978 exports exceeded imports because of price relationships reflecting the world supply-demand situation. The United States is a net exporter of fish oil, exports accounting for 92% of 1972–1976 average production. Fish oil is used principally in edible products abroad, but only in various inedible products (paint and varnish, fatty acids, resins and plastics, lubricants, and other products) in the United States. Fish meal and solubles are used in feeds primarily for animals that provide poultry, meat, egg, fish, and milk products for human consumption. This appears to be the most efficient and cost-effective way to use the various raw materials of fish meal to meet human nutritional requirements, although some of those raw materials (i.e., the fish) are also used in products for human consumption.

World production of fish meal grew at an annual average rate of about 10% in the postwar years until 1968–1970 when fishery resource-related constraints began to limit production, especially in the seven major producing-exporting countries. World production of fish meal is now about 4–5 mmt, and world exports are about 2 mmt. The catch for *reduction* into fish meal and oil by manufacturing process is about one-fourth of the total world catch of fish of 70 mmt, but the price and value of catch for reduction are quite low. For example, the U.S. landing of fish for human food in 1978 had a value to fishermen (exvessel value) of $1.7 billion and a weight of 1.4 mmt (liveweight), or $1203/metric ton (54.5 cents/pound), while landings of menhaden, the dominant raw material for fish meal in the United States, had a value of $98 million, a weight of about 1.2 mmt and an average exvessel price of about $83/metric ton (3.79 cents/pound).

Over half of the world's consumption of fish meal occurs in five countries (in order of consumption in the 1970s): Japan, Soviet Union, United States, Federal Republic of Germany, and United Kingdom. However, production and consumption patterns have changed in the postwar years.

Prices of fish meal in 1947–1978 were affected by general inflation in all producer prices, especially in 1951 (Korean War effect) and since 1973. Sharp deviations of fish meal prices from the 1947–1978 trend occurred in the form of two major price peaks (1949 and 1973) and a record postwar low in 1960. The 1949 and 1960 departures from the long-term trend were associated primarily with conditions in the markets for fish meal, whereas the 1973 departure was associated with a number of unusual national and international events in the 1972–1973 crop year. The impact of the 1972 El Niño on markets for fish meal must be considered in the context of these other factors. A broad context is necessary to evaluate the market effects of the several postwar El Niño phenomena that occurred when the anchoveta stocks off the coasts of Peru and Chile were being heavily exploited (1965, 1969, 1972, and 1976 El Niño events).

PRODUCTION AND CONSUMPTION

Production Processes

Simply stated, fish meal is a high-protein animal feed ingredient made by cooking, pressing, drying, and grinding fish or shellfish. Fish oil is mostly a byproduct of fish meal production; it is an oil that is extracted from the body (body oil) or liver (liver oil) of fish and marine mammals. Fish solubles are a water-soluble protein byproduct of fish meal production. Fish solubles are generally condensed to 50% solids and marketed as "condensed fish solubles." If prices, costs, and profitability warrant, the fish solubles may be reduced to an even lower moisture content and added to a manufacturer's output of fish meal. These three products are manufactured from whole fish specifically harvested for the purpose (usually small, oily, pelagic fish), from incidentally harvested fish not suitable or readily usable for other products, and from fish offal obtained in the processing of other products. In comparison with the weight of raw material, production of fish oil is more variable than production of fish meal, because the oil content of the raw material varies from year to year, seasonally, and among species.

Production and Export Patterns

World production and consumption of fish meal grew at an almost constant average annual rate of 10% (compound annual rate) from 590,000 metric tons in 1948 to a peak of 5.36 mmt in 1970 (Figure 1). World exports grew from 415,000 metric tons in 1955 to a peak of 3.48 mmt in 1968, an average annual rate of increase of about 18%. Much of this growth occurred on the basis of production in major exporting countries—those that are members of the Fishmeal Exporters Organization (FEO), namely Angola, Chile, Iceland, Norway, Peru, and South Africa (including production in Namibia, formerly South West Africa).

Among the major exporting countries, Peru experienced the most remarkable growth in production and exports. Peru's production was virtually zero in 1948, but reached 2.25 mmt in 1970, 42% of the world

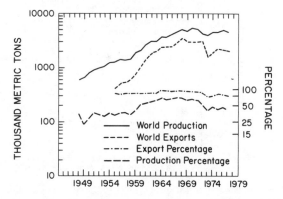

Figure 1 World production (1948–1977) and exports (1955–1977) of fish meal, and percentages of each for major exporting countries, semilog scale.

total. Apparently because of excessive fishing effort and an exceptionally severe recurrence of the complex climatological, oceanographic, and biological phenomenon known as El Niño, stocks of anchoveta off the coast of Peru were sharply reduced in the early 1970s. Consequently, Peru's production of fish meal was reduced by half in 1972 and by half again in 1973 to 420,000 metric tons, a level that Peru had surpassed in 1960. Peru's production recovered somewhat in 1974–1976, but was cut sharply again in 1977 by an El Niño phenomenon in 1976–1977. Species other than anchoveta have accounted for significantly higher shares of the catch for fish meal and oil in the 1970s in both Peru and Chile.

Despite Peru's drop, world production of fish meal averaged 4.43 mmt in 1973–1977 compared with 4.38 metric tons in 1965–1969. Japan and USSR, which are now leading producing and consuming countries, approximately offset the decline in Peruvian production between 1965–1969 and 1973–1977.

World Supply Constraints

Resource-related constraints have affected the catch for reduction into fish meal, oil, and solubles in the 1970s in numerous countries, including among the major exporters, Denmark, Norway, Peru, South Africa (including Namibia). Other factors recently affecting supplies available for export from major exporting countries include the politically related reduction in Angola's production, and the sharp decline in exports from South Africa (and Namibia) in relation to export controls and growth in domestic use by poultry and livestock. Other major producing-exporting countries are also increasing their domestic consumption.

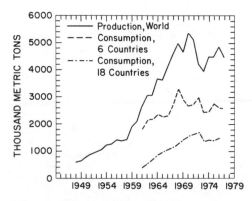

Figure 2 World production of fish meal (1948–1977), and consumption in 6 major and 18 secondary consuming countries (1961–1977).

Consumption

Six major consuming countries account for the use of more fish meal than all other countries combined, although other countries have experienced growth in consumption (Figure 2). Among the six major consuming countries, Japan and the Soviet Union have experienced a continuous upward pattern of consumption well into the 1970s based largely on domestic production (Figure 3). By contrast, the other four major consuming countries, the United States, United Kingdom, Federal Republic of Germany, and the Netherlands, experienced growth until the late 1960s and import-related declines since then. The sharpest decline in consumption occurred for the Netherlands, which is no longer a major consumer. The combined imports of the United States, United Kingdom, and Federal Republic of Germany have accounted for a steadily declining share of world exports, about 64% in 1955 and about 29% in 1977 (Figure 4).

Among 24 selected countries, consumption was generally higher on the average in 1973–1977 than in 1961–1967 (Table 1). It declined significantly in Belgium, France, Spain, and the Netherlands. On the other hand, it increased significantly in Finland, Switzerland, and Sweden; in the East European countries, except Czechoslovakia; in Chile, South Africa (and Namibia), among major exporting countries; in the Federal Republic of Germany, Japan, United Kingdom, and USSR, among major consumers; and in Mexico, Taiwan, and Thailand.

Some of the countries with significantly increased annual average consumption, comparing 1961–1967 and 1973–1977, also had increases from 1968–1972 to 1973–1977 when world exports declined 35%. These

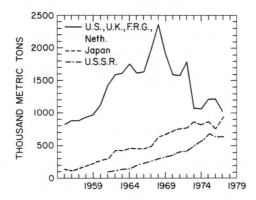

Figure 3 Consumption of fish meal in six major consuming countries, 1955–1977 or 1961–1977 (U.S.S.R.)

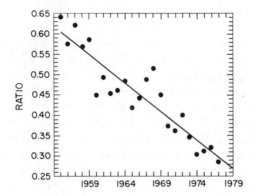

Figure 4 Ratio of imports (U.S., U.K., and F.R.G. combined) to world exports of fish meal, annual data and trend line, 1955–1977. Trend line: ratio = 1.373 − 0.01398 (T) where T = 55–77 for (t = 16.03) (t = −10.82) 1955–1977; R-squared: 0.85; Durbin-Watson: 1.46; F (1,21): 117.14.

countries include Japan and USSR among the major consuming countries, Finland and Switzerland in Western Europe, German Democratic Republic and Poland in Eastern Europe, South Africa (and Namibia) among the major exporting countries, and Taiwan and Thailand. While some of these countries increased consumption on the basis of imports and others on the basis of domestic production, any increase in the face of declining world supplies (world exports) suggests rather strong growth in demand.

ANIMAL FEEDS AND NUTRITION

The demand for fish meal derives from its cost-effective, profitable use in mixed feeds primarily for the world's poultry, livestock, and farmed fish. Fish meal is also used in feeds for mink and other fur bearing animals, laboratory animals, and animals that are kept in homes (cats, fish, and other pets). Poultry feeds account for an estimated 80% of the use of fish meal in the United States, according to sources in the fish meal processing industry.

Least Cost Feed Formulation

Both objective and subjective decision criteria are employed in the choice of feeds and feed ingredients.[1,2] The selection of fish meal and other ingredients for a particular poultry or other feed depends upon objective factors, including animal nutritional requirements, ingredient

TABLE 1. Consumption of Fish Meal in 24 Selected Countries, 1961–1977

Country	Annual Average Consumption (thousand metric tons)			Change (%)	
	I 1961–1967	II 1968–1972	III 1973–1977	I–III	II–III
Major consuming countries					
Federal Republic of Germany	225	445	326	45	−27
Japan	198	438	839	324	91
Netherlands	105	170	43	−59	−74
United Kingdon	225	403	305	36	−24
United States	447	618	440	−2	−29
USSR	166	357	599	260	68
Western Europe					
Belgium	64	100	35	−45	−65
Finland	20	42	49	148	17
France	99	112	53	−47	−53
Italy	75	112	72	−4	−36
Spain	98	158	61	−39	−62
Sweden	44	85	72	62	−15
Switzerland	30	57	84	175	48
Eastern Europe					
Czechoslovakia	49	95	48	−3	−49
German Democratic Republic	70	112	155	123	38
Hungary	27	62	51	85	−18

TABLE 1. Consumption of Fish Meal in 24 Selected Countries, 1961–1977

Country	Annual Average Consumption (thousand metric tons)			Change (%)	
	I 1961–1967	II 1968–1972	III 1973–1977	I–III	II–III
Poland	60	162	187	215	15
Yugoslavia	24	69	67	173	-4
Major exporting countries					
Chile	24	20	62	164	211
Demark	65	88	55	-15	-37
South Africa and Namibia	42	128	167	293	30
Other					
Mexico	32	91	71	124	-22
Taiwan[a]	—	(37)	60	—	62
Thailand[a]	—	(46)	68	—	48
World production	3533	4852	4433	25	-9
World exports	2215	3072	1986	-10	-35

[a]Data not available for several years, when average is in parentheses.

Source: Based on FAO data as compiled by F. C. Wright for the Annual Conferences of the International Association of Fish Meal Manufacturers (1967 through 1978).

composition and prices, and constraints that may be written into a least-cost-feed computer program. For example, a typical program for broiler feed to obtain least cost per ton of mixed feed may involve 12–14 constraints. Constraints may be included for metabolizable energy level, minimum and maximum percentages for five of the essential amino acids (arginine, cystine, lysine, methionine, and tryptophan), and for three of the essential inorganic elements (calcium, phosphorous, and sodium), the ratio of methionine and cystine, and the ratio of calcium and phosphorous.[3]

Fish meal is a nutritionally high quality, valuable feed ingredient, noted for its high content of the amino acids methionine and lysine.[4-6] These two amino acids are manufactured synthetically; therefore, depending on prices and availability, feed mixers can supplement traditional ingredients to achieve proper amino acid balance if necessary. However, if the feed will be close to least cost per ton, nutritionists may include some fish meal because of its value in addition to what may be surmised on the basis of chemical analysis of its content of amino acids, vitamins, inorganic elements, and energy; that is, they may include fish meal because of the unidentified growth factor (UGF).

The growth and feed conversion of broilers are improved when fish meal is included in the diet, according to a recent review of world literature by Pike.[7] He considered several nutritional effects:

> Most of these nutritional effects have not yet been quantified to the extent that they can be incorporated into diets formulated by computer to least cost. However, when the cost of broiler meat production is considered, the feed costs must take into account feed utilization. While diets containing fish meal may not be least-ton cost, they will generally give lowest feed costs per kg of gain because of improved utilization by the broiler.[7]

Pike reported that liveweight gains of broilers improved by 2.4% and feed conversion improved by 1.9% on the average for 12 experiments comparing diets with and without fish meal in which the diets were equated for known nutrients, including lysine and methionine.

Seemingly small percentage differences in weight gain and feed conversion are quite significant, because of the critical importance of feed costs to the broiler industry. Feed now represents over two-thirds of the cost of producing eggs, and over 70% of the cost of producing broilers and turkeys, excluding costs for processing and marketing. Other production costs include hen depreciation, chick or poult cost, labor, energy, overhead, and miscellaneous costs. Total production costs have risen about 50% during the 1970s, mostly because of the 65 to 75% increase in feed costs. The prices of inputs used in producing poultry and eggs are largely determined by forces over which the producers have little control, but they can influence the efficiency with which they use these inputs.[8]

Without the substantial improvements in productivity that have occurred, consumers would be paying much more than they do for poultry and eggs. Comparing some 1952 and 1977 data on broiler productivity for one representative state, broilers averaged 80 days of age at market time in 1952 and 3.35 pounds in liveweight, and they consumed 3.17 pounds of feed per pound of liveweight (10.6 pounds per broiler by market time). In 1977, the broilers averaged 53 days of age at market time and 4.05 pounds in liveweight, and they consumed 1.95 pounds of feed per pound of liveweight (about 7.9 pounds of feed per broiler by market time). In 1982, it is estimated that broilers will average 46 days of age and 4.00 pounds at market time and they will consume 1.80 pounds of feed per pound of liveweight (7.2 pounds per broiler).[9]

Consumption by Animals

Fish meal is being used increasingly where it has the highest value, because of the combined effects of growing world demand for animal feeds, low growth in the world supply of fish meal, and advances in the field of animal nutrition. Norwegian studies indicate that when world supplies of fish meal are relatively low and its price is relatively high with respect to soybean meal and other feed ingredients, it is most likely to be used in feeds in which effective substitution is least likely, for example in feeds for fish, and mink, and in some starter feeds for broilers, turkeys, and pigs. At lower relative prices, fish meal will be used in other feeds, for example in layer and pig diets in Europe.[10]

As already indicated, poultry feeds account for an estimated 80% of the use of fish meal in the United States according to sources in the fish meal processing industry, that is, roughly 350,000 short tons of the 1974–1978 average consumption of 430,000 short tons of fish and shellfish meal, and fish solubles. The exact usage of fish meal in a diet depends on (1) relative prices and composition of various ingredients, animal nutritional requirements, and specially written computer programs to determine least-cost feed formulation; and (2) whatever value is placed on the unidentified growth factor (UGF) for fish meal.[11] On this dual basis, some firms use 1–2% fish meal in broiler starter diets, which are fed to 0–4 week birds,[12] whereas others follow a recommendation of 2.5% in both starter and grower diets for broilers,[3,9] with 5% providing a feed viewed as a standard of excellence.[13] Limiting rates to a maximum of about 8–9% in starter diets, and 7% for growth diets prevents a "fishy" taste in the final product for human consumption.[11] The organoleptic effect on the final product relates to the level of fish fat in feeds, for which recommended maximum safe levels have been established.[5]

Farmed fish are grown under a variety of controlled conditions, such as in ponds, raceways, and other enclosures; and their nutrients may come from prepared diets, natural aquatic organisms, or both, depending on the environment. Nutritionally, they require relatively more protein and less metabolizable energy than warm-blooded animals. Among the species grown in the United States, catfish, which grow in warm water, require 30–36% protein in their diet, and trout, which grow in cold water, require about 40–45% protein, compared with 16–22% for poultry.[14–16] Feed accounts for 40–55% of the cost of production compared with 70% for poultry, and feed formulation is an important part of business management. The sharp rise in prices and the drop in availability of fish meal in 1973 prompted concern,[17–19] because fish meal is a key ingredient in feeds for farmed fish. The use of fish meal is higher than for poultry feeds (30% for cold-water fish feeds, and 10% for warm-water fish feeds), and totals at least 40,000 short tons.[20] The fish farming industry is growing rapidly, and economic information on it is increasing.[21]

Fish meal and solubles are widely used in foods for cats. In feeds for poultry, livestock, and fish, least-cost formulation is one of the key criteria for selection among various ingredients that can satisfy animal nutritional requirements. In pet foods, palatability (acceptance by the animal) is apparently more important than price or least-cost formulation, because of people's views of pets as opposed to animals for commercial production. As estimated 2.5 billion pounds of cat food are sold each year in the United States, including 23,000 short tons of fish meal.[22–24]

POSTWAR PRICE TRENDS

Price trends of fish meal in the postwar period have been influenced by a number of factors. To reduce the effects of extraneous factors, several simplifying procedures are used.

Methodology

Annual prices of several fish meals are highly correlated (Table 2). This suggests that the markets in which these fish meals are sold are related, and that the behavior of one of the prices reflects roughly what is happening in the other markets, despite differences in market location

TABLE 2. Correlation Coefficients for Prices of Fish Meal, Soybean Meal, and the U.S. Producer Price Index, 1947–1978 and 1964–1978 Annual Data[a]

Price	1	2	3	4	5	6	7	8
Coefficients Based on 1947–1978 Annual Data								
1 Soy 44	1.0000							
2 PPI	0.8157	1.000						
3 Menhaden	0.9650	0.8586	1.0000					
Coefficients Based on 1964–1978 Annual Data								
1 Soy 44	1.0000							
2 PPI	0.7309	1.0000						
3 Soy 50	0.9994	0.7140	1.0000					
4 Peruvian	0.9760	0.8070	0.9743	1.0000				
5 Hamburg	0.9792	0.7311	0.9799	0.9841	1.0000			
6 Tuna-mack.	0.9742	0.7376	0.9753	0.9823	0.9859	1.0000		
7 Japan	0.8356	0.8821	0.8288	0.8927	0.8750	0.8870	1.0000	
8 Menhaden	0.9797	0.7744	0.9787	0.9913	0.9917	0.9944	0.9028	1.0000

[a]Soy 44: Soybean meal, 44% protein, bulk, Decatur, Ill.; PPI: U.S. Producer Price Index, all commodities (1967 = 100), Bureau of Labor Statistics; Soy 50: Soybean meal, 50% protein, bulk, Decatur, Ill.; Peruvian: Imported Peruvian meal, 65% protein, bulk, at New York City, f.o.b. East Coast ports; Hamburg: Fish meal, 64–65% protein, any origin, c. and f., at Hamburg, Federal Republic of Germany; Tuna-mack.: Tuna-mackerel meal, 60% protein (1964–1975), and tuna-mix meal, 55% protein (1975–1978), bulk, at Los Angeles; Japan: Average unit value of imports into Japan of fish meal and flour, c.i.f., excluding any duties on imports in excess of quotas, converted at the average monthly currency exchange rates; Menhaden: Menhaden meal, 60% protein, bulk, at New York City, f.o.b. East Coast and Gulf ports.

and product specification. Because price data for U.S. menhaden meal are available in uninterrupted series for a longer period of time than price data for other fish meals, they are used to represent behavior of all fish meal prices, rather than, for example, the U.S. price of Peruvian anchoveta fish meal.

Prices of fish meal reflect not only the behavior of markets for fish meal and other feedstuffs, but also the general inflation in all producer prices that has occurred, especially since 1973. Inflation in all prices at the producer level in the United States may be observed in the Producer Price Index (PPI, formerly the Wholesale Price Index, WPI, 1967 base of 100) for all commodities, which is prepared by the Bureau of Labor Statistics (BLS), U.S. Department of Labor. During the 32-year period used in the analysis of fish meal prices, the index rose from 76.5 in 1947 to 209.3 in 1978, the sharpest rises occurring in 1951 (11%) and 1973 (13%), with a substantially higher average annual rate of increase in 1973–1978 (9.2%/year) than in previous periods.

Methodologically, deflation is a commonly accepted analytical procedure for removing the implicit effects of general price inflation in a historical price series to emphasize the effect of market forces related to a particular commodity. Various deflators can be used. The prices of menhaden meal and soybean meal (44% protein) used in the present analysis have been deflated by the U.S. Producer Price Index for all commodities (1967 = 100). In other words, the deflated prices for each year in the 1947–1978 period are expressed in terms of 1967 dollars.

Weekly or monthly price observations are used to obtain simple annual averages, thereby reducing both the frequency and amplitude of price oscillations.

The price of soybean meal is assumed to be a good indicator or proxy for world market conditions for major high-protein oilmeals, although it does not necessarily reflect market conditions for feed grains and other feedstuffs. Soybean meal accounts for a significant percentage of world production (52% in 1965–1974) and world trade (56% in 1965–1974) of major high-protein meals, including soybean, fish, peanut, sunflower, cottonseed, linseed, rapeseed, copra, and palm kernel meals, all on a soybean meal equivalent basis in terms of crude protein content. Fish meal accounted for 12.6% of the world production of these meals, and 17.4% of world trade in these meals (1965–1974 average percentages).[25]

The deflated annual prices of menhaden meal and soybean meal, and the ratio of these prices for the period 1947–1978 exhibit several peaks, troughs, and trends that are of interest (Figure 5). The focus will be on two major price peaks (1949 and 1973), the 1960 low, and El Niño effects.

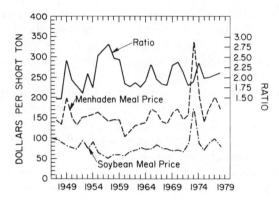

Figure 5   Deflated U.S. prices of menhaden meal and soybean meal (44% protein), and their ratio, 1947–1978.

The 1949 Price Peak

The 1949 peak in the deflated annual price of menhaden meal occurred apparently because of circumstances primarily related to the markets for fish meal apart from the markets for all high-protein meals, because the deflated annual price of soybean meal declined in 1948–1950. The *Wall Street Journal* (August 27, 1949) reported that a new feed formula using 8% fish meal rather than the usual 3% was sweeping the eastern broiler producing areas, that it allowed marketing broilers at 11–12 weeks of age rather than the 15–16 weeks required with other feeds, and that it resulted in the use of 20% less feed. Since U.S. supplies of fish meal were higher than in 1945–1948, it appears that the price increase was related primarily to increased demand for fish meal. The 1949 deflated price was not exceeded until 1973. It was 44% above the 1947–1978 trend line value for 1949, while the 1973 price was 92% above the trend line value for 1973.

The 1960 Price Drop

Prices of fish meal dropped to a postwar low in 1960, prompting an international conference.[26,27] Undeflated prices had been rather stable from 1952–1959, following the adjustment which affected all producer prices in 1951 in relation to the Korean War. Undeflated annual prices of menhaden meal had averaged $137 per short ton in 1952–1959, and had departed at most 5% from this average. However, prices of soybean meal had been declining in the 1950s resulting in the highest relative prices for fish meal in the 1947–1978 period. The ratio of prices of menhaden meal to soybean meal (44% protein) reached a peak of 2.82 in 1957, and exceeded the ratio of all other years in 1955–1959, averaging 2.59 in comparison with the 1947–1978 average of 2.125. It was the relative price level that led to increased use of substitutes for fish meal. Substitution became more feasible, because in about 1959–1960 there were sharp declines in the prices of synthetic lysine and methionine.[28]

The 1973 Price Peak

While undeflated annual prices of menhaden meal in 1977 and 1978 were relatively close to the 1973 price, that 1973 price was the record high, especially in terms of deflated prices that remove the implicit effects of general price inflation. The deflated annual price of menhaden meal in 1973 was 92% above the 1947–1978 trend line value for 1973. Unlike the previous peak in 1949 and the low of 1960, the 1973 peak in the price of menhaden meal was associated with an unprecedented peak in soybean meal prices (Figure 5).

There was a series of unusual national and international events in the 1972–1973 crop year that influenced world markets for fats and oils, feed grains, oilseeds, high protein meals, other feedstuffs, and food grains, as well as markets for food, fuel, and most other products. Briefly, grain, oilseed, and high-protein meal production failures in major producing areas of the world, the consequent rise in demand for available stocks (held mostly in the United States), including the unprecedented Soviet grain purchase, domestic monetary policy, world economic conditions, government acreage restrictions, devaluation of the dollar, and the U.S. price freeze—all occurring in the 1972–1973 crop year—contributed to the 1973 price rise. Late in 1973, fuel prices rose. The drop in Peruvian and Chilean production of fish meal was but one factor contributing to the 1973 price jump. That 1973 price jump resulted in considerable analysis, and its causes led to the World Food Conference in Rome, November 5–16, 1974.[29–36]

El Niño Effects

Several of the postwar El Niño phenomena affected world production, trade, consumption, and prices of fish meal. El Niño (The Christ Child) is so named because it often begins at Christmas time.

During an El Niño, the upwelling of cool, nutrient-rich, subsurface water is displaced by warm water, adversely affecting the fish whose life depends upon the upwelling. Boerema and Gulland[37] report that the area of distribution of the Peruvian anchoveta is practically restricted to the area of upwelling along the coasts of northern Chile and Peru. The strength of the upwelling and the width of the upwelling zone vary annually and seasonally, affecting the distribution and concentration of the anchoveta. Upwelling zones are noted for their productivity. Goulet[38] indicates that upwelling areas account for 3% of the world's total marine water surface area and an estimated 38% of the ocean's exploitable primary carbon production, although the total marine harvest depends on the trophic level (successive level in the predator-prey food chain) of the species harvested, as well as on technical and economic factors. Upwellings are caused by surface wind stress, but the cause of El Niño is not fully understood according to Newell[39] and O'Brien.[40] While correlation does not indicate causality, Allison and others[41] found that the sea surface temperature anomalies associated with El Niño are correlated with several geophysical parameters. These sea surface temperature anomalies vary in frequency, direction (warmer or colder than normal), intensity, and duration. It appears that warm water anomalies of sufficient intensity to affect fishing occurred off the coasts of northern Chile and Peru in 1957, 1965, 1969, 1972, and 1976; lower intensity events occurred in 1951, 1953, and 1963; and earlier, high-intensity events occurred in 1891, 1925, 1930, and 1940.

The economic significance of postwar El Niño phenomena varied. Production, exports, and prices of fish meal were sometimes affected in the year of El Niño occurrence, and sometimes later, in part for reasons explained by Boerema and Gulland.[37] They indicate that young anchoveta from the main (August–September) spawn are recruited to the fishery beginning in January of the following year. Peruvian fishing for anchoveta is stopped at this time, usually for 2 months, to permit new recruits to grow, thereby increasing the meal and oil yield from a ton of fish. While the adverse effects of El Niño on anchoveta are not fully known, both spawning and young fish are especially affected by the unusual environmental conditions; hence, recruitment is reduced. Thus, the impact of an El Niño on Chilean and Peruvian fish meal production might be delayed until the following year when the young recruits to the fishery would otherwise reach harvestable size. In addition, the impact of El Niño on fish meal production depends on other factors, including the intensity of fishing, the duration of the adverse conditions, the percentage of anchoveta in the catch, and managment measures. Furthermore, the impact on world exports and prices of fish meal depends on market factors, including the level of stocks of fish meal, and developments in the markets for all high protein meals.

El Niño Events of the 1950s

It appears that the El Niño events of 1951, 1953, and 1957 could have reduced the rate of growth in fish meal production in Chile and Peru. The record high ratios of fish meal to soybean meal prices in 1955–1959 (Table 3 and Figure 5) were most likely caused by declining prices of soybean meal in the 1950s and the relatively stable price of fish meal that were associated with apparently increased demand for fish meal beginning in 1949, as previously explained.

El Niño Events of the 1960s

Production of fish meal in Chile and Peru experienced low growth in 1963, perhaps because of the 1963 El Niño; production and exports dropped in 1965 and 1969 when major El Niño events occurred. Peru's production moved to a record level in 1970, but exports began a downward trend in 1969. The ratio of fish meal to soybean meal prices rose sharply in 1965 and 1969–1970, because of the significance of reduced contribution of exports from Peru and Chile to world exports (Table 3, and Figure 5).

El Niño, 1972

The El Niño event of 1972 brought higher water temperatures than previous postwar sea surface temperature anomalies, and by this time fishing capacity was substantially higher. In 1971, the Peruvian fleet was so large that the allowable catch of anchoveta was taken in 90 days compared with about 200 days in the mid-1960s and about 300 days in the early 1960s.[37] Peruvian and Chilean production of fish meal was cut in half in 1972, but Peruvian exports did not fall very much until 1973 because record high stocks of fish meal were used to offset dwindling production. Peruvian stocks of fish meal were reduced from about 800,000 metric tons in late 1971 to about 60,000 metric tons in early 1973; exports were 1.6 mmt in 1972. Monthly stocks averaged about 760,000 metric tons in 1971, 5.2 times average monthly exports compared with the 1964–1978 average ratio of about 3.5.

World exports of fish meal increased in 1972, and the ratio of fish meal to soybean meal prices fell (Figure 5). The effects of the 1972 El Niño on

TABLE 3. Selected Fish Meal Market Indicators for El Niño Effects: Combined Chilean and Peruvian Production and Exports with Year-to-Year Percentage Changes, Exports of Chile and Peru as a Percentage of World Exports, and the Ratio of U.S. Prices of Menhaden and Soybean (44% Protein) Meals, 1947–1979

Year	Peru + Chile		Year-to-Year Change		Export Ratio: (Peru + Chile) to World Exports (%)	Price Ratio: Menhaden to Soybean (44%)
	Production (mmt)	Exports (mmt)	Production (%)	Exports (%)		
1947						1.47
1948	1					1.47
1949	3		300			2.42
1950	5		59			1.94
1951	13		157			1.77
1952	18		35			1.61
1953	20		11			2.10
1954	24		22			1.74
1955	36	27	49		7	2.54
1956	44	32	23	16	6	2.69
1957	81	66	84	108	13	2.82
1958	145	117	79	77	19	2.46
1959	362	295	150	152	39	2.44

Year						
1960	612	534	69	81	50	1.84
1961	904	756	48	42	53	1.76
1962	1,215	1,134	34	50	62	1.87
1963	1,245	1,130	2	0	57	1.76
1964	1,727	1,572	39	39	66	1.92
1965	1,375	1,480	−20	−6	61	2.32
1966	1,689	1,485	23	0	61	1.95
1967	1,979	1,709	17	15	57	1.83
1968	2,157	2,254	9	32	65	1.80
1969	1,788	1,852	−17	−18	62	2.30
1970	2,444	1,986	37	7	67	2.38
1971	2,191	1,946	−10	−2	66	2.14
1972	1,011	1,699	−54	−13	56	1.79
1973	510	381	−50	−78	23	1.90
1974	1,070	725	110	90	37	2.35
1975	835	876	−22	21	40	1.98
1976	1,128	793	35	−9	37	2.00
1977	735	611	−33	−23	31	2.05
1978	1,000	762	33	20	37	2.11
1979	1,180	980	18	29	42	1.98

Source: Based on FAO data as compiled mostly by Wright[60] and Anon.[27] and price data compiled for the NMFS *Fish Meal and Oil Market Review*.[43] Some data for 1978 and 1979 estimated.

world markets were felt more sharply in 1973 when world exports of fish meal were cut in half, but that dramatic event was just one of a series of significant and unusual events contributing to the rise in world prices of all high-protein meals, as previously explained. Annual prices of fish meal rose only slightly relative to annual prices of soybean meal in 1973. In 1974, however, fish meal prices declined less than soybean meal prices, leaving the ratio of the two prices high, and comparable to ratios of 1965 and 1969. World production of major high-protein meals increased by 18% or about 10 mmt in 1974, while fish meal production increased only 15% or 0.8 mmt (all data on soybean meal equivalent basis).[42]

El Niño, 1976

Peru's production and exports of fish meal recovered somewhat from 1973 levels in 1974–1976, but fell sharply in 1977 because of the adverse effects of both heavy fishing and the El Niño in 1976. Market effects were less important, because Peru and Chile accounted for a far lower percentage of world exports than previously (Table 3). Since 1977, production in Peru has improved, but fell 10% short of the 1972–1976 average of about 750,000 metric tons in 1979. Production of fish meal in Chile has been growing more rapidly than in Peru and could reach 450,000 to 500,000 metric tons in 1979 compared with the 1972–1976 average of 157,000 metric tons. In both countries, species other than anchoveta have become more important than in the past in the manufacture of fish meal and oil, including jack mackerel, Chilean pilchard (sardines), mackerel, hake, and possibly other species, although some of these species are used in the growing human-food processing industries.[43,44]

Econometric Models and Price Analysis

The rationale for econometric analysis of the forces of price behavior has its basis in decisions in public policy and private business. Tomek and Robinson[45] provide a useful review of price analysis. Within the fisheries context, current policy problems relate to fisheries management, conservation, and development, including aquaculture.

All models of real world behavior are by definition simplifications or representations that emphasize relationships thought to be important. There are typically several parts to an econometric model of a commodity market, including behavioral equations for demand, supply, inventories (stocks), and prices. Once a model is specified, the parameters defining the relationships are estimated statistically and validated, as explained by Labys, for example.[46] Given the difficulty and complexity of building complete models for some commodities, such as fish meal, oil, and solubles, some analysts may choose to study just one set of behavioral relationships. Generally, demand relationships have received more attention than the equally important supply relationships.

For purposes of developing models of market behavior of fish meal, it may be useful to recognize some unique economic characteristics of both the fishing and agriculture industries. Hathaway has summarized five characteristics that *taken together* define agriculture's uniqueness, so far as is possible for a large, diverse industry.[47,48] Demand may be growing faster for fishery products than for agricultural products in general, as is the demand for most agricultural animal products. There appears to be a significant tendency in both industries for resources, including labor, to become economically fixed. Fisheries economists emphasize the common property nature of fisheries as a unique characteristic; see, for example, Bell's comprehensive text on fisheries economics.[49] It is generally recognized that profit and income expectations have generated excessive amounts of labor and capital in many fisheries, resulting in excessive fishing effort which sometimes contributes to the collapse of a fishery, as in the case of the Peruvian anchoveta fishery.

Besides those already mentioned in this section on price analysis, economic studies on or of relevance to fish meal include those by King,[50] Allen, Combs, and Petersen,[28] Bell and others,[51,52] Leonico,[53] Kolhonen,[34] Segura,[54] Ccama,[55] Meilke,[56] and Hansen,[57] in chronological order. Currently, Wang and Mueller,[58] and Huppert[59] are working on studies of interest.

Structural shifts in the demand for fish meal appear to have occurred in 1949, 1959–1960, and perhaps in the 1970s. In any event, consumption patterns have changed and differ among countries in the postwar period. Finally, data on production, trade, and prices are available for some countries, but data on stocks are available only for countries that are members of the Fishmeal Exporters Organization (FEO). Consumption data must be derived from production and trade data for most countries, as Wright has done for the International Association of Fish Meal Manufacturers (IAFMM).[60]

OUTLOOK

Several factors are relevant in the long-term outlook for fish meal, including (1) postwar behavior of key variables, (2) the world fisheries resource potential, and (3) prospects for fisheries development.

Postwar Developments

Production Patterns

The 1973–1977 average world production of fish meal, about 4.4 mmt, was achieved in the mid-1960s. If the only change was to restore Peru's output to its 1965 level, the world total would be 5.0 mmt, a level achieved in 1968–1971 when 20 years of rapid growth in world output ended. By contrast, world production of high-protein meals more than doubled from 40.4 mmt in 1965 to 83.7 mmt in 1979 (USDA forecast); it is projected to reach 100 mmt by 1985, according to Holz (all data expressed in terms of soybean meal, 44% protein; fish meal multiplied by a factor of 1.4452).[25,42,61,62] Because of these sharp differences in production patterns, world consumption of fish meal stabilized in the 1970s, whereas consumption of all high-protein meals continued to grow.

Consumption Patterns

Fish meal will be used increasingly where it has the highest value, because of the combined effects of growing demand for animal feeds, probable limited growth in the world supply of fish meal, and advances in animal nutrition. Substitution of other ingredients appears less likely in feeds for fish, pets, mink, and in starter feeds for broilers, turkeys, and pigs. Total and per-animal consumption data for many countries, however, will likely continue to show declines. For example, broilers are a major consumer of fish meal in the United States, and the number of broiler chicks placed in 21 states increased from 2.8 billion in 1969 to 3.7 billion in 1978, whereas U.S. aggregate consumption of fish meal and solubles fell from 583,000 to 395,000 metric tons, resulting in a decline in apparent use from about 200 to 100 grams per broiler, assuming for the sake of comparison that fish meal is consumed only by broilers. Comparing in another way, U.S. per capita consumption of edible fish products grew to a record 13.6 pounds in 1978 from 10.6 pounds in 1967, whereas U.S. per capita use of fish meal declined to 4.02 pounds from 9.24 pounds. U.S. consumption of edible fishery products has increased for many years largely on the basis of increased imports, but U.S. imports of fish meal have declined.

World Fishery Resource Potential

World Total Marine Potential

Biologists' estimates of the total potential yield of marine animals vary considerably, from about 100 to 2,000 mmt tons. Wise[63,64] indicates that the most reliable present estimate of the potential harvest of *conventional marine* species is about 120 mmt, or twice the 1973–1977 average of 62 mmt. The total world catch includes both freshwater (9.7 mmt in 1973–1977) and marine catch, and the 1973–1977 average catch for fish meal and oil was 19.7 mmt. Estimates of potential are made using two methods, "one based on extrapolation of the present catch, the other based on analyses of the theoretical flow of organic material through the food chain at different trophic levels."[65] Wise cautions that extrapolations using recent data should be interpreted carefully, because of the significant effect of the decline of Peruvian anchoveta catch (combined catch by Chile and Peru; data in million metric tons: 13.1 in 1970; 1.7 in 1973; 0.8 in 1977; and 1.2 in 1978). He further cautions that achievement of the potential assumes proper management, but believes that the "sad lessons of past overexploitation and recent extensions of national jurisdictions make the picture more hopeful now than it has been in many years."[63] Hennemuth is far more pessimistic, believing that there are several significant problems, and that the current marine catch of about 60 mmt is not likely to be increased much by the year 2000.[66]

Pelagic Species Problems

Like Hennemuth, Ursin believes there are problems with some yield-estimates based on single-species models, and reports the application of a multispecies model to the North Sea.[67] He indicates that the yield from the North Sea fisheries has increased since 1960, but that the total calculated stock size (all species) remained at about 9 mmt, even though the species composition changed. Herring from the North Sea was once the backbone of the Norwegian and Danish catch for fish meal and oil, but now Norway pout, sprat, sandeel, and blue whiting account for the bulk of the North Sea catch for fish meal and oil. Ursin also indicates that two-way predator-prey relations between cod and capelin affect the potential catch of these two species in the Barents Sea. Capelin from the Barents Sea now accounts for a dominant part of Norway's catch for fish meal and oil. The total catch of capelin in the Barents Sea peaked at about 3 mmt in 1977, but the Norwegian-Soviet Fisheries Commission recommended a significantly lower catch in 1979 (1.8 mmt) with 60% (1.1 mmt) allocated to Norway.[43]

The catch for use in fish meal and oil is relatively large (27.5% of the world total catch in 1973–1977), and it includes primarily pelagic species that are subject to cycles in abundance. Even though the Peruvian anchoveta appears to be recovering, the California sardine (i.e., the Pacific sardine off the coast of California) and the Atlanto-Scandian herring have fallen to virtual commercial extinction. The sardinelike species off the coast of Japan are now in an upswing, but the South

African and Namibian (South West African) pilchard stocks continues to require careful management (see Chapter 6).

Pelagic Species Potential Increases

Referring to conventional marine species, Moiseev estimated a potential increase in catch of about 20 mmt (i.e., bringing the world total marine catch to 82 mmt), of which about 17 mmt was for coastal pelagic species.[68] Suda estimated a somewhat larger total potential increase for all pelagic marine species (22 to 30 mmt and an overall increase bringing the world total marine catch to 93 to 110 mmt),[65] but his estimate for coastal pelagic species, about 16 to 23 mmt, is quite close to Moiseev's. Much of the potential increase for conventional coastal marine pelagics is either in temperate and cold waters of the Southern Hemisphere or in tropical waters; that is, mostly off the shores of developing countries and/or countries that are not major producers and exporters of fish meal. Suda's estimates also include oceanic (distant water) pelagic species that could be used for fish meal and oil, but the implied distant-water fishing is more costly than coastal fishing.

Prospects for Fisheries Development

Moiseev's and Suda's estimates for potential increases in the world-wide catch of conventional coastal marine pelagic species, both about 20 mmt, are about the same as the 1973–1977 average world catch for reduction (19.7 mmt). If all of that potential could be achieved and used for fish meal and oil, the world output of fish meal could be doubled to about 9 mmt. In addition, recovery of Peruvian anchoveta catch to 6 mmt (well below the 1970 peak of 12.5 mmt, but also well above the 1973–1977 average of 2.8 mmt)[64] could result in a combined increase in Chilean and Peruvian production of fish meal of about 700,000 metric tons from the 1973–1977 average of about 860,000 tons.

The doubling of world fish meal production could be stated simply as a broad planning target, like the doubling of world or national catch of fish, but there are several complex issues that could affect the achievement of any such planning target. Prices must be high enough to cover costs of production. On the other hand, if fish meal prices are too high, then demand may be affected by the development of substitutes. The Food and Agriculture Organization (FAO) has a long-term view of using the raw materials of fish meal for edible products, but very little of those materials seems suitable now, except perhaps for making fish protein concentrate (FPC). FPC faces problems of profitability in production and distribution, and, like fish meal, competition with commercially successful high protein products of agricultural origin. Fishery management authorities may face choices among fisheries for reduction, food, and recreation, not to mention the ecosystem food demands of various marine animals. These several factors make projections of fish meal consumption and production difficult.

Planning Targets and Their Feasibility

Broad biological yield estimates, such as those of Suda and Moiseev for the world, must be refined for purposes of fishery development. The species available, their size, their geographic distribution, their seasonal abundance, and their stock size and yield are the key biological factors, but some of this critical information is not known for unfished resources.[69]

Unfortunately, fisheries agency plans for fisheries development have sometimes lacked sufficient consideration of social, political and economic issues. This is true even though the plans may include estimates of capital costs and benefits (see Vondruska, pp. 259–261;[70] Bell, pp. 9 and 19–20[71]).

Fortunately, substantive economic, social, technical, and biological issues have received increased attention in some plans for development, for example, in the United States.[72] Even so, broad planning targets for doubling world or national catch[72,73] are not project specific (but see, for example, a recent FAO manual for identifying and evaluating fisheries investment projects by Campleman[74]).

The Role of Reduction Fisheries

Long-range plans and policies for fisheries development usually emphasize increased fisheries production for direct human consumption, not for reduction into fish meal and oil. This issue is complex. One factor is that fisheries for the two purposes sometimes compete, especially when stocks in a given area are exploited near maximum levels. Competition may not be at the harvesting level, but perhaps at the ecosystem level, such as in terms of two-way predator-prey relationships.[67] Also, fisheries for reduction may not show as much concern for maintaining specific stocks as fisheries for food, because they need not be as species specific, and there is the related question of distinguishing between edible and inedible fish.[75]

Dreosti objects to the view that fisheries for reduction are wasteful. He indicates that about 90% of the fish now used for reduction into fish meal and oil is either unsuitable for human food or unmarketable, leaving 10% that could be processed for direct human consumption, if economically feasible.[76] Because it "is not easy, nor inexpensive, to produce acceptable stabilized food products from small, fatty fish," Dreosti believes that it could be more economical for developing countries — whose extended jurisdiction zones contain much of the unharvested potential of small, coastal pelagic species—to convert these fish into fish meal and oil.[76] FAO, however, appears to view the use of these species for fish meal and oil only as an initial development strategy,[73] but recognizes the need to find alternative feed ingredients.[77]

Single-Cell Protein

Single-cell protein based on petroleum has been added to the long list of commercially available high-protein feedstuffs, amino acids, vitamins, and minerals, but its economic viability and ability to compete remain to be tested on a significant scale. While amino acid composition of single-cell protein can be expected to vary among organisms and culture media employed, the products evaluated by Spinelli et al.[78] differed significantly from fish meal with respect to several amino acids. However, it is possible to meet animal nutritional requirements using a variety of ingredients.

Single-cell protein production based on petroleum as a culture medium excites interest and speculation, but not much is known publicly about its economic prospects (Corkern and Dwoskin[79]). By contrast, the production of "single-cell organisms for protein concentrates, vitamins, and amino acids by fermentation process, using sugar or carbohydrates as the principal nutrient, is a well-established commercial process in the United States" (Corkern and Dwoskin,[79] p. 82). Production of single-cell protein based on petroleum is capital intensive and apparently efficient in the use of petroleum derivatives.[80] Estimated world output potential was about 1.5 mmt annually in 1975–1980,[81] and a German firm is reported to have found that single-cell protein based on petroleum is competitive with fish meal.[82] On the other hand, it is possible that rising real prices of petroleum in combination with stable or more slowly rising prices of high protein meals could limit investment in the expanded production of single cell protein based on petroleum derivatives.[61,62]

Fish Protein Concentrate (FPC)

Fish protein concentrate (FPC) is a dry flour or powder that is manufactured in much the same manner as fish meal for animal feed, but under conditions that assure safety and quality suitable for human consumption. Despite concerns about world food supplies and protein deficiencies in human diets in various parts of the world, and hopes that FPC could be used to offset those protein deficiencies, FPC is produced in only relatively small quantities. The problems appear to include low profitability, and limited consumer acceptance. FPC is not likely to displace fish meal overnight because of these problems (see Dreosti,[76] Finch,[83] and Lauritzen[84]). Fish represents the largest single cost component,[83] and the cost of fishing has been increasing, probably more rapidly than the cost of farming.[85] Both fish meal and FPC compete with products of agricultural origin.[86]

CONCLUSIONS

Fish meal is a nutritionally high-quality animal feed ingredient that experienced remarkable growth in world production and consumption in the postwar period until 1970. Since then, resource-related constraints in the major producing-exporting countries have affected the ability of these countries to help supply the growing world demand for animal-feed ingredients. Broad estimates of world potential biological yields for marine animals suggest that world annual production of fish meal could be doubled from levels of the mid-1970s to about 9 mmt, but achievement of that potential increase would have to occur from the harvest of small coastal pelagic species located off the shores of developing countries and/or countries that are not major producers and exporters of fish meal. One bioeconomic model of world living marine resources indicates a projected decline in catch for reduction purposes from 20 mmt in the mid-1970s to 15 mmt by the year 2000,[51,52] although the estimated potential biological yield of conventional coastal marine pelagic species is about 40 mmt.

The broad span of literature from several disciplines relating to fish meal suggests some problem areas, including the El Niño phenomena. There is social value in forecasting these events[87] and interest in the relationship of climate and the fisheries (e.g., Anon.[88]). The effects of El Niño events can be observed in market variables, such as prices, production, consumption, trade, and stocks of fish meal, but careful analysis is necessary because these variables are affected by developments in markets for all high-protein feed ingredients. For example, world prices of fish meal rose in 1973, following the El Niño of 1972, but they dropped in mid-1977, following the El Niño of 1976. In both cases, prices of fish meal changed primarily because of developments in world markets for all high-protein meals.

This chapter has been concerned primarily with a broad postwar perspective of developments and trends in production, consumption, trade, and prices of fish meal, but much of the adverse social and economic effects of an El Niño are felt by those people whose income and employment depend upon harvesting and processing the catch for fish meal and oil. Of course, El Niño phenomena complicate fishery management problems, but they are not the sole cause of those problems.

ACKNOWLEDGMENTS

The author acknowledges helpful comments and suggestions by Jukka Kolhonen, Mikolaj Wojnowski, Roger Hutchinson, Richard Kinoshita, and Jim Roberts, and others in the NMFS, F. Chalmers Wright and Michael Glantz. Jim Roberts provided data base and computer services. Mary Brown cheerfully provided typing through several stages of draft and final copy. The author's wife and children sacrificed family time, and provided encouragement.

REFERENCES

1 T. F. Funk and F. C. Tarte, "The farmer decision process in purchasing broiler feeds," *American Journal of Agricultural Economics*, **60**(4), 678–682 (1978).

2 Anon., "Fish meal—changing patterns in price relationships," *World Fishing*, **28**(2), 37, 39 (1979).

3 O. W. Charles, University of Georgia, personal communication (June 13, 1979).

4 I. H. Pike, *The Role of Fish Meal in Diets for Poultry*, International Association of Fish Meal Manufacturers (IAFMM) Technical Bulletin No. 3 (1975).

5 S. M. Barlow and I. H. Pike, *The Role of Fat in Fish Meal in Pig and Poultry Nutrition*, IAFMM Technical Bulletin No. 4 (1977).

6 Anon., *Available Amino Acid Content of Fish Meals*, IAFMM Technical Bulletin No. 1 (1970).

7 I. H. Pike, "Unidentified growth factors of fish meal: Possible nutritional explanations," *Feedstuffs*, **51**(15), 32–33 (1979).

8 G. B. Rogers, "Costs, prices, and productivity in the poultry and egg industries," *Poultry and Egg Situation*, U.S. Department of Agriculture PES-300, pp. 26–29 (1978).

9 L. Dansky, "The future for broiler nutrition," *Feedstuffs*, **50**(28), 21–22 (1978).

10 O. Enger, "Factors underlying the fish meal market," a paper presented at the 18th Annual Conference of the IAFMM, October 30 to November 3, New Orleans, Record of Proceedings, pp. 198–209 (1978).

11 E. H. Covell, "The success of the poultry industry—The USA," a paper presented at the 18th Annual Conference of the IAFMM, October 30 to November 3, New Orleans, Record of Proceedings, pp. 190–197 (1978).

12 W. P. Roenigk, National Broiler Council, Washington, D.C., personal communication (June 13, 1979).

13 T. D. Runnels, "Menhaden fish meal as a feed ingredient," a paper presented at the 19th Annual Fisheries Symposium sponsored by the National Fish Meal and Oil Association, March 15, Washington, D.C. (1978).

14 R. T. Lovell, "Formulating diets for aquaculture species," a paper presented at the 20th Annual Fisheries Symposium sponsored by the National Fish Meal and Oil Association, March 14, Washington, D.C. (1979).

15 L. G. Fowler and J. L. Banks, "Animal and vegetable substitutes for fish meal in the Abernathy Diet, 1973," *The Progressive Fish Culturist*, **38**(3), 123–126 (1976).

16 L. G. Fowler and J. L. Banks, "Fish meal and wheat germ meal substitutes in the Abernathy Diet, 1974," *The Progressive Fish Culturist*, **38**(3), 127–130 (1976).

17 Anon., "The feed price dilemma—from the nutritionists' point of view," and "The feed price situation—from the manufacturers' point of view," *The Catfish Farmer*, **5**(2), 22, 23, 25, 26 (1973).

18 J. W. Avault, Jr., "High feed costs spur research," *The Catfish Farmer*, **5**(2), 27 (1973).

19 R. A. Erkins, "Protein problems plague pisciculturists," *The American Fish Farmer and World Aquaculture News* **4**(5), 7–8 (1973).

20 T. Brandt, U.S. Fish and Wildlife Service, Fish Farming Experiment Station, P.O. Box 860, Stuttgart, Arkansas 72160, personal communication (August 29, 1979). Dr. Brandt provided estimates of production of various cultured species, feed conversion rates, and percentages of fish meal in diets.

21 J. Vondruska, *Aquacultural Economics Bibliography*, U.S. Department of Commerce, NOAA Technical Report NMFS-SSRF-703 (1976).

22 R. Mohrman, "Fish meal and solubles in the cat food industry," a paper presented at the 20th Annual Fisheries Symposium sponsored by the National Fish Meal and Oil Association, March 14, Washington, D.C. (1979).

23 J. Corbin, "Dietary nutrient allowances for dogs," *Feedstuffs Reference Issue*, **51**(29), 70–73 (1979).

24 Anon., *1977 Census of Manufacturers—Dog, Cat, and Other Pet Food (SIC 2047)*, Preliminary Report, U.S. Department of Commerce, Bureau of the Census, MC77-I-20D-5(P) (June 1979).

25 Anon., *Oilseeds and Products*, U.S. Department of Agriculture, Foreign Agricultural Service, Foreign Agriculture Circular, FOP 1-76 (1976).

26 Anon., "Fish meal imports prompt appeal for international talks," *Feedstuffs* **32**, 62–63 (April 30, 1960).

27 Anon., *Future Developments in the Production and Utilization of Fish Meal—Report of the International Meeting on Fish Meal;* vol. II (Appendices); and "Ensuring stable conditions in the international market for fish meal," The International Meeting on Fish Meal, FAO, March 20–29, Rome (1961).

28 G. R. Allen, G. F. Combs, and P. Petersen, "The world outlook for fish meal," Appendix B in Anon., *Future Developments in the Production and Utilization of Fish Meal*, vol. II (Appendices), The International Meeting on Fish Meal, FAO, March 20–29, Rome (1961).

29 G. W. Kromer and S. A. Gazelle, *Fats and Oils Situation*, U.S. Department of Agriculture, Economic Research Service, FOS-269 (1973).

30 Anon., *International Economic Report of the President*, transmitted to the Congress, February 1974, U.S. Government Printing Office (1974).

31 Anon., "The Economy in 1973," *Federal Reserve Bulletin 60(1)*, 1–14 (1974).

32 D. E. Hathaway, "Food prices and inflation," with comments and discussion (H. S. Houthaker, J. A. Schnittker, and others), *Brookings Papers on Economic Activity*, The Brookings Institution, Washington, D.C. no. 1, 63–116 (1974).

33 A. Eckstein and D. Heien, "The 1973 food price inflation," *American Journal of Agricultural Economics*, **60**(2), 186–196 (1978).

34 J. Kolhonen, "Fish meal: International market situation and the future," *Marine Fisheries Review*, **36**(3), 36–40 (1974).

35 J. Kolhonen, "Impact of El Niño on world fish meal trade," a paper presented at the El Niño Symposium sponsored by the 25th Tuna Conference and the 21st Eastern Pacific Oceanic Conference, October 2, Lake Arrowhead, California (1974).

36 J. A. Schnittker, "The 1972–73 food price spiral," *Brookings Papers on Economic Activity,* The Brookings Institution, Washington, D.C., no. 2, 498–507 (1973).

37 L. K. Boerema and J. A. Gulland, "Stock assessment of the Peruvian anchovy *(Engraulis ringens)* and management of the fishery," *Journal of the Fishery Research Board of Canada,* 30(12), part 2, 2226–2235 (1973).

38 J. R. Goulet, Jr., "Physical upper limits on fisheries production," in P. N. Kaul and C. J. Sindermann, Eds., *Drugs and Food from the Sea,* University of Oklahoma Press, Norman, Oklahoma, 1978, pp. 431–440.

39 R. E. Newell, "Climate and the ocean," *American Scientist,* 67(4), 405–416 (1979).

40 J. J. O'Brien, "El Niño—an example of ocean/atmosphere interaction," *Oceanus* 21(4), 40–46 (1978).

41 L. J. Allison, J. Steranka, R. J. Holub, J. Hansen, F. A. Godshall, and C. Prabbakara, *Air-Sea Interaction in the Tropical Pacific Ocean,* U.S. National Aeronautics and Space Administration, NASA Technical Note, NASA TN D-6684 (1972).

42 A. Holz, "World oilseeds and products outlook," a talk presented at the National Agricultural Outlook Conference sponsored by the U.S. Department of Agriculture, November 14, Washington, D.C. (1978).

43 J. Vondruska, *Fish Meal and Oil Market Review,* U.S. Department of Commerce, National Marine Fisheries Service, Current Economic Analysis I-34 (1979).

44 Anon., *Yearbook of Fishery Statistics,* 1977, FAO Rome (1978).

45 W. G. Tomek and K. L. Robinson, "Agricultural price analysis and outlook," in L. R. Martin, editor, *A Survey of Agricultural Economics Literature,* vol. 1 *Traditional Fields of Agricultural Economics, 1940s to 1970s,* University of Minnesota Press, Minneapolis, for the American Agricultural Economics Association, 1977, pp. 329–409.

46 W. C. Labys, *Dynamic Commodity Models: Specification, Estimation, and Simulation,* Lexington Books, D. C. Heath and Company, Lexington, MA, 1973.

47 D. E. Hathaway, *Problems of Progress in the Agricultural Economy,* Scott, Foresman and Company, Chicago, 1964.

48 D. E. Hathaway, *Government and Agriculture: Public Policy in a Democratic Society,* The Macmillan Company, New York, 1963.

49 F. W. Bell, *Food from the Sea: The Economics and Politics of Ocean Fisheries,* Westview Press, Boulder, CO, 1978.

50 G. A. King, *The Demand and Price Structure for By-Product Feeds,* U.S. Department of Agriculture, Agricultural Marketing Service, Agriculture Technical Bulletin No. 1183 (1958).

51 F. W. Bell, D. A. Nash, E. W. Carlson, F. V. Waugh, R. K. Kinoshita, and R. F. Fullenbaum "A world model of living marine resources," in W. C. Labys, Ed., *Quantitative Models of Commodity Markets,* Ballinger Publishing Company, Cambridge, MA, 1975, pp. 291–323.

52 F. W. Bell, D. A. Nash, E. W. Carlson, F. V. Waugh, R. K. Kinoshita, and R. F. Fullenbaum, "The future of the world's fishery resources: forecasts of demand, supply and prices to the year 2000 with a discussion of implications for public policy," U.S. Department of Commerce, National Marine Fisheries Service, Economic Research Laboratory, unpublished file manuscripts No. 65-1 and 65-2(1970).

53 E. S. Leoncio, *An Econometric Study of the Fish Meal Industry,* FAO Technical Report Paper No. 199, FAO, Rome, 1973.

54 E. L. Segura, "Optimal fishing effort in the peruvian anchoveta fishery," in A. A. Sokoloski, Ed., *Ocean Fishery Management: Discussions and Research,* U.S. Department

55 F. Ccama, "The demand for fish meal in the U.S., 1949–73," unpublished Master of Science thesis, Virginia Polytechnic Institute and State University, Blacksburg, Virginia (1975).

56 K. D. Meilke, "Demand for feed ingredients by U.S. formula feed manufacturers," U.S. Department of Agriculture, *Agricultural Economics Reearch* **26**, 78–89 (1976).

57 T. Hansen (University of Norway, Bergen), "Models for analyzing the demand for fish meal," *Preprints of the NATO Symposium on Applied Operations Research in Fishing,* symposium sponsored by NATO and the Norwegian Fisheries Research Council, August 14–17, 1979 (volume II, forthcoming).

58 S. Wang and J. J. Mueller, "An econometric model of the U.S. menhaden industry," unpublished draft, New England Fishery Management Council (Wang), and National Marine Fisheries Service, Northeast Region (Mueller) (1978).

59 D. D. Huppert, "An analysis of the U.S. demand for fish meal," unpublished draft, U.S. Department of Commerce, National Marine Fisheries Service, Southwest Fisheries Center (1979).

60 F. C. Wright, *Statistical Digest,* prepared for the annual conferences of the International Association of Fish Meal Manufacturers (IAFMM) (annual through 1979).

61 A. E. Holz, "Demand for soybeans and products in the 1980s," presented at the annual convention of the American Soybean Association, Atlanta, Georgia, August (1979).

62 A. E. Holz, U.S. Department of Agriculture, Foreign Agricultural Service, Washington, D.C., personal communication (1979).

63 J. P. Wise, "Food from the sea: myth or reality?" in P. N. Kaul and C. J. Sindermann, Eds., *Drugs and Food from the Sea,* University of Oklahoma Press, Norman, Oklahoma, 1978, pp. 405–424.

64 J. P. Wise, U.S. Department of Commerce, National Marine Fisheries Service, personal communication (August 1979).

65 A. Suda, "Development of fisheries for nonconventional species," *Journal of the Fisheries Research Board of Canada,* **30**(12), part 2, 2121–2158 (1973).

66 R. C. Hennemuth, "Marine fisheries: Food for the future?" *Oceanus,* **22**(1), 3–12 (1979).

67 E. Ursin, "Resource assessment: Danish computer model to assess fishing effects shows results," *World Fishing,* **28**(1), 39 and 41 (1979).

68 P. A. Moiseev, "Development of fisheries for traditionally exploited species," *Journal of the Fisheries Research Board of Canada,* **30**(12), part 2, 2109–2120 (1973).

69 J. W. Reintjes, "Pelagic Clupeoid and Carangid resources for fishery development in the Gulf of Mexico and Caribbean Sea," Proceedings of the Thirty-First Annual Gulf and Caribbean Fisheries Institute, November 1978, Cancun, Mexico (April 1979), pp. 38–49.

70 J. Vondruska, "Domestic investment in harvesting and processing sectors required to utilize fish stocks available under extended jurisdiction," in L. G. Anderson, Ed., *Economic Impacts of Extended Fisheries Jurisdiction,* Ann Arbor Science Publishers, Inc., Ann Arbor, Michigan, 1977, pp. 247–264.

71 F. W. Bell, "World-wide economic aspects of extended fishery jurisdiction management," in L. G. Anderson, Ed., *Economic Impacts of Extended Fisheries Jurisdiction,* Ann Arbor Science Publishers, Inc., Ann Arbor, Michigan, 1977, pp. 3–28.

72 E. R. Combs, Inc., *Prospectus for Development of the United States Fisheries,* a contract report prepared for the Fisheries Development Task Force, National Oceanic and Atmospheric Administration, U.S. Department of Commerce, and available from National Technical Information Service (NTIS), ATTN: Order Desk, 5282 Port Royal Road, Springfield, Virginia 22161, NTIS No. PB 298453/AS (1979).

73   Anon., *The Potential of the Fisheries to Provide Increased Food Supplies for the Developing Countries and the Requirements of Investment,* FAO Fisheries Circular No. 343, FAO Rome (1977), as cited in T. Wray, "Long term targets for development," *Fishing News International,* part 1, **18**(2), 20–21; part 2 **18**(4), 24–25 (1979).

74   G. Campleman, *Manual on the Identification and Preparation of Fishery Investment Projects,* FAO Fisheries Technical Paper No. 149, FAO, Rome (1976).

75   J. Popiel and J. Sosinski, "Industrial fisheries and their influence on catches for human consumption," *Journal of the Fisheries Research Board of Canada,* **30**(12), part 2, 2254–2259 (1973).

76   G. M. Dreosti, "Industrial fish: feed or food?" A paper presented at the 18th Annual Conference of the IAFMM, October 30 to November 3, New Orleans, *Record of Proceedings,* 1978, pp. 172–189.

77   W. Krone, "Fish meal in fish feed—a brief account of discussions on this topic during the FAO technical conference on aquaculture (Kyoto, Japan, 26 May to 2 June 1976)," FAO, Rome (1976).

78   J. Spinelli, C. Mahnken, and M. Steinberg, "Alternative sources of proteins for fish meal in salmonid diets," a paper presented at the Symposium on Finfish Nutrition and Fish Feed Technology, Heeneman Verlagsgesellschaft mbH, Besserstr. 83–91, D-1000 Berlin 42, Federal Republic of Germany (1979, forthcoming).

79   R. S. Corkern and P. B. Dwoskin, Eds., *Synthetics and Substitutes for Agricultural Products—A Compendium,* U.S. Department of Agriculture, Economic Research Service, Miscellaneous Publication No. 1141 (1969).

80   J. Spinelli, U.S. Department of Commerce, National Marine Fisheries Service, Northwest and Alaska Fisheries Center, Seattle, Washington, personal communication (September 1979).

81   J. Wells, table from a paper presented at the American Chemical Society Meeting, Philadelphia, April 9, 1975, as cited in K. S. Price, Jr., W. N. Shaw, and K. S. Danberg, *Proceedings of the First International Conference on Aquaculture Nutrition,* October 14–15, 1975, Lewes/Rehoboth, Delaware, sponsored by the Delaware Sea Grant College Program in cooperation with the U.S./Japan Aquaculture Panel, University of Delaware, Newark, Delaware, DEL-SG-17-76, 1976, pp. 235–236.

82   Anon., "Technology—protein factory harnesses bacteria," *Science,* **115**(8), 121 (February 24, 1979).

83   R. Finch, "Whatever happened to fish protein concentrate?" *Food Technology,* **31**(5), 44–53 (1977).

84   J. Lauritzen, "FPC—many openings available to realize potential," *World Fishing,* **28**(2), 41 and 43 (1979).

85   D. D. Huppert, "Economic constraints on food production from the sea," in P. N. Kaul and C. J. Sindermann, Eds. *Drugs and Food from the Sea,* University of Oklahoma Press, Norman, OK, 1978, pp. 415–424.

86   J. A. Crutchfield and R. Deacon, "The economics of fish protein concentrate," in S. R. Tannenbaum, B. R. Stillings, and N. S. Scrimshaw, Eds., *The Economics, Marketing, and Technology of Fish Protein Concentrate,* The MIT Press, Cambridge, MA, 1974, pp. 355–438.

87   M. H. Glantz, "Science, politics and economics of the Peruvian anchoveta fishery," *Marine Policy* 3(3), 201–210 (1979).

88   Anon., *Proceedings of the NMFS/EDS Workshop on Climate and Fisheries, April 26–29, 1976,* U.S. Department of Commerce, National Oceanic and Atmospheric Administration (1976).

CHAPTER 11

Peruvian Political and Administrative Responses to El Niño

Organizational, Ideological, and Political Constraints on Policy Change

LINN A. HAMMERGREN

Abstract

This chapter focuses on Peruvian policymaking for the fishing sector, and especially for the fish meal industry, from 1960 to 1979. During this period both the sectoral and national policymaking systems and conditions of the industry underwent substantial changes, leading in turn to a three-level treatment of themes of this chapter. On the first level this is a decisionmaking study of policymaking in the sector, which focuses on identifying the nature of government policy and the principal factors influencing its formulation.

Second, at the level of policy analysis, an effort is made to identify problem areas within the fishing sector and to evaluate the government's ability to deal with them.

Third, it is a comparison of the performance of the two policymaking systems in force during the period, one prior to the 1968 military coup and one following it. These two systems represent the two most popular (and extreme) alternatives proposed for policymaking related to development—an incremental bargaining system with limited government participation and a centralized planning system with direct government involvement in "strategic" areas.

Separately and in combination, these three levels of analysis produce the following arguments and conclusions. The content of government policy toward the fishing sector varied considerably, often radically, during this 20-year period in response to major changes in the political and policymaking system. It varied specifically toward those involved in the organizational and ideological components of the latter. Policy, however, remained remarkably unresponsive to changes in the sector itself, and especially in the fish meal industry. These sectoral changes were less important as sources of policy change but usually became so only when they reached crisis status—as was the case, for example, with the near disappearance of the anchoveta in 1972. Thus, certain significant continuities in fishing policy emerge under both systems, unfortunately in such critical areas as resource management and in the fragmented and often contradictory nature of individual policy decisions.

In conclusion, two additional points are made regarding the relative performance of the two policymaking systems. First, while critics of planning models may be correct in stressing the impossibility of comprehensive planning, the Peruvian experience prior to 1968 does suggest the desirability of the more coordinated development of industries based on vulnerable, renewable resources; this means a shift away from a pure bargaining model toward a planning system that incorporates a more active role for government in directing development.

Second, the study also suggests that what is commonly considered "planning" may not fill this need. Post-1968 events indicate that, while higher levels of government involvement, the creation of organizational infrastructure, and the formulation of "plans" will change the content of policy, they will not automatically make it any less fragmented or less contradictory, nor (as a consequence) will they necessarily increase the chances of preparing for or preventing crises. Such changes may, in fact, block efforts to introduce a more rational or more comprehensive base for decisionmaking (by substituting new intragovernmental sources of bias for external ones and consequently by lessening the existing degree of sensitivity to environmental fluctuations or change), as in the Peruvian case.

INTRODUCTION

Peru's fisheries in general and its fish meal industry in particular have presented recent governments with a major challenge to their ability to develop policy under rapidly changing conditions both within the industry and within the wider economic, political, social, and environmental systems. The responses of the various governments to date, and of private actors as well, have most often been inadequate and have tended to compound rather than remedy the fishing sector's problems. The lessons to be drawn from the Peruvian experience are in many instances negative ones. Only recently (since 1976) have policymakers seemed prepared to learn from as well as to rectify some of their earlier errors. Unfortunately, these errors carried with them high economic and political costs which extended beyond the fishing sector. Government and private mismanagement of the fisheries has been a major contributor to the nation's near economic collapse in the mid-seventies. The government's role in particular has further challenged its credibility as an economic planner and administrator.

Although this chapter is titled "Peruvian political and administrative responses to El Niño," the predominant characteristic of Peruvian fishing policies from 1958 to 1978 is that they have been less responsive to El Niño or to any other industry-specific condition than they have been to the more general political and economic situation. This characteristic, which gave rise to many of the sector's problems, has persisted under two distinctive policymaking systems and during periods when fishing was either a minor or major focus of government attention. Many factors contributed to the nonresponsiveness: the distractions of an attempted socioeconomic revolution, inadequate leadership, the extremely rapid growth of the fishing sector and of the government's oversight organizations, and the increasing demands on the political system as a whole. However, the crux of the problem has been the persisting fragmentation of policymaking in the sector, which has meant that decisions have frequently worked to contradictory ends, and that such apparently critical issues as resource management have been virtually ignored.

As noted above, Peru's problems with its fishing policy have persisted through two radically different policymaking systems, separated by the political transformations accompanying the 1968 military coup. Fragmented policymaking has been a characteristic associated with incremental bargaining systems like those operating in Peru prior to 1968. It has not always been seen as detrimental. Theorists have in fact argued that fragmented policymaking may allow the widest representation of interests, producing a kind of naturally balanced output.* However, the Peruvian experience with the fishing sector also suggests that both policy fragmentation and the incremental model may be inadequate for dealing with policy areas where change is rapid and where general benefit issues like resource management are not specifically represented.

The experience also suggests that a transition to a more centrally located system like that brought about in 1968, substantially isolated from external political pressures and accompanied by efforts at national and sectoral planning, will not necessarily reduce the problems of fragmentation (or, for that matter, of incrementalism). Despite the apparent opportunity for higher levels of coordination after 1968, policy remained fragmented and contradictory in its impact, this time not

---

*For a discussion of the advantages of fragmentation and incrementalism, see Lindblom.[1] For a more critical discussion of Lindblom's model and an alternative discussion of fragmented systems, see Yates,[2] pp. 33–37 and 85–119.

because of external political pressures but because of the organizational and ideological biases structured into the new policy apparatus and because of the government's increased ability to implement programs. Somewhat ironically, while the preponderance of external constraints worked against policy coherence in the pre-1968 incremental system, their elimination after that date encouraged fragmentation arising in unchecked pressures from within the bureaucracy and government.

Contrary to the often pessimistic findings in recent literature on policymaking and planning in developing areas (see for example Caiden and Wildavsky[3] and Wynia[4]), the transformation in policymaking systems did produce a marked change in policy output both in terms of content and quantity. This was primarily a consequence of increase in government power. However, even after 1968, in areas where specific programs did not exist (e.g., resource management) policymaking remained incremental and adaptive, departing little from past practices. Efforts to improve policymaking through increased government involvement and guidance may allow a more direct attack on issues ignored in an incremental, bargaining model. It will be argued, however, that their effectiveness is still subject to constraints posed by the organizational and ideological biases structured into the policy system.

POLITICAL BACKGROUND AND POLICY TRENDS

One of the key events in the development of Peru's fishing policy in the last two decades, although its impact was not immediately visible, was the 1968 coup in which the reformist, constitutionally elected government of Fernando Belaúnde Terry was replaced by the first of two military governments officially committed to effecting basic and "revolutionary" changes in the nation's economic, political, and social structure (for more general background on this period, see Kuczynski,[5] Lowenthal,[6] Stepan,[7] and Palmer[8]). Among the most fundamental changes brought by the coup were the government's redefinition of its role in shaping the country's development and its consequent reshaping of the policymaking system to one with a greater emphasis on centralized planning and government control of the economy. Although Belaúnde had been elected on a reformist platform, his government, like its predecessors followed *laissez faire* economic policies, providing support to the private sector, but allowing the latter to make most of its own decisions and so in effect to direct national development. With a relatively strong Congress, a competitive party system, and a bureaucracy that was both overcentralized and dominated by ministerial fiefdoms, the policymaking system under Belaunde tended to be fragmented, uncoordinated, and incrementalist in its orientations. Major decisions were reached only after lengthy bargaining between numerous private and governmental groups. What coherence there was with respect to policy came not as the result of planned coordination but because of the active involvement of a relatively small proportion of the population. In a country like Peru, with a 40% illiteracy rate coupled with major economic, political, and social inequities, those in a position to make far-reaching decisions were relatively few and so could rely on informal means to come to agreement.

Although some of these characteristics were preserved after 1968, the "Revolutionary Government" effected substantial changes in the basic policy process. Through its emphasis on centralized planning, it brought an end to the tradition of limited government involvement in economic and social development and at the same time altered the relative political influence of those participating in policymaking. The new government was composed of military and civilian technocrats, most of whom had a strong bias against what they perceived as the old economic oligarchy and its foreign allies (i.e., foreign investors in general, and multinationals in particular, foreign governments and international agencies, all of which, given the nationalistic and leftist perspectives of the new rulers, were seen as the perpetuators of an economic status quo that kept Peru and other Third World countries poor).

The Revolutionary Government defined itself as neither communist nor capitalist, but as a nationalistic regime committed to the acceleration and redirection of national development designed to extend benefits to the entire Peruvian population. The regime also stressed its commitment to increased popular participation, leading to a series of experiments in "cooperativization," "worker management" and to other efforts to end the decisionmaking monopoly of a small political and economic elite. (It should be noted, however, that the government was considerably more reluctant to surrender its own authority and so kept the upper levels of its own organization free of such innovations.)

Although by 1978 the military's efforts to spur economic development had clearly failed, it had changed basic socioeconomic structures through nationalization of some foreign mining interests, extensive land reform, the introduction of the "industrial community" (worker co-management) and several acts of expropriation which served to extend state control over other strategic sectors of the economy. Political structures were changed as well. While many pre-1968 institutions like the Congress and the political parties were eliminated or reduced to minimal importance, the government bureaucracy was greatly expanded and assumed a more active role in policy making. Between 1967 and 1975, total government employment increased from 7 to 11% of the work force, and the public sector share of the gross domestic product doubled from 11 to 22%.[9] This was a significant change for a country in which the central bank was privately controlled and taxes were collected by a private company until the mid-1960s. With the ouster in 1975 of the military junta's first president, General Juan Velasco Alvarado, and his replacement by the more moderate General Francisco Morales Bermúdez, the government entered a period of retrenchment (the "second phase") in an attempt to remedy economic problems stemming in part from the poorly planned, costly programs and in part from a series of unforeseen setbacks. In some cases this meant modifying or eliminating first phase programs. Despite the problems it faced, its more conservative orientations, and its promise to return the country to civilian rule by 1980, the second phase (Morales) government remained committed to a strong role in directing vital sectors of the economy.

FISHING POLICY UNDER BELAUNDE

Policy trends toward the fishing industry closely parallel these more general developments; the 1968 coup substantially increased government involvement in the sector and introduced the first real efforts at planned policymaking. Ironically, government involvement in the fishing industry (for example, investment in infrastructure) before 1968 was so limited as to become a major source of complaint by private manufacturers. By the mid-1960s they were actively demanding more government attention.

Most of these demands came from the fish meal industry. Although fish meal production represented only one of the fishing sector's activities, it accounted for about half that sector's work force, contained the bulk of its economic and political power, generated virtually all of its foreign earnings (and up to 35% of the national total), and absorbed most of its private investment.* The fish meal manufacturers also dominated the principal political lobby, the National Fisheries Society (*Sociedad Nacional de Pesquería,* SNP), founded in 1960.

Food fishing for human consumption, the major alternative, was much less lucrative and, with the exception of a few canneries, remained

---

*The inadequacy of government statistics and the relative inattention to all activities in the sector except the fish meal industry make it difficult to cite comparative figures. Those cited here are taken from a 1978 report of the Ministry of Fisheries.[10] A somewhat conflicting statistical description of the sector can be found in the planning office's 1965 report.[11]

mostly in the hands of small operators using rudimentary fishing techniques. Prior to 1968 government statements expressed some interest in developing the food fishing industry, but no more positive steps were taken. Left to their own choices, potential investors gave preference to the more dynamic fish meal industry. The industry by the mid-1960s began to cut into the already limited canning operations, the owners of which considered it more profitable to convert their plants to fish meal factories.

Throughout the 1960s, government decisions affecting the fishing sector (for the most part the fish meal industry) were made by a variety of agencies.* Each agency had responsibility for specific decisions affecting the sector, but government involvement remained regulatory within very narrow limits. Furthermore, fishing policy was not coordinated by any single office but was shaped by the diverse (and often conflicting) interests of the various agencies involved, agencies for which fishing policy was not a primary concern. Even those offices with a direct interest in the sector failed to develop a broader policy. The Ministry of Agriculture's *Servicio de Pesquería,* a logical choice for a coordinating role, tended to limit its activities to the development of fresh-water resources, leaving marine resources to IMARPE (Peruvian Marine Institute), a semi-independent research institute with mainly a scientific advisory role. The fisheries planning office (*Oficina Sectorial de Planificación Pesquera,* OSPP), whose members showed some interest in developing more coherent policies, was too small (14 members) and too new (founded in 1963) to have much effect. Consequently, the most important government decisions affecting the sector were not made by any of these agencies, but came out of the Ministry of Finance and the Congress. Those decisions involved taxation and credit policy for the fish meal industry, the two major areas of dispute between the industry leaders and the government throughout the 1960s.

The government's objectives in the sector were relatively simple. Although foreign exchange earnings were vital to the national economy, the government left the development of the fish meal industry to private groups. It still, however, viewed the industry as a major source of its own income. Throughout its history, Peru relied on export and import

*These included the Ministry of Finance; the Ministry of Agriculture through its fisheries service *(Servicio de Pesquería)*; the semi-independent research institute IMARPE, funded in part by the SNP; the Ministry of the Navy; the Ministry of Labor; and the Congress. The private *Sociedad Nacional de Pesquería* also played an important part, not only as a source of information and pressure but in its exercise of certain decisionmaking functions delegated or recognized by the state (e.g., setting producer quotas which served as the basis for export licenses issued by the Ministry of Agriculture).

taxes to generate government revenue. Thus, the fish meal industry, as a major exporter of fish meal and a substantial importer of equipment represented an important source of funds. With the extraction of tax revenue as its major active concern *vis-a-vis* the fishing industry, government actions also extended to more sporadic ventures in credit policy, resource protection and labor relations. These actions, however, were rarely initiated by the government and usually came in response to interest group pressures with private groups often taking a major role in designing policy.

Taxation and Credit

Even with respect to taxation, government policy could hardly have been considered coherent, except insofar as it involved extracting the maximum from the industry. Throughout the 1960s fish meal producers were subject to a variety of taxes, [12-19] some introduced specifically for the industry, others for more general application. As a consequence, the *Sociedad Nacional de Pesquería* (SNP) became increasingly active in protesting these taxes for the following reasons: first, the taxes imposed an inequitable burden when compared to those for other industries; second, after the industry was affected by a series of setbacks (i.e., fluctuations in resource availability and in international prices), the taxes did not account for its special problems and in effect contributed to rising costs of production as well as to increasing financial pressures; and finally, owing to the lack of overall tax policy toward the industry, tax measures tended to be contradictory in their impact, so that concessions awarded by one policy might be immediately eliminated by another.

The government's tax policy contributed to a second complaint of the industry—its increasingly heavy debt burden. The industry's rapid growth from 1958, the entrance of a number of investors with limited experience and resources, and the technological competition sparked by the boom, set the base for the heavy indebtedness of all participants. In addition, the lack of a specific government credit policy (particularly one administered through state development banks) produced an increasing reliance on short-term loans, making the industralists' situation even more precarious.* Throughout the 1960s and early 1970s, the industrialists campaigned through the SNP for a credit policy that would give them more security. While the proposal for a special development bank for

---

*An antiquated private banking system, controlled by the more prestigious elites added to the problem since many of the industralists were relative newcomers and so lacked the contacts that often proved essential to securing loans.

fishing was never realized, the industrialists did get state backing for a number of foreign long-term loans for debt consolidation.

Labor, Marketing, and Conservation

With respect to labor relations and export marketing, policy evolved in a similar fashion, with the government rarely taking the initiative but instead responding to pressures from interested actors or serving as an arbitrator between contending groups. In the case of export marketing, a drop in fish meal prices in 1960 led to an arrangement (the Paris Agreement, *Convenio de Paris*) among the private sectors of the principal producing countries, establishing national quotas to be controlled through export licenses issued by the respective governments. Following the Paris meetings, the Peruvian government officially recognized the system with a ministerial resolution which made the SNP's producer quotas the basis for export licenses. A 1966 revision of the Agreement recognized both the four commercial channels through which the industrialists exported their product and the SNP's Market Committee, which mediated among them. Thus, even more so than in the area of credit, the state held its role to giving official approval to measures introduced by the private sector and so, in effect, delegated its policymaking functions to the latter.

In the case of labor relations the situation was further complicated by the presence of numerous contending private groups, notably the SNP and the unions of fishermen and factory workers, but also including a number of other pressure groups—for example, the independent boat owners who controlled about 15% of the fleet. Labor unions in the industry were strongly organized and were generally effective with or without state involvement in securing high wages for their members. The more complex issues, however, involved the legal and contractual status of the workers and especially the fishermen (who were originally covered by special legislation and thus denied the social benefits normally provided to industrial workers). Once again, although the state tended to take a slightly more independent role, in several cases the resulting legislation was based on agreements initially worked out between the unions and the industrialists.

The final major policy area, during the 1960s, was that of resource conservation, especially efforts to limit the anchoveta catch to some level that would ensure a continuing supply. The government originally had no particular policy about this, but in 1964 both the fishermen and IMARPE noted changes in the nature and quantity of the catch, and the Institute began more intensive studies. It was on the basis of these studies that the first closed season *(veda)* was established in 1966. Although the industrialists were understandably not enthusiastic about these limits, once they were set, the SNP was active in encouraging members to respect them, and Peru's good record in resource control through the 1960s is largely a result of the private sector's cooperation.

Two aspects of the *veda* made it more acceptable to fishermen as well as to boat and factory owners. First, although it established one and later two or three closed seasons annually, it did not directly limit the size of the catch for individual operators. This produced problems later, as technological improvements along with major increases in the number of boats and factories expanded fishing and fish meal processing capacity thereby heightening competitive pressures. Until the late 1960s with only minor adjustments in the length of the open season, the catch ranged from 6 to 10 million metric tons, a range which IMARPE at that time considered to be within the margins of safety.

A second reason for the private sector's acceptance of the *veda* was convenience. The first *veda* was set in the winter months when catches tended to be smaller and fishing more dangerous. The other closed seasons were set early in the year when the predominance of juvenile fish made fishing less desirable. The closed season also provided an opportunity to make repairs on factories and boats and as such was not without practical advantages.

The minor adjustments which government involvement precipitated in the growth and development of the industry could not alter the dominant trends, nor were they intended to do so. By the mid-1960s informed members of the private sector, government, and scientific community were well aware that the industry's *apparent* health concealed a multitude of serious problems—notably its overexpanded capacity, its dependence on a single highly vulnerable resource, and the existence of pressures on further growth and expansion, which tended to make the industry's situation still more precarious. The general awareness of the problem is suggested by government reports[11,20,21] and SNP *Memorias*[12-19] for these years. (Because the SNP was the official spokesman for the industry, its discussions tend to be less critical.) Observers knew that uncontrolled, the fleet could fish out the entire stock of anchoveta in a single season, and that the existing plant capacity was *at least* double what was needed to process the allowable catch. At the same time, the efficiency of individual factories varied enormously, leading to the waste of resources and to the marginal economic status of a number of operations. It was also known, on the basis of past experience, that a bad season would bring bankruptcies and the closure of many marginal factories with the consequent unemployment of their workers. Although the problems were recognized and although it was obvious that drastic restructuring was in order, neither the government nor private groups were inclined to recommend it. Government policy was not moving in this direction but (at best) was aimed at helping those involved to adapt to the existing situation, thereby reinforcing the status quo. When tax or credit relief was given, the net result was to encourage further investments in the already overextended industry and to help marginal operations to hold on longer. Although some measures were introduced to encourage industrial efficiency (toward the end of the Belaúnde period), they too tended to spur further investment and so to increase the industry's debt burden. Finally, despite the "lip service" paid by government and private groups to the need for greater diversification within the sector as a whole, no concrete steps were taken toward that end.

This lack of substantial change even in the face of a recognized need is not surprising. It is the natural product of a policymaking system based almost exclusively on bargaining among diverse groups and interests and reinforced by the government's self-imposed minimal role and minimal policymaking apparatus. In times of crisis, such a system may be transformed into one capable of making less incremental and more decisive responses but, as the Peruvian experience suggests, it would be of little help in preventing that crisis even with prior warning.

The fragmented approach to policymaking and the absence of coordinating mechanisms both encouraged isolated demand-making and reduced the incentive for groups with dissimilar interests (e.g., the small and large producers) to enter directly into confrontation with one another. Where dissimilar or conflicting interests arose the usual tactic was to present separate demands to the government in the hopes that each would be met. A resultant consequence of this type of bargaining arrangement was the reinforcement of the prevailing optimism about the continued supply of anchoveta. The government's stance of minimal policymaking in this area meant that resource management was not an issue. In addition the conflict-avoidance strategy further diverted attention away from concern about the impact of human activities on vulnerable marine resources in favor of a free-wheeling competition for a share in the immediately available resource.

FISHING POLICY AND THE MILITARY: 1968–1972

When the Peruvian military seized the government in 1968, the fishing sector was not one of its immediate concerns. Thus, initial changes in the sector would be slow in coming. Aside from the fish meal industry, the new government did not consider fishing important to the domestic or export economy; and the industry, recuperating from one of its periodic declines, demanded no immediate attention. Furthermore, despite the industry's importance as a foreign exchange earner, its relatively brief existence and its domination by new elites isolated it from the traditional socioeconomic institutions the new government sought to overturn.* Significantly, the Ministry of Fisheries was created in early 1970, almost a year after the initial ministerial reorganization which established new offices for the government's top priority programs.

For the first year, the government ignored everything but the fish meal industry and there its actions were similar to its predecessors—rising profits provided an opportunity either to increase or to introduce taxes, which in turn led to complaints from the SNP. The new government's initial inclination to view the industry primarily as a source of revenue is suggested by such moves as a decision to earmark the proceeds of certain taxes for the newly created Ministry of the Air Force. Meanwhile, private groups in general and the SNP in particular moved cautiously, uncertain about the course of action the military would take. Viewing the government's earlier expropriations in agriculture, mining, and industry, the SNP leaders had reason to fear for their own future. Yet, the military's development stance suggested that both the industry and the sector might get some of the attention it had sought for so long. The industralists' tactic was thus to encourage dialogue and to offer their cooperation wherever possible in the hopes that whatever position was finally taken would reflect their views. In its 1969 annual report, the society applauded the creation of the Ministry of Fisheries as a positive step toward institutionalizing the sector, while noting that the government's tendency toward a greater involvement in the national economy "would not result in inconvenience for the country so long as it did not stifle private initiative."[23]

Creation of the Ministry of Fisheries

In retrospect, the creation of the Ministry itself appears as the first major step toward the development of a policy toward the sector. This policy was shaped to a large extent by the form initially given the new

---

*While some of Peru's traditional elites did enter the industry, they were generally less successful than the relative newcomers, many of whom had started from virtually nothing and hence were not "known" enemies of the new government. For a discussion of the background of the major industrialists see Abramovich.[22]

organization. Statements made before and after the Ministry's appearance suggest that resource management was not a major consideration, and that the government held an optimistic view about the future availability of fish stocks. The principal motivation was the military's desire to exercise more direct control over important sectors of the economy, especially those in the hands of strong private groups or with substantial foreign investment. Decree Law 17702 (June 13, 1969), establishing the study commission which would make the final recommendation for the Ministry's creation, stated that:

> The actual combination of organisms and activities related to the exploitation, transformation and commercialization of our icthyological wealth constitutes a fishing complex suffering from serious defects and which is oriented to the benefit of privileged minority groups while working against majority interests and the fair participation which corresponds to the State.[24]

The origins of the Ministry were not entirely ideological. There was a strong element of bureaucratic politics at work as well. The first few years of the new regime constituted an era of rapid expansion within the public bureaucracy, as administrative groups were quick to seize the opportunity to enhance their own power by having their programs promoted to higher levels of importance. With the approval of a new government that as yet accepted no limits on its ambitions or its resources, offices become divisions and divisions became ministries, with the appropriate increases in staff, budgets, and salaries. During the government's first year in power, the fishing sector had remained part of the Ministry of Agriculture, a particularly inauspicious location given the high priority assigned to agrarian reform, a priority that threatened to eclipse all other activities of that Ministry. Thus, the bureaucrats of the fisheries office had an additional incentive for seeking independent status. A critical role was played in this by the *Oficina Sectorial de Planificación Pesquera* (OSPP) which, while it had never enjoyed much power during its brief existence, had spent the years since its creation writing policy papers and in 1969 produced a series of studies calling for the "institutionalization of the sector," that is, the creation of the Ministry.[25]

Ironically, the OSPP's success in campaigning for the new Ministry meant a further dilution of its own power, a consequence of the changes in organizational size and complexity. However, it greatly increased the importance of the sector. Employment in the new Ministry immediately quadrupled from less than 200 to over 800 and by 1973 had reached

1300.* The fact that many of the newcomers had little, if any, experience in the area, did not hinder either their taking key positions or bringing their own, often naive perspectives to bear on policymaking. For example, the man chosen to head the Ministry, General Javier Tantaleán, was a sports fishing enthusiast, which he openly admitted was his only relevant background. Still, he would be the key figure in fisheries policy for the next 5 years and one of the most controversial figures of the first phase (1968–1975) of military rule.

The First Five-Year Plan

The new Ministry was both larger and more specialized in its organization than its predecessors, incorporating such new agencies as the Direction of Technical Cooperation and Training, and the Direction of Commercialization. Since these offices were created prior to the sectoral plan, their existence was instrumental in shaping the latter.† Thus, as contrasted with the rather simple plan of action elaborated by the OSPP under the Belaúnde regime[26] the Ministry's first five-year plan, reflecting all of these new inputs and its substantial resource base, was a lengthy document including over 50 specific policy objectives divided among the areas of research, extraction, transformation, industrial management, commercialization, taxation, credit, and training and social programs. Although some of these objectives were little more than statements of good intentions, several suggested major changes in the government's position toward the sector as a whole and the fish meal industry in particular, bringing it more into line with the overall goals of increased national (and State) control of the economy, redistribution of income and economic power, and encouragement of cooperative economic structures.[27] In terms of the fish meal industry the government plan noted a concern with resource conservation and increased efficiency of resource use, but it particularly stressed a commitment to decrease foreign participation and to reverse the trend of concentration of ownership into "inconvenient centers of power." The plan did not specify how the latter objective was to be achieved, but it did make it clear that the

---

*These figures are approximate and do not include employees in the decentralized agencies (e.g., PESCAPERU, EPCHAP, etc.). Owing to the controversies surrounding the original overstaffing and later efforts to reduce personnel numbers, the Ministry is reluctant to release more precise figures.

†Some, like the training office, originated in the government's commitment to such nonsector-specific themes as popular participation.

existence of "dominant economic groups" within the fish meal industry would be viewed as a problem in its own right.[28]

Despite this concern with increasing economic concentration, the five-year plan was founded on the assumption of the continued existence of a strong private fish meal industry, which would bring in foreign earnings and make a major contribution in the form of investment in infrastructure. For example, the introduction of new equipment to increase plant efficiency was programmed entirely for private investment (based on figures supplied by the SNP). Continuing private investment in the fish meal industry would allow the State to focus on what it declared its primary goal of strengthening the yet underdeveloped food fish industry. Most public investment was earmarked for the construction of ports, fishing complexes, and transportation infrastructure aimed at eliminating the major distribution bottlenecks which were hindering growth. Related programs were aimed at consumer education to encourage the population to substitute fish for other sources of protein. Much of the activity in this area, both in construction and marketing, was channeled through the newly created EPSEP (State Enterprise for Fishing Services), which at the height of its powers had complete responsibility for the sales of fresh and processed fish for direct consumption.

With the formation of the Fisheries Ministry (and in the months immediately following), the government's position toward the fish meal industry began to take more definite shape. One of the first steps was the absorption of IMARPE (formerly loosely tied to the Ministry of the Navy with SNP financing) as a "decentralized" institution (outside the ministerial hierarchy and directly responsible only to the Minister). While IMARPE retained its semi-independent status and its advisory role, it cut its ties with the SNP and weakened many of its official links with foreign scientific groups at Minister Tantaleán's insistence, on the grounds that the institute and its knowledge needed to be protected from those external influences.[29] This decision was largely symbolic and in the end was most costly to the government itself, but it is important in suggesting the extent of the government's determination to exert control while decreasing the control of private domestic and foreign actors.

A major step came in March of 1970, with the creation of EPCHAP *(Empresa Pública de Comercialización de Harina y Aceite de Pescado)*, the organization which was to take over the marketing of fish meal and oil. Fish meal production remained in private hands, but after June of that year, all sales whether domestic or foreign were to be conducted by EPCHAP. The private producers were not happy with this turn of events, although it did realize one goal of their most important members, that of establishing a single commercial channel in place of the existing four.* The relatively slight protest can thus be attributed both to dissatisfactions with the previous system and to the private producers' reluctance to antagonize the government over what they considered an essentially peripheral issue.

The transition to a state enterprise was relatively smooth, since the new organization was staffed for the most part by personnel from the largest of the private channels, that had previously handled about half of the exports. Most complaints tended to focus not on the state intervention itself but on the 3% commission charged by EPCHAP and on its practice of withholding stocks from the market in an effort to push up the price. While prices did rise over the long term, the tactic was inconvenient for the industralists who depended on a steady flow of capital to pay off their ever-increasing debts. In this case the government was willing to make concessions by reducing the commission charged and by promising to speed up various aspects of the marketing process.

EPCHAP's creation must be seen once again as less of a response to the industry's particular problems than as a result of the government's desire to extend its control over major sectors of the economy. While the previous marketing system had been criticized (even by participants), and while the single channel had obvious benefits, these facts alone do not explain the government takeover. More important to the decision was the military's basic distrust of any powerful private group, especially those with foreign ties, and its disinclination to leave major economic decisions in their hands. Beyond this, the spirit of optimism and ambition surrounding the creation of the new Ministry as well as Tantaleán's own desire to imbue it with some sort of mystique were important in encouraging this intervention.

First Effects of Government Intervention: New Programs and New Problems

Although in the case of EPCHAP the ambitiousness of the government and the Ministry ended well, the third major event of 1970 did not conclude so fortuitously. This event was the record 12.5 million metric ton catch allowed by the Ministry and briefly applauded as an accom-

---

*Interviews with representatives of the private sector (in 1978) indicated that this was one area where they harbored little resentment of the government's intervention.

plishment[30] until its longer term effects became visible. It is still being debated whether the record large catch* had a major effect on the later near-disappearance of the anchoveta, but observers claim to have noted as early as 1970 certain peculiarities in the behavior and appearance of the fish (such as a high fat content, and the stock's proximity to the coast),† that might have recommended caution, but which also made it easier to exceed the normal catch. To be fair to the military, even IMARPE did not recognize any unusual danger and had recommended a 9.5 million metric ton limit.

From the Ministry's side several other factors encouraged a larger catch. One was the inexperience of the new administrators and the inevitable disorganization following the creation of the new Ministry which meant that the catch was not closely monitored and that decision-makers were less sensitive than they would be later to the problems of resource conservation. They were not aided much by the general lack of serious concern, shared by bureaucrats and industrialists alike, about the real possibilities of destroying the anchoveta stock. The acceptance of the *veda* and the support for IMARPE's investigations indicate awareness of the danger, but this was clouded by a large dose of optimism. Virtually no one was calling for stricter conservation measures and most were willing to accept unquestioningly the most favorable evaluation of the situation.[32]

A second factor encouraging large catches was the desire of the new Fisheries Ministry and its Minister to demonstrate their competence. Having just taken control of the sector and having assumed responsibility for fish meal marketing, the bureaucrats were under pressure to demonstrate their abilities. The record catch (with the likelihood of record production and record sales) was an attractive way of doing so. Thus while inexperience and lessened control increased the likelihood of overfishing, the Ministry itself had some reasons for favoring a larger catch. There was one final consideration which suggests that, despite the prevailing "revolutionary" biases, policymakers were not entirely im-

---

*According to unofficial estimates (interviews, 1978) the catch may have exceeded 15 million metric tons. Catch estimates were always approximate and tended to be too low. See Paulik[31] for a discussion of the problems involved.

†Interviews, Lima, 1978. While these retrospective forebodings may be somewhat colored by hindsight, it is clear that recruitment in 1969–1970 was extraordinarily high which may have been a sign of some change. In any event, the reaction of the industrialists, as usual, was to take this at face value as one less reason for caution, rather than inquiring into its significance.[32]

mune to pressure from the private sector—holding the catch size to the recommended limits would have required an unusually short season. As the industrialists were already pressuring to be allowed to take full advantage of what promised to be a record year,[32–34] the Ministry let them fish out the full open season after the recommended limit had been reached, rather than setting an earlier *veda*.

Throughout 1971, since neither the government's threat to the private industry nor the threat of both to the anchoveta had become clear, debates about policy continued to center on the old issues of taxes, credits, and other charges imposed by the government. The record catch had not eased the industry's financial burdens nor had it done anything to eliminate the general problem of overexpansion. Despite its concern about increasing economic concentration, conditions in the industry forced the government to take some steps which encouraged that trend, including the extension of consolidation loan facilities to large as well as small and medium producers and the formation of a contingency fund which would aid owners in getting out of the business. The government was caught between two conflicting goals: increasing industrial efficiency and decreasing overexpansion would almost necessarily mean increasing economic concentration, since it was generally the larger owners who were the most efficient producers. While the Ministry's planning office argued that the industry was not subject to hard and fast economies of scale,[28] the largest operations did tend to have the most modern equipment, the lowest cost per ton of meal and oil produced, and the most thorough resource utilization. Overall efficiency might be increased by direct aid to the smaller and more marginal operations, but only at a higher cost and with greater organizational problems. No immediate solution to this dilemma was in sight, but in 1971 an intersectoral commission with representatives from the Ministries of Economics and Finance, Fisheries, and Industry and Commerce, and with one representative from the SNP was established to study the industry's economic and financial situation.

In 1971 another major step was taken with the publication of the organic law (Decree Law 18810) for the fishing sector which set forth the State's general and open-ended responsibility for supervising activities in most areas and made a number of more specific policy statements. One particularly controversial innovation which further diverted attention from resource management questions was the *comunidad pesquera* (fishing community), a format for introducing worker participation and, eventually, co-ownership and co-management in all private and State enterprises. The SNP immediately objected, arguing that the fishing industry was too complex and required too much specialized knowledge for this to work and suggested, as an alternative to the stipulated enterprise-wide communities, a single community for the whole fishing sector. The SNP also protested the requirement that 20% of each enterprise's profits be earmarked for the worker community (12% of which was to go into the purchase of shares), arguing that this further increased the financial burdens on the already hard-pressed industry.

The furor aroused by the industrial community also diverted attention from a simultaneous event, the introduction of a government agency, CERPER (Certificaciones Pesqueras del Perú), which would in the future have sole responsibility for quality control of fishing products, and notably of fish meal. Until this time, the function had been in the hands of private groups, the most important of which was CERTESA, and following its lead with EPCHAP the new government agency in large part adopted the staff and operations of the latter. The move itself aroused few protests, although the industrialists continued to complain about CERPER's fees, seeing this as just another means of indirect taxation. However, in light of their other problems CERPER was a minor worry and is most significant simply as one more step toward state control of the industry.

Industry protests concerning tax and credit policy, and concerning the fishing community, continued through 1971 and 1972. Although some concessions were won in the first two areas, the industrialists, through the SNP continued to claim that the government was overburdening them. Meanwhile, government activity in other areas also produced complaints, specifically about EPSEP's commercialization of food fish and either the scarcity or high price of the latter in the marketplace.[35] Although the government had a series of ambitious plans for transporting fish into the higher elevated regions (the *sierra*) away from the coast and improving distribution, the efforts were often poorly coordinated. Their attempts at price control just as often resulted in scarcities and black marketeering. The government's major construction projects moved slowly and public investment fell far below that programmed. Figures on this vary, but according to the ex-Minister Tantaleán, the projected and real amounts for *public* investment during his tenure were as follows (in millions of *soles*):

	1971	1972	1973	1974	1975
Projected	790.8	1239.0	2364.0	2454.7	2500.6
Actual	242.0	421.6	296.4	2012.8	1135.5

Private investment lagged still further behind projected figures, especially after the nationalizations.[36] The ambitiousness of many of the projects was also questionable, since several, like the large fishing complexes, would not be completed for many years and would not be expected to have an impact until much later.

Despite these problems associated with the government's first planning efforts in the sector, the situation by 1971 seemed to have reached a temporary equilibrium. Neither the government nor the industrialists were happy with the new status quo, but for the time being, neither seemed likely to change it. The government's increased involvement both in the fish meal industry and in the rest of the sector had realized a number of its preliminary objectives and left it with enough new programs (and problems) to keep it busy for the next several years. Meanwhile, political changes brought by the 1968 coup, including the elimination of the competitive bargaining system and the increase in government control had lessened the sensitivity of the policymaking system to external groups in general and, because of the government's ideological orientation, to former economic elites like the fish meal producers in particular.

Aside from decreasing the influence of the latter group, this lessened sensitivity had two additional effects. It gave policymakers more leeway in planning and implementing their programs than had been possible under the pre-1968 regime. It also further entrenched them in their narrower organizational views and discouraged attention to problems that cropped up both in individual programs and in coordination across programs. With respect to the longer term, as policymakers gained more experience with the sector, this second effect might have been partially mitigated, but before that could happen, a new factor entered to further divert their attention.

FISHING POLICY AND THE MILITARY: THE ANCHOVETA CRISIS

In the first half of 1972 events took a new turn for the worse with the drastic decline in the anchoveta stocks, a phenomenon that has been linked, although not conclusively, to the occurrence of El Niño and the earlier condition of overfishing. Once IMARPE sounded the alarm, the government took immediate steps to curtail fishing, allowing only 22,000 tons to be removed in the shortened second season. As observers had feared, the warning came too late to prevent most of the damage and the catch dropped from over 10 million metric tons in 1971 to 4.5 million in 1972 and to an all time low, up to that time, of 1.5 million in 1973. The stock was so much reduced that it was doubtful whether complete recuperation would ever occur. It was evident that over the next few years at least the industry would be operating at a fraction of its former level.

At the time of the crisis, the government had been negotiating the fifth of a series of consolidation loans for the debt-plagued industry but, under the new conditions, this was of little use. The immediate threat was that of bankruptcy for the majority of the firms and the unemployment of a large portion of the 27,000 workers directly involved. In the short term, loans were made available to the firms to pay their fixed expenses and to meet their payrolls. By late 1972, however, the Minister of Finance warned the cabinet that the emergency measures were too costly to sustain and that a more permanent solution was needed.

At this point, the fish meal crisis became a government-wide concern, removing the responsibility for its solution from the Ministry and placing it within the cabinet, and with the President and his advisors. Tantaleán continued to play a major role, although this was as much due to his personal influence within the government and with President Velasco as to his formal position. Additional background for the decisionmakers came from reports submitted by IMARPE and by an intersectoral commission (with representatives from the major banks, the Ministries of Economics and Fisheries, the SNP, and the unions).

The Decision to Nationalize

The logical step at this stage (and that toward which the reports seemed to be directed) was to use the crisis to bring about the long-needed restructuring of the industry, encouraging the closure of numerous small operations, and leaving the industry in the hands of the larger domestic and foreign companies which had the best potential for economic survival in the lean years. The unemployment problem, a major concern of the government for ideological and political reasons, could then be resolved by redirecting some of the surplus workers (and equipment) to the food fishing industry, and by having the state temporarily absorb the remainder. However the resulting concentration of economic power and the still larger role for the foreign and mixed companies, were clearly unacceptable to the policymakers. Thus, a second and more drastic policy was quickly evolved, that of taking over the entire industry and converting it to a state enterprise. This alternative, approved unanimously by the cabinet in early 1973 (although made possible only by the industry's crisis), was clearly in line with tendencies emerging within the new Ministry as well as with the preferences of the most influential groups within the government.

Once the decision to expropriate was made, the most important remaining detail was the form expropriation would take and the amount of compensation, if any, to go to owners and stockholders. Although the government had prior experience with other expropriations, notably in agriculture and mining, the situation here was complicated by the large debts of the industry, making it difficult to determine the real value of the holdings and also leaving the question of how the debts themselves were to be handled. After meetings with the SNP, and with the aid of a series of study groups involving members of the Ministry of Finance and the banks, the decision was made that the government would assume the industry's debts while paying the stockholders in 10-year bonds whatever difference remained between the assets and debits. Stockholders in companies whose debts exceeded their assets (a total of 24), would receive no payment but would at least be relieved of their debts. Thus the total cost of the expropriation to the government was $106,024,801 of which $20,895,231 represented the debts of companies without positive assets.*

Both the decision to assume the debts and the means by which the companies' holdings were evaluated have been the targets of criticism from all sides. In retrospect, it appears that the industrialists as a whole did not fare too badly and that a number actually benefited from the move which saved them from certain bankruptcy. In fact, it was rumored that, although the SNP as an organization opposed the expropriation, a number of its members, especially those with marginal smaller firms, were in favor of it from the start. There were those, however, who realistically or not, believed they could weather the crisis and who to this day regard the move as unjustified. To this can be added the group, who although may have welcomed the chance to get out of the business,

---

*The dollar amounts were calculated from the quantities cited by Malpica[37] in Peruvian soles (S/4,771,116,044 and S/940,285,414 respectively), the currency in which most accounts were paid. A separate arrangement, the Green Agreement *(Acuerdo Green)* was worked out for the five companies owned solely by U.S. firms, with the compensation made directly in dollars. The amounts are included in the figures cited. Precise figures for the cost of the expropriation vary. Those given by Tantaleán[38] are somewhat lower, but still in the same range.

felt the government undervalued their holdings.* Since so much of the process was conducted in secret, the controversy may never be resolved, but the claim by some industrialists that at least a part of the firms could have survived does appear justified. *Expropriation was clearly not the only solution, but it was most compatible with the government's political and ideological preferences.*

Whatever the outcome for the industralists and stockholders, the expropriation did not resolve the industry's problems but merely tossed them into the government's lap. It had inherited the industry's large debt, as well as its overexpanded fleet of 1256 boats, 105 factories, and 27,000 unemployed workers, all of whom would be incorporated at least temporarily into the new state enterprise, PESCAPERU *(Empresa Pública de Producción de Harina y Aceite de Pescado)*. It had also inherited the 400,000 metric tons of fish meal in stock which would help it meet its immediate obligations, but at the same time, the anchoveta catch for 1973 fell far below the originally predicted 4 million metric tons.

The government did have one advantage over the industrialists. As a single decisionmaker with ultimate authority, it had much greater flexibility in consolidating its holdings and was able to close plants, remove boats from operation, and dismiss, or, more frequently, reassign workers to other locations or jobs. These readjustments which began as soon as PESCAPERU was set up, reduced the dangers of overfishing while allowing the government to increase the efficiency of the entire operation. However, given such constraints as, for example, the government's reluctance to dismiss workers, the flexibility was still not sufficient and PESCAPERU remained an overexpanded and costly operation. Despite higher prices for fish meal and oil, the enterprises' debts and high labor costs, the subsidy on domestic sales of meal and oil, and the continuing scarcity of the anchoveta, all contributed to its problems.

Problems in the Rest of the Sector

In the meantime, the Fisheries Ministry's operations in the rest of the sector were beset by other problems. Caught up in the enthusiasm that characterized all phases of government planning in the early years of the "revolution," the Ministry was suffering from overexpansion, both in terms of its physical size and of the dimensions of the projects it was

*Evaluations of the companies' holdings began with the book figures, and were later revised by committees appointed by the government, leading in some cases to higher valuations, but usually to lower ones. For two conflicting views of the process, see Malpica[39] and Tantaleán.[40]

committed to fulfill. Its rapid growth to 1,300 employees at its peak, produced problems of coordination and control, leading to the waste, and the obvious misuse, of resources. Size and the rapid influx of new administrators, many with no background in fishing, complicated planning and led to an emphasis on a number of priorities that in retrospect were unwisely chosen. As the planners themselves would admit by 1978, in these early years most decisions were made on the basis of socioeconomic priorities with little or no concern for resource conservation.[41] Thus, on a number of occasions the Ministry's own programs encouraged overfishing and temporary declines in the stocks of a variety of species. Government sensitivity to criticism and a tendency to deny or to cover up such mistakes heightened the hostility and mistrust of nongovernment actors and reduced the willingness to cooperate on the part of the remaining private groups and the numerous labor organizations. This was not a problem limited to the fishing sector; by 1974 and 1975, the government's popularity was declining even among groups that had benefited from its early programs.

The experience of EPSEP, the Ministry's other major enterprise, offers another example of these problems as well as suggests the ways in which the administrators sometimes worked at cross purposes. EPSEP had been created to supervise one of the major programs in the sector, that of developing a major food fishing industry both for internal consumption and for export. It was to realize this through a number of subprograms, including the construction of the necessary infrastructure, an increasing responsibility for distribution and sales, consumer education, and the creation of several mixed enterprises for canning and preserving. From its earliest days EPSEP operated at a loss. The official explanation was the subsidized prices for fish consumed in the country, but mismanagement and overambitious planning also contributed to the loss. Efforts to encourage food fishing, including the redirection of a number of anchoveta operations into the area, had led to the depletion of stocks of favorite food fish in 1973, 1974, and 1975 and to the virtual disappearance of several of these from the marketplace.[41] The situation was compounded by PESCAPERU's decision to substitute other species for the scarce anchoveta in fish meal production. Complaints from the canning industry, which feared its own resource base would go the way of the anchoveta, led to a cutback in this policy, but not without some further damage to the credibility of the Ministry's planning operations. EPSEP's construction projects proved both more costly and much slower than first promised, leading to further criticisms, while its distribution and commercialization efforts so overtaxed the enterprise's abilities that by 1976 steps were begun to cut back its functions, return commercialization to the private sector and to cease many of the least profitable ventures, like its chain of restaurants. Steps were simultaneously taken to reduce EPSEP's staff whose growth had added to the operation's cost.

Labor relations was still another area in which the Ministry began to encounter problems affecting its operations in all areas, but especially in PESCAPERU. The fishing unions had always been strong, and although those directly involved in the fish meal industry had supported the takeover, the honeymoon was a brief one. Even before 1975, when the threat of job cutbacks emerged, the Ministry had created discontent as a result of its involvement in union affairs. As early as 1971 and 1972 the government began a general program of building up its own unions and popular organizations. It carried this policy into the fishing sector where its support of cooperative leaders led to charges of tampering in union politics. Not content with the support from existing unions, the Ministry encouraged the emergence of a new organization, the Revolutionary Labor Movement (MLR), which soon became a strong and controversial contender for power. The politics of the fish meal industry's two major unions, the FPP (fishermen) and the FETRAPEP (factory workers), had always been charcterized by some measure of violence, but the MLR became more closely associated with this—leading some critics to call it "Tantaleán's mafia." Thus the Ministry had created enemies with its very efforts to build support even before it gave the workers a solid complaint with the threats to their jobs.[42]

The job issue was still more serious, for it deprived the government of its most decisive advantage in dealing with labor, its ability to guarantee employment. All the workers incorporated into the public sector, either in PESCAPERU or elsewhere had entered with a guarantee of job stability. This meant, according to a general law passed in 1970 (DL 18471), they could not be fired except under very special conditions. Using its labor requirements in construction and elsewhere, the government had generally been able to resettle workers, but as the economy began to decline in 1974 and 1975, this method of decreasing unemployment became too costly.[43] The government used a series of measures to soften the blow, including incentives for voluntary resignations, but once its intentions became clear, relations with the unions took yet another turn for the worse, leading to an investigation of PESCAPERU early in 1975, sponsored by the FPP leadership, and a number of major strikes within the sector by 1976. The unofficial investigation of PESCAPERU produced one of the most controversial analyses of the government takeover charging for example that the expropriations had been a decisive benefit to the private sector, particularly to certain of its members.[39]

FISHING POLICY IN THE "SECOND PHASE": 1975–1978

At least a year before the military coup that replaced President Velasco with the former Minister of Finance, General Francisco Morales Bermúdez, it was evident that the national economy was entering a period of crisis and that a good part of this could be blamed on the government's own programs. The heavy public investments in all sectors had built up a sizable foreign debt while the restructuring of the economy and a number of unrelated setbacks (declines in copper prices, the fish meal collapse, and the rise in oil prices combined with the slim returns of the government's oil exploration programs) had cut productivity and foreign earnings. A principal reason for the coup was thus the conviction on the part of more conservative military elements that substantial changes in policy had to be undertaken. The retrenchment of the second phase and the more conservative leanings of its leaders affected the fishing sector much as it did other areas, bringing first a cutback in investments and employment and then a serious reconsideration of many of the original programs. The coup also saw the removal of Tantaleán as Fisheries Minister in part because of his close association with Velasco, but also because of his own efforts to build up an independent center of power within the sector (the MLR) and to use it as a base for increasing his influence.

Cutbacks and Reorganizations: Redefining the Government's Role

Primarily as an economic measure, but also to curb such internal power centers, a move was then made to reorganize the Ministry and the fish meal industry in particular. Some earlier steps had been taken with the 1974 transfer of EPCHAP to the Ministry of Commerce, where it assumed responsibility for the import and export of a number of commodities, and with the campaign in the same year to encourage the voluntary resignation of a number of PESCAPERU employees and to remove a portion of the boats and factories from operation. As a result of this campaign, PESCAPERU was operating with 530 boats and 51 factories by the end of 1975, although its work force still totaled about 23,000.[44] As voluntary retirements were obviously insufficient, two commissions were setup in 1975; one to make recommendations on the reorganization of the enterprise and the other to study the possibilities of short and medium-term consolidation and refinancing of its debt.

In response to the report of the first commission, the government declared PESCAPERU in reorganization in May 1975, a move that eliminated job stability for its workers. The reorganization also incorporated another major recommendation of the first commission, the transfer of the fishing fleet back to the private sector, ideally into the hands of small owners with former PESCAPERU employees receiving the first option of purchase.

Even before the announcement of the move, rumors about the commission's report and the cutbacks it recommended had caused a major strike by the fishermen which resulted in the government's decision to declare PESCAPERU in a state of emergency with the right to fire those threatening its security. By releasing only the fishermen, the government managed to keep FETRAPEP on the sidelines where it issued communiques supporting the FPP, but took no more direct action. Although the fishermen remained on strike for several months and maintained sufficient unity to prevent the massive firings threatened, the fleet was sold and by 1978 had reverted entirely to the private sector.

In January of 1977, the second stage of the reorganization began with a program to encourage the voluntary resignation of the factory workers. By 1978 these had been reduced from 13,550 to 7043 with a further reduction in the number of factories to 42.[45] The enterprise reported at the end of this period that it had "not fired any worker but had offered special benefits and facilitated the purchase of surplus holdings by those who resigned voluntarily."[46] During the same period, EPSEP also reduced its work force from 3200 to 1400 and overall employment in the Ministry itself fell to 800, once again through a program of incentives to those leaving voluntarily. A final step was taken in 1978 with the return of EPCHAP to the fishing sector and its merger with PESCAPERU.

All of these changes raised the possibility that the government might completely surrender its holdings in the industry, returning PESCAPERU to private control. By mid-1978, the remaining workers were actively, if peacefully, campaigning to prevent this move, taking their case not only to the military but to civilian political leaders[47] with an eye to the promised return to civilian rule by 1981. The military government denied any such intentions. In part to avoid problems with the unions but perhaps out of conviction as well, the SNP officially disclaimed any interest in such a move. Given the persisting scarcity of anchoveta, the labor problems, and the fact that PESCAPERU had only recently begun to reduce its outstanding debt with the treasury, and considering the government's past history of changing its policies without warning, the enterprise posed little temptation over the short run. Furthermore, private industrialists had found a back door to the fish meal industry through a government regulation ordering canning operations to convert waste materials to the latter product. The government's monopoly really extended over the anchoveta only and so long as that species remained scarce, thus private entrepreneurs could truthfully disclaim their interest.

The Limits of Reform: Remaining Constraints on Government Action

By 1978, the situation in the sector as a whole was hardly what the government had projected but there were signs of progress. While no one believed that the anchoveta would return to their earlier abundance, there were optimistic, if not entirely realistic, predictions of an early return to 4 million metric ton catch norm. Furthermore, as the government began to cut back its own involvement, private investment was, once again, attracted to the sector. However, this attraction was based not so much on the future of the fish meal industry as on the opportunities in a variety of other areas, notably canning and food fishing. There were even those who were willing to see the anchoveta crisis as a disguised blessing, in that it had forced attention to the need for diversification and to the dangers of basing an industry on one resource, while pushing investors into a search for other opportunites.

Nonetheless, numerous problems remained. Despite the official emphasis on diversification in all sectors since 1968, the accomplishments had not been great. In the period from 1960 to 1976, food fishing accounted for only 1.5–12.9% of total tonnage extracted and annual growth had averaged 3.7%.[48] In the last 6 years (1970–1976) the growth rate had risen to 10.6%, but from 1973 on this was due less to an increase in national fishing capacity than to a series of agreements allowing foreign fleets to fish Peruvian waters. The slow growth hinged on several factors; infrastructure (fleet and land complexes) remained inadequate, internal demand had not increased adequately, and private investment had virtually disappeared from 1973 to 1977. More importantly, the government while still intent on building up the sector gave no indication of coming to grips with the need to plan the balanced exploitation of marine resources. The result was a continued tendency for selective reliance on a few species and consequent threats to their stocks.

The extent to which the drive for diversification frequently ended at the planning level is sugested by the fate of the old PESCAPERU fleet. The 504 boats sold to private owners were in theory the foundation of a new food fishing industry, but the cost of coverting them was too high for many of the new owners. Thus, it appeared that most would remain dedicated to anchoveta and sardine fishing, selling their catch to PES-CAPERU for the production of fish meal. Other observers expressed a related fear that in the face of the continued scarcity of anchoveta and of pressures from the new private owners and remaining PESCAPERU workers, the government might allow the overexploitation of a substitute species, most probably the sardine, leading in turn to its disappearance and a smaller industry crisis. While government planners were aware of the danger and could cite the earlier crisis as an example, the question remains as to whether decisionmakers within or outside the Ministry would heed their warning and hold firm despite the complaints and possible suffering of the pressure groups involved. Unfortunately, past experience, the demands of an entire economy in crisis, and the lingering tendency to view resources as inexhaustible in one form or another, do not argue for the more conservative position.

CONCLUSIONS

By 1978, some 20 years after the beginning of the anchoveta boom, and 6 years after the industry's collapse, a number of conclusions could be drawn about Peru's experience both in the industry and in the wider fishing sector. While the fish meal crisis contributed heavily to the country's subsequent economic problems, the experience was not an entirely negative one. When the industry was thriving, it provided funds for growth in other areas, and perhaps more importantly, both its rise and fall considerably increased interest in developing Peru's fishing economy. Thus, even the errors made by all three administrations may serve as a basis for further development. There are some signs that in both the government and the private sector there is an awareness of the past mistakes and a willingness to learn from them.

It is also clear that despite the special problems posed by such environmental factors as El Niño, Peru's experience in the fishing sector was not a unique event in that country and is indicative of some more general weaknesses evident in other policies of the Revolutionary Government as well as of its predecessors. Peru's history has been one of the unmanaged growth and subsequent collapse of a number of industries (e.g., nitrates, rubber, guano, oil) of which fish meal is only one of the most recent. Even in areas where the rise and fall has been less dramatic (e.g., copper) or in other types of programs (the agrarian reform conducted by the Velasco government), we find similar instances of a failure to prepare for any but the most favorable circumstances, a tendency to pursue single objectives with little or no regard for their impact on other goals, and, within single programs, an inattention to the kind of details that may mean success or failure even over the short run. This attitude toward policymaking, summarized by one Peruvian as "letting God look after Peru,'' can be traced at least partially to simple inexperience, a result of the long tradition of limited government involvement and still more limited capabilities. During the pre-1968 period, the constraints on government action giving rise to these tendencies also cushioned their potentially disastrous impact. However, after 1968 with the government's sudden shift to an activist role and with such unpredicted setbacks as El Niño, that potential was rapidly realized, not only in the fishing sector, but in the entire effort at directing national development. While both at the sectoral and national levels, the military successfully introduced a number of radical and long overdue changes, its performance at managing these changes and their impacts has been far less positive.

This evaluation of the Peruvian experience leads in turn to some more general conclusions about styles of policymaking in countries experiencing or desiring rapid change. The different situations confronting the pre- and post-1968 governments make it difficult to compare their performance directly, but it is evident that if the post-1968 system had its drawbacks, many of the crises it mismanaged had their roots in the inadequacies of the earlier incremental approach to policymaking. While critics of comprehensive planning models may be correct in stressing the impossibility and impracticality of real-world planning, the Peruvian case does suggest the need for some kind of a more comprehensive effort to direct and program development efforts both nationally and sectorally, and to create the kind of organizational infrastructure necessary to bring this about.

It also suggests that, while planning and management obviously depend on the creation of organizational infrastructure, organization-building alone does not guarantee that either will be done, or if done, will be done well. If the government prior to 1968 could not have planned or managed the sector's development due to lack of institutions, the post-1968 government did not do noticeably better despite the proliferation of government offices. The bureaucratic explosion in fact worked against both planning and management goals in that it hampered coordination and introduced extraneous and perhaps inappropriate priorities based on the narrow perspectives of individual offices. The plans elaborated for the sector were really little more than budgets, but they lacked the latter's base in a realistic assessment of resources and capabilities. Furthermore, they often bore little relation to the decisions actually taken—as in the case of the first 5-year plan which had counted on heavy private investment.

The new organizations were also plagued by a number of shorter-term problems—inexperience, a lack of coordinating mechanisms, and insufficient internal control which might have been remedied had the crisis not struck so early. The personalistic leadership style of Tantaleán, a more general characteristic of the first-phase government, further complicated both coordination and control, for what he could not directly supervise often went unattended. Tantaleán would later justify his actions as a means of building an *esprit de corps* within the Ministry,[49] but in an organization the size of the one he headed, less attention to mystique and more emphasis on management structures might have been more appropriate.

Thus, the end result of increased government involvement in policy-making and the accompanying organizational explosion was that, despite the outwardly monolithic structure, the pattern of fragmented decision-making toward a number of frequently contradictory ends persisted. More recent efforts to reorganize and rationalize the bureaucracy have removed some of the problems produced by sheer size. However, the crucial issue of priorities remains, as does the temptation to simply give way to the most immediate and vociferous demands, whether these originate within the bureaucracy itself or in the political environment.

ACKNOWLEDGMENTS

Research for this study was financed by a Fulbright-Hays Lecturing and Research Fellowship to Peru.

GLOSSARY

CERPER *Certificaciones Pesqueras del Perú* (Peruvian Fishing Certification), government agency founded in 1971 to do quality control of fishing products and of fish meal in particular.

EPCHAP *Empresa Pública de Comercialización de Harina y Aceite de Pescado* (Public Company for the Marketing of Fish Meal and Oil), government agency founded in 1970 to replace commercial outlets and so to establish a state monopoly on the sale of fish meal and oil (and later of a number of other non-fish products).

EPSEP *Empresa Pública para Servicios Pesqueros* (Public Company for Fishing Services), government agency established in 1970 to direct development of the food fishing industry and which at the height of its powers had complete responsibility for the sales of fresh and processed fish for direct consumption.

FPP *Federación de Pescadores del Perú* (Federation of Peruvian Fishermen)

FETRAPEP *Federación de Trabadores Pesqueros del Perú* (Federation of Peruvian Fisheries Workers), Major federation for land (factory) workers in fish meal industry.

IMARPE *Instituto del Mar del Perú* (Peruvian Marine Institute), Semi-independent research institute originally attached to the Ministry of the Navy and funded in part by the SNP. In 1970, absorbed into the newly created Ministry of Fisheries.

INP *Instituto Nacional de Planificación* (National Planning Institute), founded in 1963 by an interim military government but rose to prominence in the early 1970's with the military government's emphasis on national economic planning. Responsible for drawing up the national five year plans and coordinating the work of the sectoral planning offices.

MLR *Movimiento Laboral Revolucionario* (Revolutionary Labor Movement), Labor organization centered in the fishing industry and rumored to have been founded or at least supported by Minister Tantaleán.

OSP *Oficina Sectorial de Planificación* (Sectoral Planning Office), any of the planning offices set up within each ministry to direct the formulation of a sectoral plan in cooperation with the INP. In this case the fisheries planning office.

OSPP *Oficina Sectorial de Planificación Pesquera* (Sectoral Planning Office for Fisheries), name for fishing's OSP when prior to 1970 the sector was administratively still a part of The Ministry of Agriculture.

PESCAPERU Government company set up to run the nationalized fish meal industry. Originally controlled both fleet and factories but returned the fishing boats to private ownership in 1976 and 1977.

SERVICIO DE PESQUERÍA Fisheries Service, office of the Ministry of Agriculture prior to 1970 responsible for overseeing fresh water and marine fishing (although it emphasized the former and virtually ignored the latter).

SNP *Sociedad Nacional de Pesquería* (National Fisheries Society), private interest organization formed in 1960 to represent those with business in the sector and to provide common services to them. Dominated by the fish meal industry until the latter's nationalization.

REFERENCES

1   C. Lindblom, "The science of muddling through," *Public Administration Review*, 79–88 (Spring, 1959).

2 D. Yates, *The Ungovernable City,* MIT Press, Cambridge, MA, 1979.

3 N. Caiden and A. Wildavsky, *Planning and Budgeting in Poor Countries,* Wiley, New York, 1974.

4 G. Wynia, *Politics and Planners,* University of Wisconsin, Madison, WI, 1972.

5 P. Kucynski, *Peruvian Democracy under Economic Stress,* Princeton, 1977.

6 A. Lowenthal, Ed., *The Peruvian Experiment,* Princeton, 1975.

7 A. Stephan, *The State and Society,* Princeton, 1978.

8 D. Palmer, "Peru: Authoritarianism and Reform" in H. Wiarda and H. Kline, Eds., *Latin American Politics and Development,* Houghton Mifflin, Boston, 1979.

9 Ibid, p. 208.

10 Perú, Ministerio de Pesquería, Oficina Sectorial de Planificación Pesquera (OSPP), "Diagnóstico del Sector Pesquero" (unpublished working paper), 1978.

11 Perú, Ministerio de Agricultura, Servico de Pesquería, "Política Económica de la Pesquería del Perú," 1965.

12 Perú, Sociedad Nacional de Pesquería (SNP), *Memoria,* 1962–1963.

13 Perú, Sociedad Nacional de Pesquería (SNP), *Memoria,* 1966–1967.

14 Perú, Sociedad Nacional de Pesquería (SNP), *Memoria,* 1968.

15 Perú, Sociedad Nacional de Pesquería (SNP), *Memoria,* 1969.

16 Perú, Sociedad Nacional de Pesquería (SNP), *Memoria,* 1970.

17 Perú, Sociedad Nacional de Pesquería (SNP), *Memoria,* 1971.

18 Perú, Sociedad Nacional de Pesquería (SNP), *Memoria,* 1972.

19 Perú, Sociedad Nacional de Pesquería (SNP), *Memoria,* 1973.

20 Perú, Ministerio de Agricultura, Oficina Sectorial de Planificación Pesquera (OSPP), "Diagnóstico Económico Social de la Pesquería," 1963.

21 Peru, IMARPE, "Panel of Experts' Report on the Economic Effects of Alternative Regulatory Measures in the Peruvian Anchoveta Fishery," IMARPE Report No. 34, 1970.

22 J. Abramovich, *La Industria Pesquera en el Perú,* Universidad Federico Villareal, Lima, 1973.

23 SNP, *Memoria,* 1969, p. 16.

24 J. Tantaleán Vanini, *Yo Respondo,* Manturano-Gonzalez, Lima, 1978, p. 210.

25 Perú, Ministerio de Agricultura, OSPP and Servicio de Pesquería, "Argumentos para la Insitucionalización del Sector Pesquero" (unpublished mimeo), 1969.

26 Perú, Ministerio de Agricultura, Servicio de Pesquería, 1965, pp. 157–66.

27 Perú, Instituto Nacional de Planificación (INP), *Plan de Perú, 1971–75:* Plan del Sector Pesquero, Vol. III, 1971.

28 Ibid., p. 163.

29 J. Tantaleán Vanini, *Yo Respondo,* Manturano-Gonzales, Lima, 1978, p. 210.

30 *El Comercio* (Lima), Jan., 2, 1971.

31 G. Paulik, "Anchovies, Birds, and Fishermen in the Peru Current," 1971, reprinted in this volume.

32 SNP, *Memoria,* 1970, p. 19.

33 IMARPE, 1970, p. 10.

34 Paulik, p. 157.

35 *Ultima Hora* (Lima), December 2, 1971.

36 Tantaleán, p. 218.

37 C. Malpica, *Anchovetas y Tiburones,* Runamarka, Lima, 1965, p. 32.

38 Tantaleán, p. 152.

39 Malpica, *passim.*

40 Tantaleán, pp. 148–72, 306–30.

41 Perú, Ministerio de Pesquería, Oficina Sectorial de Planificación (OSP), "Evaluación Socio-Económica del Sector Pesquero," 1978, p. 13.

42 J. Burneo et al., *Empleo y Estabilidad Laboral,* DESCO, Lima, n.d., p. 181.

43 *Ibid.,* pp. 177–83.

44 Perú, PESCAPERU, *Memoria,* 1975.

45 Perú, PESCAPERU, "PESCAPERU," 1978, p. 7.

46 *Ibid.,* p. 8.

47 *El Comercio* (Lima), October 17, 1978, p. 6.

48 Perú, Ministerio de Pesquería, OSP, 1978, pp. 24–26.

49 Tantaleán, p. 206.

CHAPTER 12

The Impact of El Niño on the Development of the Chilean Fisheries

CÉSAR N. CAVIEDES

Abstract

Since their inception in the late 1950s the industrial fisheries of northern Chile have developed in a highly speculative fashion. Without regulations, the catching and conservation of the fish were left to the discretion of entrepreneurs, so that the catches depended more on the demands of the international market than on the size of the fish population. Under these circumstances, any ecological disruption led to serious reductions in a fishing stock that had been exploited almost to the limits of its capacity.

The oceanic and ecological effects of the 1972–1973 El Niño were felt in the waters of northern Chile with more intensity in 1973 than in 1972, coinciding with the troubled last months of the Allende administration and its overthrow by the military junta. With the drastic political change in the country there was also a new economic orientation that emphasized high production as a measure to expand exports, but lacked concern for the conservation of marine resources and for the social cost of increased production. In line with these trends, there occurred a concentration of ownership of the fishing companies and an increase in the catch of jurel (*Trachurus murphyi*), which had become a substitute for the dwindling stock of anchoveta in the early 1970s.

Viewed in this perspective, the oceanic and ecological anomalies of El Niño 1972–1973, although less dramatic than in Peruvian waters, have led to deep and long-lasting consequences for the marine resources and the coastal communities of northern Chile.

INTRODUCTION

On the western coast of South America, Chile shares with Peru one of the major fishing grounds of the world in terms of biological productivity, that of the Peru Current. While Peru's catches ranged from 52 to 84% of South America's total landings between 1970 and 1978, Chile was the minority partner with 10 to 19% of the landings during the same period.

The fisheries located in the Great North, between 18°S and 27°S (Figure 1), attract 60% of the fishing effort of Chile and account for one-third of the nation's employment in marine resource exploitation. In terms of fish processing capacity, this region is dominant, containing 79% of the industrial fisheries personnel and installations of the nation. It is not surprising, then, that in good years the catches of the North make up to three-quarters of the country's landings, whereas in years of low production they may decline to only half.

Because of their concentration off the coast of northern Chile, the industrial fisheries are vital for the coastal communities where fishing facilities and processing plants are established and they further contribute, through exports, to the health of the national economy. However, variability in the yields of the northern industrial fisheries adds an element of uncertainty that is readily reflected in the regional as well as the national economy.

The fisheries of northern Chile, as those of neighboring southern Peru, are sensitive to environmental disturbances. Any major oceanic El Niño phenomenon in these waters has been accompanied by a decline in the yields of the northern Chilean fisheries. 1972–1973 was not an exception to that rule, as will be reported in the following pages. While the ecological effects of that El Niño in Chilean waters were not as drastic

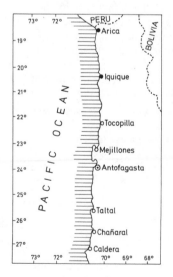

Figure 1 The cost of northern Chile.

as in northern Peru, they occurred along with turbulent sociopolitical events in Chile: the overthrow of the socialist government of President Salvador Allende and its replacement by a military junta. The political change marked the beginning of a phase of clear neocapitalistic features in the development of Chilean fisheries, whose characteristics and effects are worthy of closer examination.

THE EL NIÑO PHENOMENON IN CHILEAN WATERS

Off the coast of northern Chile, the Peru Current is divided into three branches that run parallel to the straight coastline. Between 300 and 400 miles from the coast, the cold offshore branch of the Peru Current flows northward. It borders on a warm current to the east—the Peruvian Countercurrent—moving south at varying distances of between 180 and 270 miles from the coast. A third belt of cool waters contiguous to the littoral consists of a complex of minor currents, some of them oriented equatorward, others southward.[1] Cold minor currents that flow in the immediate vicinity of the coast are strengthened by local centers of upwelling that contribute to making the coastal fringe of water the richest fishing zone. This is where all the activity of the northern Chilean fisheries is concentrated. Separated from the rest of the eastern tropical Pacific by the two outer branches of the Peru Current, only those oceanic anomalies that can be felt in the coastal belt have a real bearing on its biomass.

The last major anomaly that was thus noted was El Niño in 1972–1973. During the summer months of 1972 (January, February, March), sea surface temperatures along the coast of northern Chile rose slightly above normal. In the port city of Arica, the highest sea surface temperature was recorded in January, 22.8°C (monthly average of 19.9°C), whereas at Antofagasta, with a monthly average of 20.4°C, the highest value was 24.4°C. Still, those record highs were scattered through the summer months and did not indicate a continuous temperature anomaly.[2] Similarly, the upwelling centers evinced a pulsating strength during those summer months, revealing, thus, that variations in the intensity of the coastal upwelling, and not an overlapping by warm water as along the coast of northern Peru,[3] characterized the coastal waters of northern Chile during that particular year.

That 1972 was really an anomalous year was shown by the oceanic conditions along the coast during early winter. The May sea surface temperatures at Arica oscillated around 18.5°C, and around 17.7°C at Antofagasta. This meant a positive variation of 1.3 and 1.4 degrees centigrade, respectively, over the monthly average for each station. Further south, at Caldera, the sea surface temperature of May was about 16°C, 1°C above the monthly average, whereas at Valparaíso the temperatures of that month stayed very close to the monthly mean of 13.2°C. Apparently, oceanic conditions were upset only in the segment of the Chilean coast north of 30°S.

During a cruise of the Chilean Oceanographic Institute of the Navy in the winter of 1972 (August, September) a strengthening and widening of a warm-water flow in the south direction was observed between 100 and 150 miles from the shoreline. This enhancement of the southward flow was coupled wih a weakening of coastal upwelling and with a slackening of the coastal flow northward. The offshore branch of the Peru Current was also weakened.[4] Only at the beginning of spring did the coastal waters become somewhat normalized, but strong insolation still prevented a cooling. In December 1972, the waters off Arica, Antofagasta, and Caldera were nearly 2°C above the average of normal years (18, 18.4, and 17°C, respectively). January 1973 brought an even more dramatic warming of the coastal waters. Arica and Antofagasta registered record high sea surface temperatures of 23.6 and 23.9°C, respectively, during several days, and at the end of the month Arica averaged 21.7°C and Antofagasta 22.1°C. Mean values were 1.8°C higher than those of normal years. For the rest of the summer the upwelling centers of the Chilean North remained inactive and only by mid-April 1973 did the sea surface temperatures return to their normal levels.

In summary, the temperature anomalies of the coastal waters in northern Chile proved to be more acute in the winter months of 1972 and through the summer of 1973 than in the summer of 1972 (January–March), as had been the case along the coast of northern Peru. In spite of this, the higher sea surface temperatures occurred in phase with a general warming of the southeastern Pacific area[5] and with a weakening of the coastal upwelling.

The unusual oceanic conditions in the waters of northern Chile did not influence the yield of the pelagic fisheries until the winter of 1972, and became more pronounced only in the summer of 1973. In fact, while January 1972 alone yielded 44% of all the anchoveta caught in 1972 (367,913 tons), in the three summer months of 1973 the volume of anchoveta made up only 30% of the meager 191,795 tons fished in 1973. The landings of other cool-water species like sardines and hake had declined as well.

Inversely, the catches of tropical-water fish increased remarkably: landings of swordfish peaked at 410 tons (the highest amount of the decade), doubling the 1963–1972 average. Less spectacular, although equally significant, were the relatively greater landings of tollos (genus *Mustelus*), bonito *(Sarda chilensis)*, and pejegallo (genus *Callorhynchus*), all of which are warm-water fish that are seldom caught during normal years. The landings of tuna were not as great as expected during warm-water years. Here, it must be noted that fishermen specializing in large tropical or subtropical species seldom venture beyond 100 miles and, therefore, do not encounter the swarms of tuna from the high seas. Also, the reduction in the fisheries yield that was caused by the oceanic anomalies of the 2 years mentioned earlier occurred in 1973, and not in 1972 as it did in Peru and the rest of the eastern Pacific. The same applies to the return to normal oceanic conditions, which had begun by April 1973; it was reflected in an increase of the volume of fish, not in the same year, but in 1974 (Figure 2). This recovery was followed, however, by a slight decline in 1975, which was related to a new, but lesser weakening of the coastal upwelling along the littoral of northern Chile. This time, sea surface temperatures rose by only 1°C, particularly during the summer months. The catches of anchoveta and sardines again approached the meager levels of 1973, whereas the landings of jurel continued to increase on a trend that began in 1970, and accounted for 35% of all the commercial catches made in northern Chile.

The oceanic anomalies of 1975 were not as pronounced as those of 1972–1973. This is also substantiated by the fact that in 1975 much less swordfish, bonito, and tollo were caught than in the preceding years. In 1976 normal water conditions along the northern coast of Chile were reestablished, and the upwelling centers returned with great strength

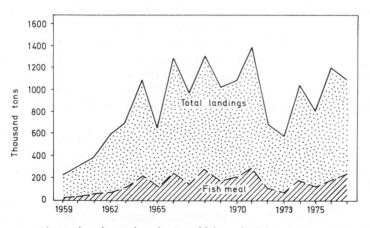

Figure 2 Volume of catches and production of fish meal. Chilean fisheries, 1959–1977.

with the pelagic fisheries reaching their highest yield of the 1970s. No effects of the reported warming of the southeastern tropical Pacific were noticed off the coast of northern Chile.[6]

From the developments that have been outlined above it appears that only during major El Niño events that extend over wide areas (such as those of 1957 and 1972–1973), when oceanic conditions of the whole southeastern Pacific are greatly disturbed, do the northern Chilean waters and their fisheries reflect the anomalous circumstances. In other years apparent El Niño conditions in the triangle formed by the Galapagos Islands, Ecuador, and Northern Peru appear to be only regional conditions without implications for the rest of the coast of South America which remains under the influence of the coastal branch of the Peru Current.

THE DEVELOPMENT OF THE NORTHERN CHILEAN FISHERIES

Industrial fisheries on the northern coast of Chile were initiated in the late 1950s, shortly after they had begun in Peru. In northern Chile the establishment of industrial fisheries is to be understood in the framework of general plans for the development of the North, whose economy, until then, had hinged almost exclusively on mineral resources. During the mid-1950s a boost was given to the languishing activities of the city of Arica by declaring it a duty-free port and making it the center of the Chilean motor vehicle assembly industry. Both the active trade of sumptuaries and the incipient car industry brought to Arica a degree of development which its sister city of Iquique—a former flourishing nitrate export center—could not emulate, and which it greatly resented. Between 1958 and 1960, to halt the critical decline of Iquique, legislation was passed that allowed the establishment of fish processing plants and the duty-free importation of the necessary machinery. With the emphasis on free enterprise and economic liberalism that characterized the administration of President Jorge Alessandri (1958–1964), Iquique was invaded by profiteers and speculators from Santiago who were eager to make fast profits from low investments and tax-free revenues.

In the early 1960s the Barrio El Colorado in Iquique (Figure 3) developed into a bustling industrial area with up to 24 fish processing plants in operation. From a dozen old-fashioned fishing boats in 1958, the fishing fleet of Iquique grew to 119 iron-hull trawlers by 1966. In the port area of the city, trawlers were built under license from a firm in Oregon. In this way, a booming economy based on the marine resources injected new vitality into a city that had been on the verge of collapse.[7]

Figure 3   Fish processing plants of the Barrio El Colorado, Iquique.

Attracted by the prospect of easy employment and good salaries, a mass of unqualified laborers from the hinterland of the North came to the city quickly filling the jobs available.[8] The personnel in the upper ranks of the processing plants consisted mainly of natives of Santiago and Valparaíso (in Central Chile). Thus, although the establishment of the fishing industry vitalized the economy of Iquique, the social benefits for the original population of that city were curtailed by the influx of outsiders.

The success of Iquique was to be emulated by other cities of the North. To diversify its economy, endangered by an early shut-down of several auto assembly plants, Arica also entered into the fisheries business. Thus, the industrial center of Playa de los Gringos, south of the city, came into existence. In 1966 the fishing fleet operating at Arica consisted of 41 trawlers delivering raw fish to seven processing plants. In the city of Antofagasta the fishing boom was less noticeable, as only three fish meal plants with a supporting fleet of 16 trawlers were opened in the La Chimba sector.

It is noteworthy that the development of fisheries in northern Chile, unlike that in Peru, occurred chiefly in the major cities of the region and that the smaller coastal settlements seldom benefited from the establishment of fish processing plants. This is a peculiar aspect of the development plans proposed for the North. Regional growth has usually been envisaged for the benefit of large urban centers as a means of reducing the existing levels of unemployment and of taking advantage of cheap labor in addition to the existing urban infrastructure. However, that policy invariably worked to the disadvantage of smaller coastal communities by contributing to their further depopulation and decline. For example, fish processing plants that were originally established in small fishing villages, such as Pisagua (three plants) and Patillo (one plant), did not last long, and by the beginning of the 1970s they had either closed down or moved their facilities into one of the major cities. Only in secondary urban centers like Tocopilla and Mejillones could small fish-processing enterprises survive, although operating at much lower levels than in Iquique, Arica, or Antofagasta.

In Iquique, the major fishing center of the country, the labor force employed in marine resources exploitation comprised 3547 workers in 1976 or about 19% of the economically active population. Assuming that an average family consists of four persons, the number of people depending directly on fisheries could be estimated at about 15,000. If the families of personnel employed in activities associated with fisheries, for example, transportation, marine equipment, and machinery servicing, are also taken into account, the population in Iquique related to these marine activities rises to about 20,000 persons or 30% of the inhabitants.[9] By comparison, in the competing port of Arica in the same year only 1494 laborers, merely 5% of the economically active population of that city, were employed in fisheries. If the comparison is extended to other communities in Northern Chile, it shows that the large city of Antofagasta (127,000 inhabitants) has only 2% of its active population in marine resources industries. The percentage is higher, however, in the smaller towns of Tocopilla (4.7%), Mejillones (6%), and Taltal (7.6%) due to the fact that among a comparatively lower number of economically active people the fisheries workers amount to a relatively higher percentage.[10]

The proportion of the population involved in marine resource exploitation attests to the extremely important role played by Iquique in the framework of industrial fisheries in Chile. The dominance of that city was further enhanced by the value of its fisheries production. In 1976 this value reached US $52.6 million, which was nearly one-half of all fisheries production in Chile. The monetary value of Arica's production totaled only $34 million.[11] These figures support the fact that approximately two-thirds of the nation's industrial fisheries installations are located on the coast along the 290 miles between Arica and Antofagasta and that this area produces nearly 75% of Chile's income derived from fisheries.

RECENT TRENDS IN CHILEAN FISHERIES

The uncontrolled and speculative growth that characterized the early development in the industrial fisheries of Chile could not withstand the first serious crisis it was faced with, namely the temporary reduction in the fish population caused by a moderate oceanic anomaly in 1965. Greedy entrepreneurs who, after only a few years of operation, had recovered and multiplied their initial investments, liquidated their assets to the benefit of larger firms and were happy to withdraw from the business. This temporary crisis initiated the trend toward the concentration of the industry in a few hands, a trend that was to culminate in the 1970s. In Iquique alone the number of independent industrialists was reduced by more than half. In the other fishing ports of the North the squeeze was less dramatic, as the establishment of fish processing plants had less of that speculative character that it had in Iquique.

The fish population decline in 1965 had few lasting effects, however. From 1966 until 1971 the Chilean fisheries yields oscillated from 1 million to 1.3 million tons per year and peaked at 1.4 million tons in 1971 (Figure 2). In the meantime, the government of the country had politically shifted from Christian Democracy to the socialist Popular Unity. Neither of these progressive administrations made a major effort to control the rather independent but highly hard-currency-yielding economic sector. Measures to prevent overfishing or to protect endangered species were not even taken. Apparently, the fishing industry was much too productive to be curbed in its freedom of operation, and when the oceanic anomaly of 1972 began to develop and the fish population declined, the government of the Popular Unity was too beleaguered by political opponents and inflation to react to the impending crisis in the fisheries in the way that the Peruvian government did.[12]

Low productivity, social unrest, and political instability discouraged independent operators to such an extent that they did not oppose the involvement of government officials in their enterprises. In any case, the takeover of some fish processing plants and fishing fleets by the government proved to be an unglamorous venture. In 1973 the political and economic situation in the country deteriorated while the crisis in fisheries worsened so that the management of that economic sector increasingly became a burden to the government. After June, the landings of anchoveta, sardines, and jurel fell to almost one-tenth of the volume of fish caught in normal years. The 1973 total reached only 581,417 tons, which was even less than the 1965 total (Figure 2). This was the most serious recession since the inception of the Chilean industrial fisheries.

The critical situation in the fishing sector reflected, without doubt, the institutional crisis that existed in the country as much as it reflected the impact of El Niño on biological productivity in coastal waters. Both crises passed by the end of the year, as the government of the Popular Unity was forcibly ousted in a military coup. Meanwhile normal oceanic conditions returned along the coast of northern Chile. Still, there were long-lasting effects: (1) a trend that began to develop a few years earlier, namely resorting to jurel as one of the main species of industrial fisheries was maintained; and (2) the ultimate concentration of activity in that industry in a few major companies was established.

THE RISE OF JUREL AS ONE OF THE MAIN SPECIES IN CHILEAN FISHERIES

As in the Peruvian fisheries, the anchoveta *(Engraulis ringens)* had become the main species in the Chilean fisheries. From 1959 to 1965 anchoveta accounted for nearly 80% of the volume of fish utilized in the production of fish meal and fish oil. Sardines *(Sardinops saga)* were next, but at levels seldom surpassing 10% of the total. The remainder was composed of other species such as ajugilla *(Scomberesox sp.)*, pejerrey *(Odontestes sp.)*, and mackerel *(Pneumatophorus peruanus)*.

From 1965 to 1969 the proportions of the landings of anchoveta and sardines remained almost constant (Figure 4). Then suddenly in 1970 jurel *(Trachurus murphyi)*, until then accounting for less than 2% of the

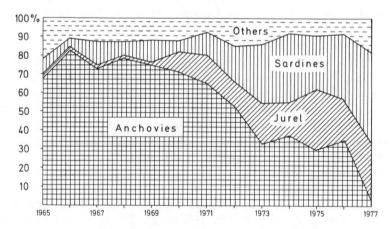

Figure 4 Main species exploited by the fisheries of northern Chile. Percentages of the total volume of catches, 1965–1977.

catches, rose to 10% becoming the second most important species in the fisheries of northern Chile. The rise of jurel, paralleled by a dramatic increase in the landing of sardines since 1972, was accompanied by a decline in anchoveta catches. One might explain the decline in anchoveta landings by fatigue of a fish stock that had been intensively exploited for more than 15 years (it had been exploited continuously since 1959) and that had suffered severely from ecological disruptions related to the El Niño 1972–1973. The Chilean fisheries were fortunate in that, as the anchoveta stock decreased, jurel and sardines replaced it.

Jurel is a species found mainly in subtropical waters with temperatures between 10° and 18°C and a salinity of between 34.1 and 34.5 per thousand.[13] The stock shares an ecosystem with anchoveta and sardines on which it feeds and whose seasonal migrations it follows. In its adult stage, reached after 15 months, the jurel measures between 30 and 60 cm. Spawning occurs during the transition from spring to summer (November–December) in cold-water areas, usually at 30 to 50 miles from the coast.[14] It is also at that distance from the shoreline where the highest concentrations of jurel have been detected. Yet, a tradition has been established by fishing fleet operators not to catch beyond a narrow coastal band of 25 to 30 miles and to direct their fishing effort at the vicinity of the major fishing ports of northern Chile: Arica (38–64%), Iquique (26–59%), and Antofagasta (2–9%). During the years of maximum jurel catches (1973, 1975, 1977), the major amount of this fish was caught in late summer and early autumn (March to May), at the time of year when the stock is growing. Further, it has been observed that among the samples fished during summer there was yet a large proportion of females in the phase of reproduction and a high percentage of young, relatively small fish.[15]

These two elements of judgment, the intensity of the fishing effort in a few coastal areas and sustained fishing activities during spawning and maturing periods, show the indiscriminate manner in which the exploitation of the jurel stock was carried out from the time it became a main pillar in the Chilean fishing industry. A preliminary report in 1977 on the state of the jurel fisheries produced by the Universidad del Norte (Chile), pointed to the possible dangers of overfishing, but was considered nothing more than an informative piece of literature and its warnings passed by without much notice.[16] So far, the fishing of jurel has been pursued according to the availability of fish in the vicinity of the coast and the demand for raw fish from the fish meal plants, without consideration of regulatory measures to prevent extinction of the species. Thus, once again, it was left to nature to take care of the recovery of the stock. After the first fishing semester of 1978—when the fishing season for jurel was almost over—the port of Arica registered an increasing trend in the landings of anchoveta (31% of the total), which meant a certain relief in the demand for jurel in the following months. Nevertheless, during the mentioned semester, jurel still made up 40% of all landings.

The Concentration of Ownership of Industrial Fisheries

Uncontrolled proliferation had been the dominant characteristic of the early development of the industrial fisheries in northern Chile. Although the decline in landings in 1965 brought about the first step toward conglomeration of smaller enterprises, the healthy state of the fishing population in the following years enabled medium and large industries to subsist comfortably. But in 1973, when an ecological and a political crisis coincided, the trend toward concentrating the enterprises in a few hands was strengthened.

Several circumstances were responsible for this development. Increased costs of maintenance and an escalation of salaries, during the years of the Allende administration, had made the fishing industry less attractive to industrialists. Beleaguered by rising production costs, lack of foreign currency for the purchase of equipment and dismayed by continual labor unrest, independent operators were readily inclined to sell their establishments to larger companies or to let the Chilean state intervene and operate them under such adverse conditions. For larger concerns the operation of fish processing plants under these conditions was not as critical as for smaller entrepreneurs, since by owning several plants in different parts of Chile, they were able to meet international demands in spite of frequent worker unrest. It is important to mention here that in 1971, the highest production of fish meal and fish oil for the period between 1966 and 1975 was achieved (Figure 2).

When the Chilean government was taken over by the military in 1973 and many of the socialist measures undertaken under Allende were revoked, the policy of the military government regarding fishing industries was to leave them under the control of the private sector and to provide tax incentives to those companies to inspire them to increase exports. Only enterprises that had been created and patronized by the state-owned Corporation for the Development of Production (CORFO) remained under fiscal ownership. Two of these enterprises in Iquique, the Tarapacá Fishing Company and the Iquique Fishing Establishment, had provided sufficient arguments in favor of the neocapitalistic policies of the military government to leave the fishing industry in private hands. With about 50% more personnel than the privately owned companies, their efficiency lagged behind by 60%, both in terms of volume and value of production (Table 1).

TABLE 1. Fishing Companies of the Three Major Ports of Northern Chile

Port	Company	Vessels	Storage Capacity (all vessels)	Processing Capacity (tons/hour)	Personnel	Value of Production (1,000 US$)
Arica	Pesq. Coloso	11	1,540	60	312	6,306
	Pesq. Indo	10	1,400	140	356	12,416
	Pesq. Eperva	5	900	50	257	11,962
	Pesq. Guanaye	5	700	50	144	3,499
	Chilemar	5	720	55	103	b
Iquique	Pesq. Coloso	14	2,940	70	368	14,224
	Pesq. Tarapacá	12	1,810	100	535	6,670
	Pesq. Guanaye	7	1,510	40	224	5,587
	Pesq. Iquique	10	1,440	105	412	5,841
	Pesq. Eperva	7	1,400	70	253	11,035
	Pesq. Indo	7	1,220	60	226	8,972
	Corpesca	a	a	23	37	b
	Pesq. De Lucchi	a	a	27	29	b
Antofagasta	Pesq. Guanaye	7	1,510	70	142	6,487
	Pesq. Zladar	4	800	50	82	3,600

[a]Does not own trawlers. Fish is supplied by independent fishermen.
[b]Value of production not reported.
Source: *Anuario Estadístico de Pesca*. Servicio Agrícola y Ganadero. Division de Protección Pesquera, Santiago, 1977.

In line with the new direction of the Chilean economy introduced by the economic advisors of the military during 1974, the diverse fish processing establishments in northern Chile came under the control of one of four major corporations: Coloso Fishing Company, Eperva Fishing Enterprises, Indo Fishing Industries, and Guanaye Fishing Enterprises. All of them own fish processing plants and fishing fleets in the three major ports of northern Chile—Arica, Iquique, and Antofagasta. Only a few smaller companies resisted the consolidation drive: three in Iquique and one each in Arica and Antofagasta.

Also, to increase efficiency, there was a reduction in the number of fishing vessels and personnel. While there were 123 fishing units operating in the waters off northern Chile in 1970, there were only 106 in 1975. This reduction in the number of units did not mean, however, a substantial decrease of the total fishing and storage capacity; it remained almost the same: nearly 18,000 tons of raw fish. The maintenance of the fishing effort was due to the replacement of some of the smaller vessels of 100 to 140 tons that were common in the early 1970s, by high capacity fishing units of 200 and 300 tons of storage capacity.[17] No doubt such innovations made Chilean fisheries more efficient, but they also posed greater threats of overfishing endangered stocks. Yet, as mentioned above, the major concern was not overexploitation but to gain higher profits from the greater processing capacity and increased exports. Even less consideration was given to the social costs associated with increased efficiency. For example, of the 7408 individuals employed in the fishing fleets, fish processing plants and dockyards in northern Chile in 1970, the number had decreased to 5220 individuals in 1975, and 85% of those were employed in the ports of Arica and Iquique.[18] Thus, while in 1975 production reached levels similar to those of 1970, the installed processing capacity of the Chilean fishing industry had shrunk by nearly 5% and the labor force had been reduced by 29%. Obviously, the approach toward productivity had been radically altered in the course of those five years.

FISHERIES IN THE PRESENT CONJUNCTURE OF CHILEAN ECONOMY

In the framework of an economic model based on free enterprise, high profits to attract investment, and increased exports with an emphasis on raw materials and semimanufactured products, fisheries have become a respectable element of the export-oriented economy of the country. In 1975 the total value of the production reached US $107,210,500 and accounted for approximately 7% of the value of Chile's exports.[19] By comparison, in 1973 exports from this sector of the economy had amounted to $28,450,000 and in 1971—a successful year for the fishing industry—they had reached no more than $50,340,000.

The relatively high value of the production had been achieved by stressing efficiency and by diminishing the costs of production—especially through labor reduction—in order to make Chilean products competitive on foreign markets. Competitiveness, although never critical in the international marketing of fish meal and fish oil, has not been based so much on the price or quality of the product as on the reliability and speed of delivery, and there Chile has been among the first five suppliers of the world. Thus, while certain countries made the decision not to buy from Chile for political reasons (namely not to entertain commercial relations with a regime that had come to power by force of arms) others were reluctant to abandon the Chilean source of supply. In any case, the ecological crisis of 1972–1973 created an inflated demand for fish derivates and forced the major fish meal and fish oil consumers to purchase the Chilean product, and at prices that were much higher than in previous years ($323 per ton in 1975 as compared with $90 per ton in 1965). Among the old customers of Chile, the German Federal Republic, the United States, and Japan continued to buy fish meal after the military coup and even increased the volume of their purchases, whereas the United Kingdom and the Netherlands severed their commercial relations, but they had not been major buyers anyway.[20] Favored by the situation on the international market, the Chilean government and the producers—interested in improving the balance of payments—tried to meet the foreign demand by increasing production, lowering cost, and shortening the time of delivery. In all these aspects they tried to surpass other competitors. An estimate of the total value of production, as compared with the amount of investment, revealed that in 1976 production was 2.5 times higher than assets,[21] a figure that demonstrates the efficiency of the fishing industry reached under the present political and economic circumstances.

Little is known about the potential of the marine resources in the coastal waters of northern Chile. Even less is known about the limits of exploitation of the biomass. From superficial observations of the fish population it appears that the fisheries could still be expanded beyond the present spatial limits of 25 miles, provided a reliable control of the stock fluctuations and a monitoring of the activities of industrialists could be guaranteed. If this were the case, the basis for a rational and profitable fishing industry in the waters of northern Chile could be developed. At the present time, however, emphasis has been put on efficiency with almost total disregard for the biological substratum of the industry, an approach that poses a real threat to the further expansion and health of the fishing industry.

Considering the fisheries in context with the overall development of coastal northern Chile, the management of the industry, because of its highly speculative character, still leaves much to be desired. Investments, equipment, and infrastructure appear to be oriented toward economic gains within a short period of time and toward maximizing profits with low levels of reinvestments. Under these premises, the future of the human population that depends on the marine resources is jeopardized every time international demand for such poducts decreases, or the prices of these commodities decline, or ecological imbalances affect the fishing stocks. Until now, the fish processing industry of northern Chile has concentrated almost exclusively on the production of fish meal and fish oil which had been selling well in the 1970s. It would be worthwhile to recommend, nevertheless, the development of canneries and fish-freezing plants whose products sell at much higher prices and the demands of which do not change as capriciously as those for fish meal and fish oil. Furthermore, experiences from the past, particularly ecological disruptions that have brought fish stocks near collapse, strongly suggest that the overall development of the coastal communities of Chile's North should not be based on the resources of the sea alone and that development alternatives should also be explored. Unfortunately, until now plans for the development of the North have rested too heavily on the fisheries—because of their quick returns—without seriously considering the fluctuations of the fish population. Great pressure has been applied to the living marine resources without proper knowledge about the limits to their exploitation. In other instances, particularly those that depend on the internal policies of the country, the exploitation of the sea's resources has been carried out with over-concern for the economy of the nation and little concern for the precarious nature of the equilibrium to which the fish stocks in the Peru Current are subjected.

In this respect it is important to observe that the three recent democratic administrations—the liberal Alessandri, the reformist Frei, and the socialist Allende—as well as today's military government have viewed the renewable resources of the sea as inexhaustible and that the fisheries have been considered by each government as the mainstay for the development of the northern littoral. This is obviously the easiest approach for a government to take because it does not require a demanding search for other approaches to development. Such governmental complacency causes great havoc among the economic interests involved in the fishing industry whenever an ecological disruption depresses the yields of the fisheries, but the strategies for the development of the North are not altered.

It appears then that the experiences gained from El Niño 1972–1973, even though that event coincided with the occurrence of significant political and economic changes both in Chile and Peru, did not serve to change substantially the economic approaches toward an important living marine resource for both nations and did little to alter long established attitudes of environmental disregard.

ACKNOWLEDGMENTS

The author is indebted to Professors Jorge Bastén and Patricio Contreras (Antofagasta), Mr. Patricio Labarca (Iquique), Mr. Humberto Sanhueza (Arica), and José Cañón (Santiago) who kindly provided access to their sources of information. Field research was conducted with the support of a fellowship from the Social Sciences and Humanities Research Council of Canada, which is gratefully acknowledged.

REFERENCES

1 N. Silva and H. Sievers, *Informe preliminar sobre los resultados de la operación Marchile X-ERFEN, 1* (1 al 28 de julio de 1976). Comité Oceanográfico Nacional, Instituto Oceanográfico de la Armada de Chile, Santiago, 1978.

2 Original records from the Centro Nacional de Datos Oceanográficos (C.E.N.D.O.C.), Instituto Oceanográfico de la Armada de Chile.

3 W. Wooster and O. Guillén, "Characteristics of El Niño in 1972," *Journal of Marine Research,* **32**, 387–404 (1974).

4 H. Sievers and N. Silva, "Masas de agua y circulación en el Océano Pacífico sudoriental. Latitudes 18°S–33°S. Operación Oceanográfica Marchile VIII," *Ciencia y Tecnología del Mar,* **1**, 7–67 (1975).

5 F. R. Miller and M. Laurs, "The El Niño 1972–1973 in the Eastern Pacific Ocean," *Inter-American Tropical Tuna Commission Bulletin,* **16**, 403–448 (1975).

6 K. Wyrtki, E. Stroup, W. Patzert, R. Williams, W.H. Quinn, "Predicting and observing El Niño," *Science,* **191**, 343–346 (1976)

7 R. Salinas, "Un ejemplo de valorización pesquera del litoral norte de Chile. La industria de la harina de pescado en Iquique," *Boletín de la Associación de Geógrafos de Chile,* **1**, 9–15 (1967).

8 J. Bähr, Migration im Grossen Norden Chiles, *Bonner Geographische Abhandlungen,* **50**, 122–124 (1975).

9 Instituto Nacional de Estadísticas, *Características básicas de la población (Censo 1970),* Tarapaca, Santiago, 1970.

10 Instituto Nacional de Estadísticas, *Características básicas de la población* (Censo 1970), Antofagasta, Santiago, 1970.

11 S.E.R.P.L.A.C., Diagnóstico visual del sector pesquero de la Primera Región, S.E.R.P.L.A.C., Iquique, 1976 (Report for internal circulation).

12 C. Caviedes, "El Niño 1972. Its climatical, ecological, human, and economic implications," *Geographical Review,* **65**, 493–509 (1975).

13 R. Méndez and S. Neshyba, "Relación entre las variaciones mensuales de temperaturas superficiales y la distribución de los recursos pesqueros de sierra, jurel y sardina, en el Pacífico Sureste," *Revista de la Comisión Permanente del Pacífico Sur,* **5**, 139–146 (1976).

14 J. Jusakos and P. Contreras, *Proyecto Jurel. Informe Global, Junio 1975–Julio 1977,* Departamento de Pesquerías, Universidad del Norte, Antofagasta, 1977, p. 42.

15 Ibid, pp. 1–2.

16 Ibid, p. 17.

17 Ibid, p. 16.

18 División de Pesca y Caza, Servicio Agrícola Ganadero, *Anuario Estadístico de Pesca, 1971,* Ministerio de Agricultura, Santiago, 1972, p. 60. (See also S.E.R.P.L.A.C. report.)

19 División de Protección Pesquera, Servicio Agrícola y Ganadero, *Anuario Estadístico de Pesca, 1976,* Ministerio de Agricultura, Santiago, 1977, p. XVIII.

20 División de Protección Pesquera, pp. XVII–XVIII.

21 S.E.R.P.L.A.C. *Diagnóstico visual del sector pesquero de la Primera Región,* S.E.R.P.L.A.C., Iquique, 1976 (Report for internal circulation.)

CHAPTER 13

Panel of Experts' Report (1970) on the Economic Effects of Alternative Regulatory Measures in the Peruvian Anchoveta Fishery*

INSTITUTO DEL MAR DEL PERU

Editor's Abstract

This report was written at the end of 1969 by a panel of fisheries economists convened by IMARPE. The authors state their awareness that pelagic species such as the anchoveta seemed particularly susceptible to reduction of fish stocks as a result, in part, of excessive, uncontrolled fishing. They call for strict control on the amount of fishing, given the best available scientific advice. The report, alluding to the 1969–1970 anchoveta catches, notes that some of that excess was allowed in order to meet economic difficulties that existed at that time in certain sectors of the industry. Several constraints on the maintenance of a healthy fishery, such as excess fleet and processing plant capacity, are discussed in some detail with suggestions for methods for the reduction of fleet and plant capacities, including the expansion of the food fish industry. Brief reference is made to the possible impact of intensive fishing pressures on the guano birds' primary source of food.

THE ANCHOVETA STOCK

Scientific studies by the Instituto del Mar, the FAO Stock Assessment Panel and others have shown that the anchoveta stock can, with proper fishery management, sustain an annual catch of around 9.5 million tons (the estimates range from as low as 8.5 million tons to as high as 10.5

---

*Reprinted with permission from the Instituto del Mar del Peru, Lima, Peru.

million tons). With efficient handling and processing methods, these catches could produce about 2 million tons of fish meal, worth at present prices ($170 per ton) about $350 million.

The first objective of any system of regulation and management of the anchoveta fishery must be to ensure the continuation of this substantial contribution to the Peruvian economy. The second should be to obtain this harvest without excessive cost and to arrange that the benefits from this fishery in terms of employment, profits, etc., are distributed in the most desirable manner.

Experience in other fisheries has shown that excessive, uncontrolled fishing can reduce fish stocks to a level at which only very small catches can be taken. Fisheries on pelagic species such as the anchoveta seem to be particularly liable to these effects. Strict control of the amount of fishing, based on the best available scientific advice, is therefore essential to maintain the prosperity of the industry.

In the 1969–1970 fishing season, the anchoveta catch of 11 million tons was about 1.5 million tons in excess of the estimated greatest sustainable catch referred to above. This is some three times greater than the excess of half a million tons suggested by the Stock Assessment Panel as reasonable for a trial period (on the understanding that it would be stopped or reversed as soon as any danger signals appeared).

It is highly probable, therefore, that unless the year-class recruiting to the exploited stock in the second half of the 1969–1970 season (i.e., at the beginning of 1970) was an unusually strong one, this large catch will have reduced the stock, leaving relatively few survivors from the 1970 recruitment to support the fishery at the beginning of the 1970–1971 season; that is, in September–December 1970, and leaving a reduced spawning stock, which may give more recruitment later in that season, that is, at the beginning of 1971. This means that the allowable catch during the 1970–1971 season will probably have to be set below 9.5 million tons and even more certainly below 11 million tons. Some of the excess catch allowed in the 1969–1970 season (particularly the extra 300,000 tons for small factories) was to meet economic difficulties in certain sections of the industry. However, such action does not help to provide a long-term solution to these difficulties; indeed, if they are not solved, it may be increasingly difficult to apply sufficiently drastic restrictions if these should appear necessary to safeguard the anchoveta stock. For instance, if the recruitment at the beginning of 1971 is average, the allowable catch for the 1970–1971 season might be set at 9 million tons (i.e., below 9.5 million tons because the 1970 recruits were heavily fished in the first half of 1970); such a limit could be accepted by most of the industry, without much difficulty. However, if, as is possible, the 1971 recruitment is poor, then the allowable catch in 1970–1971 might have to be much lower. The industry would be able to operate without distress under such a catch limit only if its present economic condition is improved.

Changes in the Amount of Fishing

In this report, the limit to the amount of fishing that is desirable on biological grounds is generally expressed in terms of the catch (the sustainable yield). A better measure would be directly in terms of the fishing effort. Although the best measure of fishing effort, taking into account the effect of technological changes, has not been precisely determined, it appears likely that the true fishing effort has been increasing in recent years, and the dangers of allowing the fishing effort to increase beyond the level giving the maximum sustainable yield have already been pointed out. In fact, there are good reasons for suggesting that a level rather lower than that giving the maximum sustainable would be an even better objective. First, since other things being equal, the cost of fishing is roughly proportional to the fishing effort, a 20% reduction in effort below that giving the maximum sustainable yield should enable costs to be reduced by this amount or more, but, because catch changes little with changes of fishing effort in its vicinity, would be likely to reduce yield by only 5% or less. Second, increases in fishing cause changes in the pattern of fishing; as fishing intensity increases the abundance of the larger, older fish decreases and the catches contain increasing proportions of small fish, which are taken progressively earlier in the first year of life. This is shown by the following recent events in the anchoveta fishery:

1 Between 1962 and 1968 the proportion of recruits less than 12 cm in length in the catches taken in the months January to March (the main recruitment season) increased from 14 to 62%. Detailed data are not yet available for later years, but it is known that the catch in March 1969 was a record high, which was broken by even bigger catches in January and March 1970, which might imply an even higher proportion of small fish in these years;

2 In each of the months of September, October, and November 1969 (i.e., the season when the older, larger fish predominate in the stock), the catch was lower than in any full month of fishing since 1965.

Such uneven distribution of fishing within the season is clearly undesirable on practical grounds, especially in requiring, together with the
shortening of the fishing season described in a later section, an unneces-
sarily high capacity in vessels and processing plants to deal with the
comparatively short period of peak landings. The concentration of
catches in a very short period also makes the proper scientific measure-
ment of the stocks more difficult. The allowable catch set for a particular
season should take into account, among other factors, the strength of the
year-class recruiting during that season. This cannot be determined even
approximately until about March–April which, even with the present
length of season, gives little time for making necessary adjustments to
the allowable catch and the closing date for the season. With a seasonable
pattern of fishing like that taking place in 1969–1970, when at the end of
the season some 1% or more of the allowable catch was being taken
each day, differences in the closing date of only a few days can make big
differences to the stock.

Protection of Small Fish

The Stock Assessment Panel proposed a closed season at the beginning
of the year as a means for protecting the small fish. While supporting
this view, the present Panel notes that this period is normally one of high
catches. Therefore, the closure at this time tends to reduce the efficiency
of the fishery, and the possibility of other measures to protect the small
fish should be considered. Some of these were described to the Panel,
but none seemed to be immediately acceptable. For instance, closure of
certain sea areas or certain ports, rather than the fishery as a whole
would, in principle, offer more discriminatory protection of the *peladilla*.
It would, however, be difficult to enforce, and without proper enforcement would be largely ineffective.

Prohibition of landing catches containing more than a certain percentage of small fish was also mentioned. So long as it remains difficult for
the fisherman to determine the sizes of fish in a shoal before setting his
net, it seems that such a measure would chiefly increase the wastage
through the dumping of small fish at sea. However, in the early years of
the fishery, the absolute (as well as the relative) numbers of small fish in
the catches were low, suggesting that the fishermen can achieve some
degree of discrimination between large and small fish, e.g., by going to
areas where larger fish had been caught on previous days. The Panel
believes that if more were known about the relative distribution of
different sizes of fish, it might be possible to frame regulations that
protect the small fish without interfering with the efficient exploitation of

the larger fish in the early part of the year. Studies on this distribution should, therefore, be made, for example, by intensive sampling of catches, together with detailed records of when and where each catch was taken.

The Panel also noted that a closure in the summer (January–March) period has only become necessary in the last few years, during which the proportion of small fish taken in these months has greatly increased. For instance, in 1967 Schaefer made calculations up to 1966, which suggested that at that time there would have been no benefit from any closure. This gives another reason for proposing that an ultimate objective of management might be to reduce the effective fishing effort as discussed in the previous section.

Catch Statistics

The statistics of catches or sustainable yield quoted above are in terms of reported "nominal catch." It has been noted by the Stock Assessment Panel (and reported to the present Panel) that the actual catches taken by the fishery exceed the reported catch, perhaps by upward of 20%. Although this underreporting does not affect the conclusions of either Panel, provided a consistent use is made of the "nominal catch" statistics, its existence must make the enforcement of some otherwise desirable forms of management difficult. It is highy desirable that reliable statistics of the actual catch should become available. It should also be noted that this relatively high degree of underreporting indicates a substantial underutilization of fish removed from the sea. This is discussed in a later section in relation to increasing the efficiency of the plants.

THE PRESENT MANAGEMENT OF THE FISHERY

Since 1965 the regulations in the anchoveta fishery have been aimed, by setting an annual total allowable catch, at restricting the annual catch to the annual removal that the stock can withstand as estimated by the scientists. With some reservations concerning the 1969–1970 season (noted in the previous section), these regulations have been successful in achieving this aim. The measures used have included:

1 A restriction of fishing to 5 days per week.
2 A closed season (usually lasting about a month) in the period January–March when small fish are abundant.

3   A closed season later in the year, the starting date of which is set
at the time when the allowable catch is reached.

In addition, the number of licenses to operate processing plants has
been limited, but this has had no effect on controlling the amount of
fishing.

While these regulations have been imposed over the major parts of the
fishery based on the ports in the central and northern parts of the
country, the much smaller fishery in the southern part has been exempt
from some of them. In particular, fishing has been allowed out of Ilo at
times when ports elsewhere have been closed. Since there is evidence
that the anchoveta stock in this area may be separate fom those
elsewhere, such differential regulations are not unreasonable, but they
have resulted in some years in the desirable allowable catch for the
fishery as a whole being exceeded. This is because the closing date for
the season has been set when the total catch has reached this figure,
rather than at a somewhat earlier date to allow for the estimated catch
taken at Ilo.

As indicated above, in the 1969–1970 season the total catch exceeded
by 1.5 million tons the estimate of maximum sustainable yield of 9.5
million tons estimated by the Stock Assessment Panel. This suggests that
under the present system of management, it is becoming difficult to keep
the amount of fishing within the limits necessary to ensure the long-term
well-being of the industry. The pressures to exceed these limits are likely
to become greater in the future. It is estimated that the fleet likely to be
operating in the 1970–1971 season will, if conditions are average, reach
a catch of 9.5 million tons in 130–133 days. This implies that, if there is
a 2 months closure within the period December–March to protect the
small fish, the season should close (if the quota is set at 9.5 million tons)
in the middle of May. This would mean a total closure of 5½ months
between 1 September and 31 August 1971. Such a long closure and short
fishing season are likely to create problems for at least the marginal
operators, and result in considerable pressure to extend the season.

This may be optimistic, since it assumes that the state of the fish
stocks will permit a catch of 9.5 million tons. The excessive catch in
1969–1970 may well mean that the allowable catch will be less than 9.5
million tons. In any case, it is certain that catches consistently in excess
of the sustainable yield will quickly reduce the stock to a point where,
even without any restrictions in fishing, only a small catch could be
taken, and very severe restrictions in fishing would be essential to rebuild
the stock. If this should occur at a time when part of the industry is still

in financial difficulties, and when fish meal prices are high, it may be very difficult for the Government to maintain the necessary restrictions. Under these circumstances, which could arise in the next few years under the present system of management, the collapse of the anchoveta stock, and the industry based on it, could easily occur.

PLANT AND FLEET CAPACITY

It is clear, therefore, that further measures to regulate the anchoveta fishery are necessary, both to safeguard the future of the resource and to increase its economic efficiency, especially by reducing costs. The potential reduction in costs, if the economic optimum were achieved, is large and can be estimated. The optimum could be achieved if the catching and processing capacities worked at full efficiency throughout the year (i.e., with no closed season, a 7-day working week, etc.). As shown in detail in later sections of this report, under unrestricted, year-round fishing both fleet and processing plant capacity could be roughly halved, implying an approximate halving of the fixed costs of each of them. At present, each accounts for about 25% of the total costs of the industry, so that, if the economic optimum system of management were achieved, total costs could be reduced by about a quarter, that is, by $50 million per year. The distribution of this benefit between fishermen, shareholders, and government could depend in part, on the method used to reach the optimum but could of course be changed by fiscal policy. Social and other considerations may, however, mean that the objective of management should not be to achieve maximum monetary gains, but to distribute the potential benefits in socially desirable ways. Nevertheless, as the status quo will not maintain itself and as new regulations are necessary to safeguard the future of the industry, a favorable opportunity presents itself to achieve substantial gains by taking effective action now, and to distribute them in accordance with the aims of government policy as a matter of deliberate choice.

The Problem of Excess Processing Plant Capacity

The present plant capacity is substantially in excess of that required to process the recommended allowable catch of 9.5 million tons per year, or any greater volume that is likely to be recommended in the foreseeable future in the light of better knowledge of stock size or of any increase in stock abundance that may occur. It is especially so if, as a result of the recent high fishing intensity, it is necessary to reduce the allowable catch below 9.5 million tons for 1 or more years. This is shown by the figures in Table 1, resulting from the Panel's calculations for different types of regulatory systems. The calculations show that, even for the 5-day fishing week and 8-month year, between a quarter and a third of the licensed plant capacity is surplus.

Surplus processing capacity encourages plant owners to expand their fleets in order to obtain larger shares of the allowable catch, thereby permitting them to improve profitability through the wider spread of plant overhead and fixed costs that higher plant utilization brings with it. However, while such a course of action may prove economically beneficial for those expenses, at least in the short term, the economic conditions of the industry generally will have deteriorated. This is because the introduction of new capacity is not, in the short-term at least, accompanied by the elimination of a comparable amount of less efficient capacity (provided only that fish meal prices are not unduly depressed).

The failure of market forces to make an adjustment to restore equilibrium lies in the low proportion of total costs (and of fish meal prices even when depressed) accounted for by direct or variable costs. This, in fact, applies to both the processing and catching sides of the industry. This is true even of processing plants (and also fishing vessels) of relatively low efficiency. In the short-term, only direct costs must be covered for a plant (or a vessel) to continue in operation, and, given the durability of the capital equipment, the short term can run into, chronologically, an extended period, particularly if the industry's creditors are patient.

Largely because of differences in efficiency, there are wide differences between processing plants in the direct cost per ton of fishmeal produced (and between vessels in the direct cost per ton of fish landed). Although the average direct cost of any given plant (or fishing vessel) varies with its degree of utilization, such variations in direct costs are, in general, much less than the differences between plants (and between vessels) at any given level of total plant (or fleet) capacity. Consequently, reductions in fish meal prices would be far more effective in squeezing capacity out of commission than the same proportionate reductions in the degree of utilization of total plant (and/or vessel) capacity, resulting, for instance, from expansion of capacity in the face of an unchanged catch limit. Since the new arrangements in Peru for the marketing of fish meal have brought greater stability in fish meal prices to Peruvian producers, the squeezing of capacity as a direct consequence of price reduction is likely to diminish

TABLE 1. Factor Capacity (tons/hour) Necessary to Handle the Anchoveta Catch Under Different Conditions

Regulations in Force	Based on the Maximum Possible Performance		Based on Observed Performance of Factories			
			Processing More than 10,000 Tons Meal per Year		Processing More than 15,000 Tons Meal per Year	
	tons/hour	%	tons/hour	%	tons/hour	%
No veda	3600	45	—	—	—	—
2 months' summer veda; 7-day week	3900	49	4830–5440	60–68	4480–5040	44–63
4 months' veda; 7-day week	4300	54				
4 months' veda; 5-day week	5100	64	6040	75	5600	70

in the future, thereby maintaining the condition of excess capacity in the absence of a deliberate policy of capacity reduction.

Just as increases in processing capacity stimulate increases in fishing fleet capacity so also increases in fleet capacity stimulate increases in plant capacity but, because of the structure and orientation of the Peruvian fishing industry, the strength of the latter is much less than the former. Nevertheless, there is a feedback, which in the absence of direct intervention of some sort, tends to induce a progressive deterioration, or, at best, a highly unstable equilibrium. Furthermore, the greater the surplus of capacity within the industry generally, the greater the pressure for increases in the annual allowable catch. Thus, in years of good fishing and/or high fish meal prices, requests will understandably be made for fishing to continue after the catch limit has been reached, to avoid wasting the opportunities provided by the favorable conditions; also in years of poor fishing and/or low prices, different but equally understandable requests can be made for continuation to enable the industry to cover its losses. Once the allowable catch as recommended by the scientists has been exceeded as a result of such pressures, a precedent has been set for similar steps in later years.

The growth in plant capacity in Peru was halted by the introduction of licensing but not before a substantial excess capacity had been created. Similarly as regards fleet capacity, if the recently announced commitment to the principle of limiting it is implemented and if it is limited to that which already exists and is on order (and the Panel foresees difficulty in achieving this), there would still be a substantial excess in fleet capacity. Even if it was considered expedient to forego the cost saving derived from a seven-day week, and say, a 10-month fishing season, and continue to operate the industry within its present framework of a 5-day week and an 8-month season, excess capacity in both plant and fleet would remain significant, at about 25% in both plants and vessels.

The core of the industry's problems undoubtedly lies in excessive plant capacity: it was on this side of the industry from which the impetus to expansion arose. It is recognized that the capacity problem cannot be effectively tackled without, at the same time, dealing with the burden of debt associated with it. The Panel understands the 2nd Commission has determined that, apart from stock financing, the written-down value of the industry's fixed assets is very little more than the industry's short-term debts, and only two-thirds of its short and long-term debts taken together. Therefore, although some reports indicate the balance sheets of many companies are in a perfectly healthy state, it seems clear that many others must be totally insolvent, although some of them may be operating at a profit at the present time.

The 2nd Commission has rightly placed heavy emphasis on improvements in technical efficiency on both the catching and processing sides of the industry, as well as on restructuring the industry's debt. Improvements in technical efficiency together with some alleviation of the pressure of the short-term debt (whether by mean of further moratoria, conversion into long-term debt, increased share capital, debt cancellation or any other means) will obviously help matters a great deal. Indeed, technical improvements which would raise processing yields appear to offer the greatest source of gains to the industry at the present time. The Panel believes, therefore, that the Commission is right in stressing the need for technical improvement and for greater long-term credit facilities for this purpose, among other things. In view of this, it is tempting to conclude, given the maintenance of present restrictions on the total catch and on plant capacity and the introduction of a scheme to restrict expansion in catching capacity, that the continuation of firm fish meal prices will produce profit margins that will be sufficient to generate the finance needed to return the industry to health on capital as well as on current account.

It must be stressed, however, that, while such technical improvements would benefit the financial situation, they would inevitably increase the temptation and pressure to exceed the allowable catch specified by the scientists. It is essential, therefore, that technical and financial improvements be accompanied by measures to reduce significantly the excess capacity in both plants and fleets, thus improving economic efficiency to ensure that the economic benefits do not fall too far short of what is attainable and to avoid the risk of a collapse of the resource. If this is not done, the ability of the industry to provide greater rewards for its workers, owners, and the state, will be much less than it is capable of being, as will its ability to keep in the van of technological progress, to provide knowledge, skills, and capital for the development of ancillary industries (for example, the food fish industry), and hence to add to the dynamism of the Peruvian economy generally. It will also reduce the industry's ability to withstand financial setbacks, resulting from a drop in fish meal prices or other causes.

Technical efficiency and the proper structuring of the industry's capital are important but are not enough; a high standard of economic efficiency is also needed if the industry is to have the strength, not only to meet years of adversity in fishing and in international marketing conditions, but also to make, if not the maximum, a high contribution to the welfare of the Peruvian people. In a world short of capital, it is particularly important for a country as hungry for development as Peru to ensure that progress is not retarded by a misallocation of the capital that is available. As already stated, on the Panel's calculations, the excess capacity in the industry's fixed equipment is of the order of about 25%. Since the original cost of these assets was not far short of $250 million, the capital denied to other parts of the economy may therefore be said to be about $60 million, which could be employed to add say, $20 million a year to the Peruvian national income. These are conservative estimates.

There are other consequential advantages of increased economic efficiency. First and foremost, an economically healthier industry will utilize its raw material more efficiently; with an input of 9.5 million tons, every 1% improvement in fish meal yield would add about 95,000 tons to annual fish meal output without any change in input or, at a conservative price of $150 per ton, $14.25 million to its value. The average fish meal yield in 1969 was about 17.5%, with individual plant yields varying from less than 15 to just over 21% (as Table 2 shows). It is not too optimistic to aim at raising the average yield to, say 18.8% (the average of the 52 plants with the highest yields) for all plants. If this were done, the annual fish meal output, with input maintained at 9.5 million tons, would be increased by 123,500 tons, and value of output (at $150 per ton) by $18.5 million annually. Improved yields of oil would add further to the value of output and the potential here is relatively much greater, perhaps three times as great, but the same proportionate improvement as in fish meal would produce (even at prices well below prevailing levels) nearly $5 million a year on this account. Thus, these two sets of yield improvements together would add $23.5 million to the annual value of the industry's output. The degree of improvement indicated above, especially in fish meal yield, is very moderate and gross gains in the value of output nearer $50 million are attainable. The scope for improvement in the yield of fish meal is especially fruitful (especially if inputs have been underrecorded so that the real yields are lower than those quoted here). The net gain would, of course, be somewhat smaller, as higher yields would bring some additional costs.

The whole of these gains cannot be said to stem directly from the concentration of production into fewer, more efficient plants; a substantial part of them comes from raising the technical efficiency of plants generally, whatever their number (and hence is in principle attainable without the elimination of any excess capacity). However, if a reduction of licensed capacity was applied by closing predominantly factories with relatively low outputs, having average yields of about 16%, as economic criteria would appear to demand, then the diversion of input to other factories would amount to between 10 and 15% of the total (see Tables 2 and 3), say 12.5% or about 1.2 million tons out of 9.5 million tons. If their throughput were processed by factories averaging the 18.8% re-

TABLE 2. Number, Output, and Licensed Capacity of Plants in Relation to Yield (Output of Meal as a Percentage of Recorded Input of Fish) in the 12-Month Period that Ended December 31, 1969

Percentage Yield	Plants[a]			Output			Licensed Capacity		
	No.	% of Total	Cum. %	'000 Tons	% of Total	Cum. %	Tons/hr.	% of Total	Cum. %
Under 15.0	6	4.7	4.7	7.0	0.4	0.4	269	3.4	3.4
15.0–15.4	3	2.4	7.1	14.0	0.9	1.3	185	2.3	5.7
15.5–15.9	8	6.3	13.4	67.6	4.2	5.5	502	6.3	12.0
16.0–16.4	8	6.3	19.7	67.4	4.2	9.7	472	5.9	17.9
16.5–16.9	13	10.2	29.9	167.5	10.4	20.1	886	11.1	29.0
17.0–17.4	19	15.0	44.9	260.3	16.2	36.3	1109	13.9	42.9
17.5–17.9	20	15.7	60.6	296.5	18.5	54.8	1385	17.4	60.3
18.0–18.4	7	5.5	66.1	120.3	7.5	62.3	538	6.8	67.1
18.5–18.9	7	5.5	71.6	139.3	8.7	71.0	442	5.6	72.7
19.0–19.4	5	3.9	75.5	92.4	5.8	76.8	414	5.2	77.9
19.5–19.9	4	3.1	78.6	171.2	10.7	87.5	506	6.4	84.3
20.0–20.4	3	2.4	81.0	71.0	4.4	91.9	260	3.3	87.6
20.5–20.9	3	2.4	83.4	61.3	3.8	95.7	188	2.4	90.0
21.0 and over	3	2.4	85.8	56.0	3.5	99.2	192	2.4	92.4
Subtotal	109	85.8	85.8	1591.8	99.2	99.2	7348	92.4	92.4
Not known	10	7.8	93.6	11.9	0.7	99.9	337	4.2	96.6
Not working	8	6.3	99.9	—	—	99.9	277	3.5	100.1
Total	127	99.9	99.9	1603.7	99.9	99.9	7962	100.1	100.1

[a]Includes any plant registered in 1969.

381

TABLE 3. Number, Capacity, and Production of Plants in Relation to Annual Production of Meal During 1969

Production '000 Tons Meal	Plants			Output			Licensed Capacity		
	No.	% of Total	Cum. %	'000 Tons	% of Total	Cum. %	Tons/hour	% of Total	Cum. %
Nil	9	7.1	7.1	—	0	0	350	4.3	4.3
0.1– 1.9	20	15.6	23.4	18	1.2	1.2	866	10.8	15.1
2.0– 3.9	6	4.7	28.1	21	1.3	2.3	370	4.6	19.7
4.0– 5.9	6	4.7	32.8	31	1.9	4.2	278	3.4	23.1
6.0– 7.9	16	12.5	45.3	111	6.9	11.1	765	9.5	32.6
8.0– 9.9	8	6.2	51.5	71	4.4	15.5	367	4.5	37.1
10.0–11.9	12	9.4	60.9	133	8.3	23.8	725	9.0	46.1
12.0–13.9	5	3.9	64.8	65	4.1	27.9	358	4.4	50.5
14.0–15.9	3	2.3	67.1	45	2.8	30.7	172	2.1	52.6
16.0–17.9	11	8.6	75.7	185	11.5	42.2	774	9.6	62.2
18.0–19.9	5	3.9	79.6	96	6.0	48.2	370	4.6	66.8
20.0–21.9	4	3.1	82.7	82	5.1	53.3	269	3.3	70.1
22.0–23.9	4	3.1	85.8	93	5.8	59.1	348	4.3	74.4
24.0–25.9	2	1.6	87.4	50	3.1	62.2	200	2.5	76.9
26.0–27.9	2	1.6	89.0	55	3.4	65.6	168	2.1	79.0
28.0–29.9	5	3.9	92.9	143	9.0	74.6	432	5.3	84.3
30.0–39.9	5	3.9	96.8	168	10.5	85.1	581	7.2	91.5
40.0–49.9	4	3.1	99.9	179	11.2	96.3	487	5.9	97.4
Over 50	1	0.8	100.7	59	3.7	100.0	207	2.5	99.9
Total	128	100.0	100.0	1602	100.0	100.0	8087	99.9	99.9

ferred to above, then fishmeal output would rise by 33,600 tons, which at $150 per ton, would amount to $5 million. A corresponding improvement in oil yields would bring the gain up to about $6.33 million, the most that can be expected as a direct result of reducing licensed plant capacity by 25%. It would, however, greatly improve the climate within the industry, by creating the attitudes necessary for progressive technical and economic improvement. Furthermore, the indirect effects of eliminating excess capacity (e.g., by removing the temptation to exceed the allowable catch), are highly important, but the Panel thinks it right to refrain from placing magnitudes upon them and, hence, from bringing them into the narrow reckoning of advantages and disadvantages.

Any reduction of plant and vessel capacity could be associated with a redistribution of plants to make the geographical pattern of processing capacity fit more closely the pattern of availability of fish on the grounds. This needs examining in the light of present and prospective shipping facilities and of other factors, none of which the Panel had time to study. But the Panel is confident, from what it has seen and heard, that there are substantial economies to be obtained in this direction. Further substantial gains seem possible by utilizing new unloading methods which would reduce the waste of both fish and free floating fish oil during discharge from vessels. However, as the Panel had no time to investigate these aspects of the matter, no magnitudes are placed on the potential gains obtainable from these. Nevertheless, they are sufficiently important not to be overlooked.

As already indicated, however, whatever the prospects in these directions, significant gains can be derived from reduction in plant capacity through savings in the costs of processing the permitted catch. The Panel understands that the average cost of production of the industry generally is at present about $120 per ton of fish meal, which, however, is subject to wide variation. The concentration of output in fewer plants would of itself bring substantial savings almost regardless of whether the plants enjoying the higher production were the most efficient producers. The total costs of the plants closed down would be saved at the expense only of the additional (i.e., variable) costs of processing the throughput transferred to the retained plants. (In other words, extremes apart, the average costs of almost any plant are nearly bound to be higher than the marginal costs of almost any other plant.) Quite clearly, however, the magnitude of the saving would depend on the extent to which the eliminated capacity were confined to high cost producers.

As indicated above, if the savings were predominantly to be found among plants with low outputs then a 25% reduction in licensed capacity would affect nearly 1.2 million tons of raw material. From the figures available, to the Panel, it seems that the average cost per ton of fishmeal produced by these plants is at least one sixth higher than that for the industry generally, say $140 for each of the 190,000 tons produced by them (their average yield being 16%, giving a total cost of $26.6 million). If the 1.2 million tons of raw material were processed by plants enjoying the 18.8% yields referred to earlier, then their output would increase by about 225,000 tons. The variable costs of these plants would probably be about $80 a ton, so that the additional costs of producing this further 225,000 tons of fish meal would be about $18 million, or $9 million less than the cost incurred by the plants closed down. It follows, therefore, that with the additional value of $6.33 million of output arising from improved yields of fish meal and oil, there would be a gain in profit of over $15 million a year, following a 25% reduction of licensed capacity, which accounted for 12.5% of output. This, viewed conservatively, represents more than a 25% increase in total annual profits over the level estimated for the present arrangements (i.e., 9.5 million tons of fish processed with a yield of 17.5% to produce 1,622,500 tons of meal at a cost of, say, $120 per ton and sold at $150, would produce a profit of $30 a ton or almost $50 million in total).

The Problem of Excess Fleet Capacity

As indicated previously, the Panel believes that the present capacity of the anchoveta fleet is too large and is still increasing rapidly in size. Unless it is reduced, serious problems will arise in the near future. In fact the fishery seems to be following the pattern of other fisheries (e.g., those for salmon and halibut in the north Pacific, whales in the Antarctic and yellowfin in the eastern tropical Pacific) in which the main limitation on excess fishing has been by a closed season, where the fishing season became progressively shorter. Using reasonable estimates of new vessel construction, the fleet in 1970–1971 would, under average conditions, be able to take 9.5 million tons in 130–133 days, which would amount to a 6.5 month fishing season. Without additional controls, future seasons would become even shorter, since anyone wishing to keep his plants in full use during the fishing season must increase the number of vessels, or increase the fishing capacity of the existing ones, to maintain his share of the limited catch.

The degree of excess fleet capacity will depend on the type and amount of controls imposed on the fishery. The most effective use of the fleet will be made, and the fish harvested at least cost, if the controls are kept to a minimum, that is, with a closed season of 1–2 months in the period

December–March, no forced closure in the middle of the year, and no control on the number of days fishing per week. However, there are social benefits to be gained from some of these additional controls, and a pattern of an 8-month fishing season and 5-days fishing per week seems to have wide support in the industry. Most calculations on the desirable size of fleet have therefore been made on this basis. They show that the capacity of the fleet existing in June 1970 could be reduced by some 20–30%.

Such a reduction would clearly be of benefit to the industry in reducing costs. The reduction would be equal to the total costs of the vessels leaving the industry, less the additional variable costs incurred by the remaining vessels, since these would rather spend more time at sea and hence handle more fish.

If the cost structure of the fleet were uniform (i.e., the variable costs were the same for all vessels), the reduction in costs would be equal to the fixed costs of the sampled vessels. Data for a sample of vessels, examined by the Panel, suggest that the fixed costs for the fleet as a whole (i.e., depreciation, insurance, annual refit, etc.) constitute some 47% of the total, which amounts to about $45–50 million. The removal of 20–30% of the fleet would, therefore, imply a saving of $9–15 million. This, however, is likely to be a conservative estimate because:

1   It represents the saving based on the fleet situation in 1970 instead of, more appropriately, in 1971 or 1972 when, if no control on fleet size is imposed, the excess capacity will be greater.
2   It is likely that the variable costs per ton of fish landed will be lower for the vessels remaining in the fleet than for those removed.

In addition, fleet size and fixed costs could be still further reduced if, instead of a 5-day week, 8-month season fishery, fishing were allowed over a longer period, for example, a 7-day week and a 10-month season.

It must be emphasized that in implementing a policy of fleet reduction, in the long-term interests and operating efficiency of the industry, it is necessary to allow the replacement of old and inefficient vessels by newer and more efficient ones and to encourage the use of new technological advances. This could be readily achieved without increasing the fishing capacity above the desired level if an appropriate tonnage of old vessels were scrapped for each new vessel constructed. The amount of tonnage to be scrapped for any given number of new vessels constructed can be determined reasonably reliably from data on the fishing capacities of vessels of different sizes in the fleet. Relevant data for the 1968–1969 season of the average catch during that season for vessels in successive size (hold capacity) categories, expressed as the catch per ton of hold capacity, are given in Table 4. They show that this index increases fairly steadily with increase in vessel size up to about 200 tons capacity, and thereafter decreases. This decrease for the largest vessels is mainly due to the decrease in the average number of trips made by them; many of them were completed during the 1968–1969 season and fished for only part of it. A better measure of the relative capacity of these vessels is therefore the catch-per-ton of hold capacity per trip, which shows less difference between the 200 and 300 ton vessels.

The scrapping ratio for a given size of vessel is then given by the ratio of the catch-per-ton hold capacity of the old vessels likely to be scrapped (mainly in the 70–199 ton range) to that of vessels of other sizes. These ratios are given in the last column of Table 4, except for the largest vessels, which did not fish a full season; the values for vessels over 160 tons hold capacity are in the range 1.2–1.4 and average approximately 1.3.

Thus, if for every ton of new capacity, assumed to be vessels over 200 tons, 1.3 tons of vessels under 120 tons were scrapped, the total catching capacity would be the same as if there had been no new construction and the old vessels had continued fishing at the same level of efficiency. The real fishing capacity of the fleet, as a whole is, however, likely to increase from one season to the next with the introduction of new technological developments. This increase is difficult to estimate. The best available measure is the factor of 1.2 used by the Instituto del Mar to correct for the increase in efficiency (essentially the increase in catching capacity per gross ton of vessel) between 1960 and 1969. This period is about equal to the effective lifespan of a vessel, so that if the increase in efficiency is constant, a new vessel might be expected to have on the average, during its fishing life, a catching capacity some 20% in excess of that of the vessel it replaces, in addition to the effects of any difference in size. Thus, it seems likely, on the basis of the data available, that an increase in the scrapping ratio by 20% to 1.56 would be necessary to stabilize the effective catching capacity of the fleet, provided the pattern of scrapping were uniform. However, such a procedure would, in effect, tend to penalize that section of the fleet being scrapped and rebuilt on the basis of the presumed capacity of the rest of the fleet, and so might discourage new building. To avoid this, it might therefore be considered more desirable to correct for the effects of changes in efficiency by other means, for example, withdrawal of part of the fleet without replacement.

While, in theory, the desired reduction in fleet capacity could be achieved in a scrap and rebuild program by setting an artificially high

TABLE 4. Catching Capacity of Vessels in Relation to Hold Capacity During the 1968–1969 Season

Hold Capacity	Number of Vessels	Average Catch (tons)	Catch per Ton Hold Capacity (tons)	Average Number of Trips per Vessel	Catch per Ton Hold Capacity per Trip (tons)	Catch per Ton Hold Capacity Relative to That of 70–119 Ton Vessels
0– 69	7	2,349.8	39.16	61.4	0.638	0.818
70– 99	285	4,181.1	49.20	100.8	0.488	1.032
100–119	320	5,072.9	46.12	104.0	0.443	0.968
120–139	262	5,903.6	45.41	112.9	0.402	0.949
140–159	123	7,506.5	50.04	115.5	0.433	1.046
160–179	155	10,817.7	63.63	126.4	0.503	1.335
180–199	91	10,987.2	57.83	124.3	0.465	1.213
200–229	69	14,140.5	65.77	135.4	0.486	1.380
230–249	13	9,341.0	38.92	83.4	0.467	—
250–299	50	12,627.1	45.92	104.7	0.439	—
300	16	11,090.4	33.61	80.4	0.418	—

scrapping ratio (e.g., by requiring the scrapping of 600 tons of old tonnage for a license to build one new one of 300 tons) such a policy would distort the normal economic considerations used in planning new construction, and would be likely to induce delays in the scrapping of old vessels in the expectation that the ratio would be decreased as the reduction of the fleet capacity was achieved.

The Panel considers, therefore, that the approach to fleet reduction should be kept separate from the problem of establishing the scrapping ratio and that a withdrawal of vessels, over and above those involved in a scrap and rebuild program is necessary to achieve the desired reduction in total fleet capacity of up to 30%.

In relation to the problem of reducing fleet capacity, the Panel discussed briefly the question of the independent vessel operators in relation to those operated by factories. The Panel was not able to obtain sufficient data to determine the relative efficiency of the two groups of vessels. It appears that the factory fleets have advantages in terms of better financing, and often better maintenance, and supply of spare parts. There is also an advantage to the factory owners in providing some better guarantee of supply of raw material. However, in other fisheries, for example, the South African pilchard fishery, which has many features in common with the Peruvian anchoveta fishery, the independent vessels proved much more efficient than factory-owned vessels. This may be due to the greater interest in the vessel, its catches and its upkeep shown by the skipper-owner. As a result, South African factory owners encourage the purchase of vessels by independent owners.

Further study of the relative efficiencies of the different groups of vessels in the Peruvian fishery seems desirable. Until it is known what is the desirable composition of the fleet in Peru on economic or social criteria, that is, all factory-owned vessels, all independent vessels, or (as seems likely) some mixture of the two, it would seem undesirable to introduce regulations that either explicitly, or in effect, tended to eliminate the independent operators.

METHODS OF REDUCING CAPACITY

Reduction of Plant Capacity

Although a reduction of plant capacity should bring large economic benefits it may not be easy to implement. Considerations weighing against a deliberate policy of capacity reduction stem largely, if not wholly, from short-term rigidities. These tend to increase as wait-and-see attitudes are adopted and as jockeying for position takes place as soon as the possibility arises of any such policy being introduced. The rigidities are such that they virtually take out all hope of a voluntary scheme of capacity reduction. A compulsory element is essential.

Compulsion may be used in a variety of ways. The Panel sees no purpose in examining all possible ways of doing so since any scheme is in the last resort acceptable or unacceptable on grounds of equity, including social and political considerations on which the Panel is not competent to pronounce. No attempt is therefore made in this report to recommend the adoption of any specific measure but the principal elements of some possible ones, including some of their main advantages and disadvantages, are considered.

One possible indirect method of controlling, and, if set low enough, reducing plant capacity, not involving any formal compulsory scheme of capacity reduction would be by setting plant *input* quotas. Such quotas possess the advantage of giving each plant an added incentive to maximize the value of its output from each ton of input at the minimum cost; plant operators could be given a degree of certainty about the future which would permit them to plan their operations more efficiently; tendencies toward obtaining as much as possible as quickly as possible (with all that this implies for the length of the fishing and processing season) would be diminished and the administration of plant and vessel licensing or other regulatory measures might be avoided. Such a scheme does, however, have a number of disadvantages, as follows: (1) there are the initial problems of determining the quotas and, perhaps also, of their periodic revision; (2) detailed daily inspection systems would be needed if widespread opportunities for evasion were not to exist; (3) competition between plants would be virtually eliminated and this would adversely affect the dynamism of the industry; (4) the dangers of industrial stagnation would become very real unless there were to be periodic reviews of the quotas which in themselves would give rise to serious problems (apart from the fact that the greater the frequency of revision the greater the erosion of that element of certainty which many businessmen would regard as one of the primary virtues of a plant quota system, and the greater the attention that plant operators would give to the means of enlarging their quotas to the detriment of industrial efficiency—unless a method was determined of proportioning quotas to relative efficiencies); and (5) there would be a strong tendency for all plant owners (or if transfers were permitted, all companies) to proportion their plant and catching capacities to one another as well as both to their plant quotas. This would tend to promote the retention by some companies of excess fleet capacity to ensure that there would be available to them, in difficult fishing years, sufficient fishing capacity to catch all the fish to which the quotas entitled them at the times they could use it; other companies would also retain surplus fleet capacity in the hope that they would be able to purchase some of the plant quotas from others, perhaps those who could not use them as effectively. (Clearly, the disadvantages of plant quotas would be magnified if restrictions were placed on their transferability, though nontransferability would greatly reduce the problem of excess capacity.)

Despite these disadvantages, plant quotas offer a good prospect of eliminating the strong tendency toward continuing expansion of capacity and of achieving a fairly stable equilibrium in the total capacity of both plant and fleet. Although this might suggest that, with a plant input quota scheme, no restriction would need to be imposed on plants and vessels as such, there are dangers in leaving capacity free from all restrictions; for example, it may generate the requisition of plants and vessels at the outset in order to bring pressure for, or on the unwarranted expectation of, the granting of plant quotas.

An alternative scheme would be by setting plant *output* quotas, which would have effects very similar to those discussed above for input quotas. They would, however, have the advantage of very easy and simple enforcement, particularly where practically the whole of the output is marketed through a single channel as is planned for the future in Peru. On the other hand, in addition to the disadvantages described above for input quotas, output quotas might encourage wasteful use of raw material by individual plants and so lead to reduced production for the industry as a whole. It may be remembered that maximum yield only accidentally coincides with maximum profits. Since, by definition of a catch quota, the ultimate scarcity in this sphere is the raw material, this fact should be highlighted to ensure that it is treated with proper economic respect; an input quota at least does this, but an output quota does not.

As regards the implementation of a formal scheme involving a reduction in plant capacity, the Panel was impressed by the large measure of agreement it met among the producers on the merits of considering the imposition of a levy on the industry to purchase the excess capacity that it is required to withdraw. As indicated in a previous section, the excess capacity amounts to between one quarter and one third of the total, which means that a reduction in capacity of about 2000 tons per hour from the present 8000 tons per hour is required. If this amount of reduction was the aim, it would mean that the levy on the retained capacity would need to be three times the rate of compensation for the withdrawn capacity (whether the compensation covered the surrender of licenses only or the licenses and the plants with which they were associated is a question of subsidiary importance).

The Panel did not ascertain what an appropriate rate of compensation would be, but $12,000 per ton of licensed capacity may be used illustratively in the belief that this may be at, or even above, the upper limit of the range that would need to be considered. It follows that, at this price, the withdrawal of 2,000 tons of capacity would cost $24 million, or $4000 per ton of retained capacity. Alternatively, if the levy were imposed on the industry's output instead of its retained capacity (and assuming output to be 1.7 million tons a year), a levy of $14 per ton would raise the required sum in one year, while $7 per ton would raise it in two years and correspondingly lower rates if the yield rose beyond the 17.5% suggested elsewhere in this report. None of these seems excessive in relation to the improved margins that would be enjoyed by the retained capacity.

If such a scheme were to be introduced then the offer might well be made to all plant owners, but, as it is unlikely that voluntary acceptance would produce the required capacity, there would need to be compulsory purchases of all those plants that failed to satisfy some criterion or combination of criteria. These could include a minimum annual volume of output on average during, say, the last 3 years (e.g., in the year ending December 31, 1969, just over half the plants, together accounting for nearly 40% of licensed capacity but only 15% of total output, individually failed to produce 10,000 tons—see Table 3). As a plant with comparatively low output is not necessarily inefficient (there are many small plants that are run very efficiently), there could be an exemption for any plant that, over the same period, averaged some minimum annual output per ton of licensed capacity (e.g., in the year ending December 31, 1969, the range ran fairly smoothly from nil to about 400 tons, 42 plants failed to achieve 150 tons and together these accounted for 31% of capacity but only 10% of output—see Table 5). The Panel was tempted to suggest a third criterion, a minimum fish meal yield, but was given to understand that input figures are insufficiently reliable for the purpose. Perhaps, therefore, it would be preferable to grant an exemption to plants that only marginally failed to satisfy the second criterion, and which did not have stickwater plants for at least, say, half of the period concerned, on condition that such plants were installed by some specified date. It might also be useful to add the condition that all plants must possess approved weighing scales, which are subject to official periodic inspection (so as to improve the accuracy of input and hence yield figures, among other

TABLE 5. Number and Capacity of Plants in Relation to the Annual Output of Meal per Ton of Licensed Capacity

Output of Meal/ Ton Capacity	Plants			Capacity		
	Number	% of Total	Cum. %	Total	% of Total	Cum. %
Less than 50	17*	14.4	14.4	776	10.1	10.1
50– 99	9	7.6	22.0	642	8.4	18.5
100–149	16	13.6	35.6	862	11.2	29.7
150–199	16	13.6	49.2	1106	14.4	44.1
200–249	14	11.9	61.1	963	12.5	56.6
250–299	23	19.5	80.6	1602	20.9	77.5
300–349	11	9.3	89.9	815	10.6	88.1
350–399	5	4.2	94.1	439	5.7	93.8
700 and more	7	5.9	100.0	478	6.1	99.9
Total	118	100.0		7683	59.9	

*Not including nine plants without production and one for which data on capacity was not available.

things), as well as other conditions designed to raise efficiency and the quality of the industry's products. In an increasingly competitive world, high standards of quality are of growing importance.

Reduction of Fleet Capacity

The problems involved in controlling or achieving a reduction in fleet capacity are essentially the same as those for plant capacity and they can be tackled in a similar way.

One possible method is to restrict the carrying capacity of the fleet attached to any factory to a certain proportion of the processing capacity of the factory; a limit on the hold capacity of 1.4 times the licensed daily factory input was suggested to the Panel in discussion. While this could apply to individual factories or to all factories belonging to the same company or group taken together, its application to individual factories would prevent the deployment of fleets along different parts of the coast in accordance with the changing distribution of fish. This practice on a company or group basis, however, has already enabled some companies to markedly improve their operating efficiency and to reduce costs.

While such a scheme has merit, the restriction of all fleets to the same proportion of the processing capacity (even allowing movement between factories belonging to the same group) would seem to introduce an undesirable degree of rigidity into the fishery. The existing proportion varies quite widely between different plants (from under 0.50 to over 1.80; see Table 6); since some of this variability is undoubtedly due to real differences in the optimum fleet capacity in different ports, due to the pattern of distribution of the fish-distance from port, etc., the application of the same ratio to all parts would lead to over-capacity at some ports and undercapacity at others.

Furthermore, the situation under such a system, whereby the construction of vessels by some companies was prohibited while others were allowed or even encouraged to build, would seem likely to create difficulties of enforcement. For example, vessels might be built nominally to fish for one company, but actually fish for another. This difficulty might be overcome by allowing deliveries of fish only to the plant for which the vessel is licensed, but that would introduce a rigidity into the system that seems undesirable. The efficiency of the industry is undoubtedly increased by the ability of vessels to deliver to other factories when their own cannot accept fish, for example, due to temporary breakdown, or oversupply.

Another disadvantage is that the scheme offers no immediate promise of a reduction in the fleet. Rather, unless the qualifying ratio of fleet to

TABLE 6. Ratio of Fleet Hold Capacity to Daily (24 hour) Licensed Plant Capacity of Same Company (Data for April 1970)

Ratio of Hold Capacity of the Fleet to Daily Plant Capacity	Number of Companies (or groups)	Hold Capacity of the Fleet		
		Tons	%	Cum. %
Under 0.50	3	2,395	1.1	1.4
0.50–0.59	2	1,700	1.0	2.4
0.60–0.69	5	5,301	3.2	5.6
0.70–0.79	5	10,550	6.3	11.9
0.80–0.89	6	11,717	7.0	18.9
0.90–0.99	10	24,545	14.7	33.6
1.00–1.09	4	9,915	5.9	39.5
1.10–1.19	4	13,085	7.8	47.3
1.20–1.29	4	11,915	7.1	54.4
1.30–1.39	4	28,865	17.3	71.7
1.40–1.49	3	16,010	9.6	81.3
1.50–1.59	6	14,866	8.9	90.2
1.60–1.69	2	8,250	5.5	95.7
1.70–1.79	1	2,320	1.4	97.1
1.80 or above	3	4,540	2.7	99.8
Total	62[a]	166,974	99.8	

[a]Includes 99 plants.

plant capacity is set very low, the capacity will be encouraged to expand beyond the present excessive level through new construction by plants with ratios below the qualifying ratio. In this connection, the Panel noted that the ratio of 1.4 seems high. The data in Table 6 show that 72% of the fleet capacity belonged to companies having a ratio of 1.4 or lower. Thus a restriction to 1.4 would, initially, affect only 28% of the fleet.

A second possible method of control, closely linked with the system of factory input quotas, would be, instead of limiting the carrying capacity, to set catch quotas for individual vessels or groups of vessels. Vessels or groups of vessels could be licensed to catch a fixed tonnage of fish, or better still, since the total allowable catch can vary from year to year, a fixed percentage of the total allowable catch.

The advantages of a catch quota allocated in this way to vessels are several. First, if properly enforced, the limitation of total catch to the desired level is immediately achieved without the need for additional regulations such as an overall catch quota, or closed season. Second, within the set total for a group of vessels, each owner can adjust his operations, or size of fleet, type and size of vessel, etc., to give what he believes to be the most efficient system, giving the prescribed catch at the least cost. This eliminates the need for the complex system of scrapping ratios referred to earlier, and also should maintain the efficiency of the industry as a whole. Third, for those vessel operators who are also factory operators, the existence of a fairly well assured catch quota, and the associated reduced likelihood of surplus from other operators, reduces some, although not all, of the incentive to build up plant capacity to take advantages of any temporary peaks in catches.

Against this, the application of catch quotas to vessels, or groups of vessels, has disadvantages. The first, and most obvious, is the difficulty of determining the initial allocation. Whatever period is chosen some operators will have had, or may reasonably claim to have had, unusually poor success. Allocation purely on the basis of catches made during any one period will therefore be held by these operators to be unjust. (However, allocation on any other basis, for example, hold capacity, would be unjust in favoring those who in the past did not use this capacity to the full.)

An allocation scheme that virtually guarantees a given catch will also reduce some of the pressure toward greater efficiency, and will, in the long run, tend to reduce the competitive position of the industry in the world market. Both these disadvantages could be reduced if licenses (for the capture of a certain percentage of the annual quota) were freely transferable between companies. Then the more efficient operators, who make the best use of their licenses, would be in a position to offer the less efficient an attractive price for their quota. The possible dangers of concentrating the fleet in the hands of a near-monopoly could be avoided by making sales or transfer of licenses subject to government approval, or by the government acting as a broker, buying licenses from the less efficient, and selling to the more efficient.

A successful scheme of allocating catch quotas to vessels, or groups of vessels, would mean that the license for a given quota would have considerable capital value. That is, the benefit in reduced annual costs arising from the limitation in fleet size would accrue, in the form of capital appreciation, to those holding licenses. This would clearly improve the stability of the industry, and facilitate the repayment of some of the large outstanding debt. It would also be possible to divert a proportion of this benefit to other uses by charging license fees. For instance, if reduction of costs was 10% of the present fixed vessel costs (which could be achieved by a quite moderate decrease in fleet capacity), the savings might be around $5 million a year; a significant proportion of these could be diverted to uses other than the benefit of the vessel owners, by charging a license fee of, say, $2 to $3 per ton.

As mentioned above, the allocation of catch quotas to vessels, or groups of vessels, has very similar features to the allocation of input quotas to factories. The Panel believed that either or both systems could have considerable long-term advantages, but, as discussed in more detail in relation to factory input quotas, the practical problems of introducing such systems and enforcing them efficiently in the present situation in Peru are large. Therefore, while not suggesting a transferable catch or input quotas for the present, the Panel felt that they should be borne in mind for the future.

It should be emphasized that whichever method is used to control the fleet capacity, a reduction from its present level is required, which raises similar problems to those considered for the reduction of plant capacity.

One measure that would reduce fleet size, and that the Panel believes should be introduced for other reasons, would be to insist on statutory standards of seaworthiness for all vessels. They would be subject to annual surveys, and those failing to reach the necessary standard would have their licenses permanently rescinded. While this would remove a number of old, inefficient and unsafe vessels, further reduction would almost certainly be needed. This reduction would be in addition to the scrapping of vessels to qualify for new construction discussed in a previous section. Since any scrap and rebuild program would enhance the value of the old tonnage, a reduction program would require some degree of compulsion, compensation, or both.

The reduction program could be handled in the same way as was suggested for plants by the imposition of a levy on the vessels remaining in the fleet to provide compensation grants for the vessels withdrawn. These would be open to all, but compulsory for all vessels that failed to achieve a certain standard fishing performance. The levies imposed under such a scheme could be substantial since, at present, the total allowed catch is maintained at more or less a fixed level so that the removal of vessels from the fleet would mean that the total catch was distributed among the remaining vessels with little addition to their costs.

The Panel considers that the most suitable performance standard in such a scheme would be the tonnage landed per season. Relevant data are shown in Table 7. These show that a limit of 4000–4500 tons per year would be necessary to achieve the 25% reduction required to reach optimum fleet size operating 5 days per week over an 8-month season.

In the absence of detailed information on the pattern of costs and earnings of these marginal vessels, the Panel could not estimate a figure which would be a reasonable compensation for withdrawal. This figure

TABLE 7. Number and Total Landings of Anchoveta Vessels, as a Function of the Vessels' Landings During the 1968–1969 Season

Landings ('000 tons)	Number of Vessels			Total Landings			Hold Capacity		
	Number	%	Cum.%	Weight ('000 tons)	%	Cum.%	Tons	%	Cum.%
0– 1.5	153	10.7	10.7	72	0.7	0.7	15,934[a]	8.2	8.2
1.5– 3.0	94	6.7	17.4	213	2.1	2.8	1,640[a]	6.0	14.2
3.0– 4.5	141	10.0	27.4	530	5.3	8.1	15,782[a]	8.1	22.3
4.5– 6.0	231	16.3	43.7	1216	12.3	20.4	26,449	13.7	36.0
6.0– 7.5	247	17.4	61.1	1670	16.9	37.3	29,788	15.4	51.4
7.5– 9.0	174	12.1	73.2	1422	14.5	51.8	23,439	12.1	63.5
9.0–10.5	92	6.6	79.8	898	9.1	60.9	14,692	7.6	71.1
10.5–12.0	80	5.7	85.5	901	9.0	69.9	13,731	7.1	78.2
12.0–13.5	76	5.5	91.0	970	9.8	79.7	14,804	7.6	85.8
13.5–15.0	55	3.9	94.9	784	7.9	87.6	11,198	5.8	91.6
15.0–16.5	31	2.2	97.1	488	4.8	92.4	6,552	3.4	95.0
16.5–18.0	22	1.6	98.7	380	3.8	96.2	5,240	2.7	97.7
18.0–19.5	12	0.8	99.5	225	2.2	98.4	2,775	1.4	99.1
19.5–21.0	4	0.3	99.8	81	0.8	99.2	866	0.4	99.5
21.0–24.0	2	0.14	99.9	44	0.4	99.6	390	0.2	99.7
24.0–25.5	1	0.07	100.0	25	0.3	99.9	296	0.1	99.8
Total	1413			9918			193,576		

[a]With corrections for vessels whose capacity is not known.

should take into account the present value of the vessels, the increased value of a license caused by a scrap and rebuild program, the improved catches per vessel achieved if the number of vessels are reduced within a fixed overall quota, and any levy that might be applied to finance compensation.

Some guide to the first of these is given by a limited set of data made available to the Panel concerning sales of anchoveta vessels into the food-fish fleet. These averaged around $20,000 for 7-year old, 100 ton capacity vessels. A compensation price of $200 per ton was therefore acceptable for at least some owners. The compensation need only cover the opportunities foregone by withdrawing from the anchoveta fishery, and should therefore be less than $200 per ton to the extent that the vessel could be sold for use elsewhere, or as scrap.

One obvious opportunity for some surplus anchoveta vessels is the food-fish industry. However, the Panel had no good information on the extent of the food fish resources, and the number of vessels that these resources could support, and until the potentialities of the food fish stocks have been determined, substantial uncontrolled diversion of vessels from the anchoveta fishery must run a risk of merely diverting the problem of surplus capacity from one fishery to another.

For the purposes of illustration a figure of $100 per ton might be used for the compensation in withdrawing vessels from the anchoveta fishery. The capacity of vessels landing less than 4500 tons in 1968–9 was 43,000 tons; at $100 per ton the compensation would be $4.3 million. These vessels together landed 815,000 tons of fish, so that if they were removed from the fishery, the catches of the remaining vessels could be increased by this amount, worth, at $11 per ton, $9.0 million. Of this increase in grossing, about half would be accounted for by increased variable costs, giving a net benefit to the remaining fleet of about $4.5 million. The required total compensation is therefore about equal to the annual net benefit. The application of a levy over two or three seasons, at the rate of, say 5–10 soles per ton of fish landed, would therefore be sufficient to raise money for compensation, and leave a net benefit to the remaining vessels.

The compensation could, alternatively, be based on past catches. This might be more realistic, since most of the vessels concerned have been fishing only part time. Their real effect on the stock, and the benefit from their removal is less than that suggested by their hold capacity. The total compensation should presumably not be affected by the method of calculation. In the example above, figures of $4.3 million compensation for a fleet which landed 815,000 tons were mentioned. This would imply a rate of compensation of around $4 per ton landed.

GUANO BIRDS

On several occasions the question of the relation between the guano birds and the anchoveta fishery was mentioned to the Panel. The abundance of guano birds—cormorants, gannets, and pelicans—along the Peruvian coast fell precipitously in 1965–1966 to the lowest level ever recorded. This sharp decline is thought to be associated with the El Niño phenomenon; a mild El Niño occurred in the 1965–1966 summer. Since 1965–1966, the populations have failed to return to anywhere near their previous abundance. This behavior is in direct contrast to their behavior after severe reduction in the past. According to the most recent census available (1968–1969), the total number of birds is 5.4 million. The factors acting to hold the population at such a low level are not known with certainty. The lower size of the anchoveta stock, caused by current fishing intensities and also perhaps contamination by DDT and other persistent pesticides may be implicated. Although no direct evidence exists that DDT compounds are interfering with reproduction of the guano birds in Peru, DDT has been linked with rapid declines in similar species of fish-eating birds in other parts of the world. The well-documented demise of the brown pelican in the California current system, which is oceanographically similar to the Peruvian current in that both are eastern boundary current upwelling systems, is but one example.

Using Jordan's (1964) average figure of 430 grams of anchoveta consumed per day by birds gives an estimated annual consumption of 0.85 million tons of anchoveta by the present bird population. Recoverable guano production per bird is thought to be 45 grams per day; thus the sustainable annual production of guano is approximately 90,000 tons. At a price of $56.00 per ton the annual gross value of the guano is $5.05 million. Thus the gross value of each ton of anchoveta in terms of guano is $5.94, as compared to a gross value of $30.20 in terms of meal (assuming 5.3 tons of anchoveta per ton of meal and meal price of $160.00 per ton).

If 50% of the fish denied to the birds were caught by Peruvian fishermen, then the loss of $5.94 per ton of guano would result in the addition of $15.10 per ton to the value of meal output. Assuming similar rates of return from the trade in guano as that in fish meal, if the industry's marginal costs of catching and processing one ton of fish into meal are less than the fish meal price to an extent which at least is as great as the cost of exterminating the bird population to the extent required to save two tons of fish (including an appropriate rate of return on the cost of extermination) then it would be worthwhile to incur such costs regardless of the price of guano.

It must be kept in mind that the assumed product values are probably fixed in favor of meal but, far more important, this approach to the matter ignores the threat it presents to the very existence of a bird population that has already reached perilously low levels. Furthermore, the ecological consequences of eliminating the bird populations are not understood. In other animal populations, control of predators has failed to give the expected increase in the prey species, for example, because predation takes more of the diseased individuals, and in their absence disease increases. Until more is known about the interactions between the bird and anchoveta populations, any type of predator control program to reduce the birds would be extremely hazardous.

CHAPTER 14

Prediction of Environmental Changes and the Struggle of the Third World for National Independence

The Case of the Peruvian Fisheries

MATTHIAS TOMCZAK, JR.

Abstract

Before 1972, more than a third of Peru's export earnings came from the catch of the anchoveta for fish meal production. El Niño is a natural phenomenon that causes the cold, nutrient-rich surface waters off the Peruvian coast to be replaced by warm water poor in nutrients, causing the anchoveta to disappear from the region. A major El Niño occurred in 1972–1973; it was accompanied by a drop of anchoveta catches from above 10 to below 2 million metric tons and by a "state of emergency" for the Peruvian fishery.

This chapter deals with the effects of major changes of natural conditions in a Third World country on its efforts to gain control over its own resources. In order to answer this question, the degree of imperialist control over Peruvian economy is evaluated. It is shown that the Agrarian Reform of 1969, which was designed by the national bourgeoisie to improve the conditions of an independent capitalist development of the country, was initially supported by the Peruvian masses because of its anti-imperialist character, and that it was this mass support that, in contrast to earlier efforts, enabled the government to implement the reform. It is pointed out that the reason for this support is found in the government's attitude towards Peru's traditional population, the Quechua and the Aymara peasants: for the first time, the government gave in to a number of their demands on communal land ownership and national identity. It is shown that as a result of mass support, imperialist control over Peru's economy weakened during 1968–1973, after which period unity between government and the masses broke, and the imperialists regained complete control over the country's economy. This development is traced back to the mounting pressure of the imperialists on the Peruvian government and to the government's fears that the struggle for national liberation could turn into a struggle for social revolution which caused the government to turn against the masses. Beginning in 1974, the situation deteriorated rapidly, and in 1978 imperialist exploitation of Peru's resources was at a higher level than ever.

To evaluate the importance of the El Niño of 1972–1973, the structure and development of the Peruvian fishery is analyzed. Three groups of workers of different historical and social background can be distinguished, the traditional fishermen who catch fish for human consumption, the anchoveta laborers, and the workers in the fish meal factories. It is shown that at the impact of the El Niño, a split occurred between the last two groups which could be exploited by the imperialists and the government in order to inflict a heavy defeat on the working class. However, it is also noted that the way the government reacted to the El Niño was a consequence of the political line that it had adopted by then and that the victory of the imperialists over the mass movement of the period was not determined by the occurrence of an El Niño. It is concluded that a possible El Niño forecast would affect some details of Peru's struggle for national liberation but would be insignificant for its outcome, which solely depends on the degree of unity in the country against imperialism.

INTRODUCTION

A large part of this chapter does not deal with El Niño or with the Peruvian fishery explicitly, as suggested by the title of the book. It will, however, be seen that all of it is relevant to the Peruvian fishery and that it is impossible to assess the impact of adverse natural conditions such as the 1972–1973 El Niño on the struggle of a Third World country for independence without a clear understanding of the degree of imperialist exploitation still present and of the various contradictions inside and outside the country. These contradictions are many, and it is always possible to study a particular one without relating it to the basic contradiction between imperialism and the peoples of the world and, by this method, avoid naming imperialism as the cause for the catastrophic consequences that the breakdown of the anchoveta fishery (after 1971) had for the Peruvian masses.

On the international level, there is the contradiction between Peru as a country of the Third World and the companies and governments of the imperialist countries, particularly the superpowers. On the national level, there are the contradictions between the Peruvian government and the masses, between the Peruvian capitalists and the workers, between the workers and the peasants. In today's world situation, Peru's basic contradiction, which is an antagonistic one, is the contradiction on the international level, which means that all contradictions on the national level are of secondary importance and can make way for unity against imperialism, some of them only temporarily, others permanently. It is the aim of this chapter to look at the actual development of these contradictions with respect to Peru since 1968 and to investigate how they were affected by an event which was outside the range of control of any of the forces involved, the El Niño of 1972–1973.

THE STRUCTURE OF THE PERUVIAN ECONOMY

In this section some figures on the economy of Peru are given to illustrate the fact that Peru is a country of the Third World. This concept implies the existence of a First and Second World, which can be briefly described as follows: the U.S. and the U.S.S.R.—the two superpowers— constitute the First World and aim at world hegemony, including complete control over its resources. The other imperialist countries form the Second World; they try to increase their imperialist influence in all parts of the world which brings them into contradictions not only with the Third World but also with the superpowers, although they cannot move independently of them. The remaining countries form the Third World and struggle for liberation from foreign domination. While the Third World countries progressively overcome their differences in the course of their struggle, the contradictions between the individual countries of the Second World and between the superpowers inevitably grow, due to increasing competition and growing contradictions within the First and Second Worlds which, if properly exploited, allow the Third World to win individual battles and to decrease the degree of economic dependence.

The economic structure of any nonsocialist country of the Third World is characterized by imperialist control over the larger part of its economy, the largest share of this control being exercised by one of the superpowers, either through direct investments or through foreign debt. Historically, the economy of all Third World countries was based on agriculture, and most of the resources plundered by the imperialists were agricultural products. The structure of the society corresponded to this situation, power being exercised by the imperialists through the local landlords. As a result, the traditional system of agricultural production—a large variety of products for local consumption, with a high degree of self-supply within the communities—was replaced by a system of large cash crop plantations on the landlords' properties and of increasing import of foodstuffs. This situation still persists. At the same time, however, during the last decades, imperialist interest shifted gradually toward plundering of mineral resources and an increase of foreign investment in manufacturing. This has been accompanied by the development and growth of the proletariat in the Third World countries and by the necessity for a change in the social structure, commonly called the Agrarian Reform, the main purpose of which is to guarantee cheap labor by making enough peasants available to the labor market. Since this can only be achieved, in a capitalist economy, by ruining the peasants' existence as peasants, and since in the Third World countries peasants constitute the majority of the population, it is the peasants who form the solid basis for a most determined and ever-growing struggle to end imperialist control over the country. They take up the fight for an Agrarian Reform, that ends exploitation by the landlords and provides them with the land to grow the kind of products needed in the country, thus reducing agricultural imports. Obviously, this kind of Agrarian Reform is completely opposed to the Agrarian Reform envisaged by the imperialists.

This very brief outline of the economic struggle of a nonsocialist country of the Third World can clearly be exemplified with the economic situation of Peru. The traditional structure is reflected in Tables 1 and 2. It is seen that in 1961, 81% of the cultivated area was occupied by only 1% of the agricultural units which were the large *haciendas* for the production of agricultural export commodities: cotton, sugar, and coffee (Table 2). Cereals, meat, and other basic food had to be imported. Still

TABLE 1. Land Ownership in Peru, 1961

Size Group (Hectares)	Agricultural Units		Total Area of Size Group	
	Number	%	Hectares	%
< 5	734,272	83.5	1,055	5.7
5–10	76,829	8.7	482	2.6
10–200	59,604	6.8	1,967	10.7
> 200	7,267	1.0	15,101	81.0
Total	878,667[a]	100.0	18,605	100.0

[a]Includes 695 agricultural units without group classification.
Source. Reference 1.

TABLE 2. Agricultural Exports and Imports of Peru, 1961–1965[a]

	Exports (Mil. Soles)			
Year	Cotton	Sugar	Coffee	Others
1961	2182.8	1721.1	610.4	389.8
1962	2659.9	1459.5	648.5	247.2
1963	2507.8	1739.0	685.6	499.6
1964	2487.0	1715.8	999.7	516.2
1965	2389.8	1005.2	778.3	421.4

		Imports (Mil. Soles)				
Year	Cereals	Animal Products	Oils, Fats	Fruit	Misc.	Others
1961	1025.1	546.9	134.1	67.4	359.7	22.3
1962	967.1	800.8	115.2	80.8	395.8	34.1
1963	856.9	1013.4	34.7	115.9	425.8	36.1
1964	1243.5	887.9	88.7	128.0	390.2	40.3
1965	1489.4	1091.8	231.5	162.4	531.2	45.4

[a]Cotton: includes derivatives. Sugar: includes derivatives. Animal products: meat, dairy products, nonedible fats, skins, wool (llama). Oils, Fats: of vegetable origin. Miscellaneous: sugar derivatives, coffee, tobacco and others.
Source. Reference 2.

in 1975 half of the milk produced in Peru (to a large extent by foreign capital) was based on imported milk powder; virtually all wheat was imported.[3] The gradual change of the imperialists' interests is evident from Table 3 where it can be seen that between 1919 and 1929 the two principle agricultural export commodities fell back to third and fourth place behind oil and copper. The two World Wars resulted in a heavy cutback of foreign industrial investment in Peru, and in 1945 sugar and cotton again ranged far ahead of oil and copper.[5] Since then, however, the tendency of the beginning of the century was taken up in greater strength than ever before; from 1952 to 1971, the percentage of direct U.S. investments in Peruvian mining never fell below 50% of the total direct U.S. investments in Peru and sometimes reached as much as 64.4%. Direct U.S. investments in Peru's oil industry ranged between 10

TABLE 3.   Principal Exports of Peru, 1887–1929 (Peruvian pounds)

Year	Cotton	Sugar	Wool	Copper	Oil[a]
1887	60,000	280,000	110,000	—	—
1900	326,000	1,456,000	297,000	621,000	185
1909	1,207,000	1,159,000	394,000	1,214,000	152,000
1919	6,635,000	8,311,000	1,632,000	4,920,000	2,320,000
1929	5,154,000	3,337,000	1,052,000	6,672,000	8,698,000

[a]includes derivatives.
Source. Reference 4.

and 25% of the total before 1965 and again after 1969. Direct U.S. investments in Peruvian manufacturing increased steadily from about 7% of total U.S. investments to a level of 15%.[6] The leading role of the United States in the control of Peru's economy is reflected by Table 4 which includes some Second World investments for comparison. Table 5 documents the degree of foreign control in manufacturing.

These figures show that a gradual change in the Peruvian economy occurred as a response to changing imperialist interests. Agricultural products which until 1950 made up the larger part of all exports, were gradually replaced (as export commodities) by minerals and oil; in 1960 their export share had declined to 36.1%. Agricultural imports, on the other hand, remained steady at a level of 17–22% of all imports.[9] In the following years the agricultural share of export products declined further, and was replaced by fish meal, until the catastrophic El Niño event in 1972–1973. After 1964 agriculture even showed an absolute decline, as increasing areas of land were taken out of production. Peru's capability to feed its people depended totally on the selling of its resources to the imperialists for the purchase of even the most basic foodstuffs.

The only successful strategy to overcome this situation is the development of an agriculture for internal consumption rather than for export. Once the need for agricultural imports no longer exists, a diversified export structure can be developed which will enable the country to build up the capital for the development of a modern industry. To apply this strategy, the productive forces of the peasants have to be set free by expropriating the landlords and distributing the land to the peasants. The key question is the Agrarian Reform. It occurred in Peru in 1969. One of the questions to be answered in this analysis is why it did not result in the liberation of Peru from imperialist domination, and to what extent the El Niño of 1972–1973 influenced the outcome.

TABLE 4. Foreign Investments in Peru, 1965–1968

Year	Total (Mil. US-$)	USA (%)	Canada (%)	U.K. (%)	Switzerland (%)	Japan (%)	FRG (%)	Others (%)
1965	822.7	75.0	1.4	9.7	2.9	0.2	0.4	10.4
1966	841.1	72.0	1.5	10.0	3.4	0.5	0.5	12.1
1967	901.0	54.7	14.6	5.1	5.1	1.5	2.0	17.0
1968	870.1	52.2	12.3	5.5	5.4	3.2	2.4	19.0

Source. Reference 7.

TABLE 5. Foreign Capital in Peruvian Industry, 1969

	Peruvian Companies		Foreign Companies	
	Number	% of prod.	Number	% of prod.
I. Nonferrous mining	4	2.1	4	97.9
Iron and steel	13	76.7	3	23.3
Oil refineries	4	85.6	5	14.4
Fishing	83	75.6	21	24.4
Sugar refineries	16	100.0	0	0.0
II. Dairy products	11	39.6	2	60.4
Mill products	58	58.6	8	41.4
Breweries	5	83.6	2	16.4
Tobacco industry	1	12.4	2	87.6
Spinning, weaving, finished textiles	140	58.4	19	41.6
Printing industry	127	97.4	3	2.6
Industrial chemical products	30	26.6	18	73.4
Other chemical products	90	33.6	48	66.4
Concrete	2	13.3	3	86.7
Motor vehicles	15	5.5	15	94.5
Plastics	76	85.8	5	14.2
Total industry activity	2153	56.0	242	44.0
Percentage of total industrial production:				
Group I		18.7		11.8
Group II		17.2		18.9
Others		20.0		13.4

Source. Reference 8.

THE VELASCO GOVERNMENT AND THE AGRARIAN REFORM

The Peruvian government was the first to declare, in 1947, sovereignty over the maritime economic zone of 200 nautical miles. Since then, successive governments have consistently maintained this position in order to secure national control over marine resources.* After the Velasco government came to power in 1968, Peru's anti-imperialist position strengthened considerably. In the United Nations, the Foreign Minister stressed the need for the Third World to close its ranks against imperialism,[10] and in the organization of the Nonaligned Countries in

---

*The Foreign Minister of the Velasco Government, Mercado Jarrín, confirmed Peru's determination to defend the 200 nautical mile economic zone in a special conference for the foreign diplomatic corps on 11 May 1970.

which Peru had been a very active member, he reaffirmed his country's dedication to common interests.[11] In the Special Latin American Coordinating Commission, ECLA, Peru put considerable effort into the setting-up of a joint Latin American platform for negotiations with the European Community.[12] It tried to obtain stable and just prices for its copper as a member of the Intergovernmental Council of Copper-Exporting Countries (CIPEC) in cooperation with the other member states Chile, Zaire, and Zambia. In 1971, when Ecuador arrested several tuna fishing vessels of the United States for fishing within 200 nautical miles from the coast, Peru defended this decision against several acts of harrassment by the United States in the Organization of American States (OAS).[13] It can thus safely be said that Peru has a long record of anti-imperialist politics and that it became, after 1968, one of the important countries in the allied front of the Third World against foreign domination.

Given the high degree of imperialist control over Peru's economy and political structure, a government devoted to such a political stand is easily overthrown unless it is backed by mass support in the country. Masses make history, and whether or not a government relies on the masses and draws on their support for its anti-imperialist measures is the key question for its success. Solidarity and growing unity of the Third World is the other important factor, but the fate of any individual government determined to fight imperialism is decided on whether or not it has mass support.

The Velasco government, like all earlier Peruvian governments, was not a workers' and peasants' government in the sense that it represented the working class or the peasants. It was a bourgeois government engaged in a heavy battle to gain independence for Peru, and to this extent it was supported by the masses because it represented their desire to overcome imperialist dependence. The president's annual Message to the Nation partly served the purpose to raise and consolidate this support:

> First of all, to struggle for the sovereignty of Peru and its "definite emancipation"—as is declared in the Manifesto—signifies, in essence, to struggle for political self-determination of the Peruvian State and for national control of the country's resources. Because, sovereignty is the potential of a nation to determine its proper destiny without foreign interference and the right to be master of its proper riches. Therefore, sovereignty implies the prevalence of the national interests over the foreign interests. And since imperialism is nothing other than foreign domination over the economy and politics of a country, the struggle for sovereignty necessarily is the struggle against imperialism.[14]

As has been discussed above, the key question in any Third World country is the Agrarian Reform. The bourgeois character of the Velasco government will become particularly clear when looking very briefly into its Agrarian Reform.

The Agrarian Reform of the Velasco government was declared by law issued in June 1969, Decreto Ley 17716, and was announced in a presidential speech the same day which ended with the words:

> To the man of the land we now can say in the immortal and libertarian voice of Tupac Amaru: Peasant: "No longer the landlord will consume your poverty."[15]

According to the Agrarian Reform Law, the area of any agricultural unit was limited to a maximum size, depending on the region of the country and the type of agricultural activity.[16] Any land that exceeded these limits, and all land which was "idle or deficiently exploited,"* became available for distribution among the peasants. Peasants could obtain land either as cooperatives, as peasant communities or as individual families and had to purchase the land on a 20-year loan at fixed interest (Article 83).† The former owners were paid for the expropriated land by the state, in bonds of three classes which carried interest between 4 and 6% and are redeemed over a period of 20 to 30 years. The important bonds are those of classes A and B because they are accepted as cash by the State Development Bank at any time when used to finance up to 50% of an industrial enterprise of which the bond holder secures the other 50% himself (Article 181).

The above regulations are significant in showing the bourgeois character of this Agrarian Reform. Its aim was to stimulate the development of a nationally owned industry while at the same time releasing the productive forces of the peasants. Technical aid and credit facilities were set up for peasants in order to further assist this process, and the law stipulated that the order of preference for these aids should be: cooperatives, peasant communities, agricultural societies, small and medium proprietors (Article 91). Simultaneously, the government took control of a number of foreign companies by expropriation. During 1969–1974 a total of 29 foreign companies in the manufacturing industry were taken

---

*"Land is considered idle if, in spite of being apt for agriculture, it was not the object of organized exploitation; deficiently exploited where bad use is made of the natural resources which determines its destruction, or where the income from the main crop of the land is less than eighty percent (80%) of the average income of the zone." (Article 16)
†Article 87 stipulates that the granting of land is declared void if the peasants fall behind the repayment schedule for two successive years.

over by the government, of which 21 operated in fish meal and fishing, 3 in concrete production, 3 in paper and pulp industry, 1 in the printing industry, and 1 in oil refining.[17] The fish meal industry is a particularly outstanding example to which we will return later.

It can be anticipated from the theoretical analysis in the previous section that this strategy was bound to fail because it treated the development of a national self-reliant agriculture as a corollary to the growth of a national industry rather than the main target of the anti-imperialist process. The reason, of course, is that the national bourgeoisie is more concerned with the improvement of its own industrial profits than with agricultural development. Nevertheless, the Decreto Ley 17716 can justifiably be called a revolutionary act of Peru's bourgeoisie: It smashed the power of the feudal agricultural oligarchy and replaced it by the power of the bourgeoisie, it prevented the deterioration and destruction of the soil by the *haciendas*,* and stopped the decline in agriculture. As can be seen from Table 6, however, it did not result in a dramatic increase of agricultural production for foodstuffs, and the Gross National Product (GNP) was increasingly supplied by industry. Velasco's Message to the Nation on 24 June 1969, which announced the Agrarian Reform Law, is clear proof of the government's awareness of its own political line when it states that the Law

TABLE 6. Agricultural Production in Peru, 1968–1974 (in 1000 metric tons)

Year	Rice I	Papa II	Maize III	Wheat IV	Total, I–IV	Total agricultural production as % of GNP
1968	286	1708	526	113	2633	
1969	444	1821	585	127	2977	16.4
1970	587	1930	615	125	3257	16.4
1971	591	1968	616	122	3297	15.6
1972	436	1712	589	140	2877	15.0
1973	440	1712	590	140	2882	14.2
1974						13.8

Source. Reference 18.

*"Deficient exploitation" as mentioned in the law includes increase of salt content because of insufficient irrigation, lack of maintenance of irrigation systems, erosion, and overgrazing as well as all forms of partial use of the land only. (See Alejandro Robles Recavarren, *Diccionario de la Reforma Agraria Peruana*. Lima, 1975, p. 60.)

is also a law of impulse for the Peruvian industry the future of which depends decisively on creating a growingly larger inner market of high diversified consumption.[19]

According to the government's 5-year plan 1971–1975, only the *increase* in demand for food should be satisfied by increased agricultural production, "in spite of the politics of substitution of imported food products by the increase in internal production."[20]

The government did not deny the bourgeois character of the Agrarian Reform. Foreign Minister Mercado Jarrín told the Conference of the Nonaligned 1970 at Lusaka:

> The society we are building will no longer consist of an elite of large land-owners, an insignificant professional middle class and a large mass of workers and illiterate peasants. Today there will be industrial managers, small and medium proprietors, large professional and wage-earning groups, workers and literate peasants.[21]*

The picture of a modern capitalist state comes out quite clearly. The problem is that building a modern capitalist state today gives only temporary relief from imperialist control. The key question is independence from agricultural imports.

The Velasco government was not the first one to promote an Agrarian Reform in Peru. We mentioned already that the imperialists, too, have a certain interest in an Agrarian Reform and studies had been made in the United States on how to achieve a new land distribution without sparking an anti-imperialist turmoil.[22] Earlier governments had introduced agrarian reform laws but failed to implement them. When the Velasco government

---

*It is worth noting that this change from a feudal agrarian to a bourgeois industrial society coincides with the politics recommended for the Third World by the revisionists. This is one of the reasons why the Velasco government repeatedly had to deny allegations that it tried to free the country from U.S. domination in order to sell it to U.S.S.R. domination. In fact, revisionists held posts in state administration even before the Velasco government. Gustavo Espinoza R., coauthor of *El Problema de la Tierra*, who extensively quotes Marx, Lenin, and East European scientists, and who has been Prime Engineer of the *Junta Nacional de la Producción Alimenticia*, is just one example. The activities of the revisionists cannot be treated in detail in this analysis which deals with the major contradiction in Peru: the struggle between the Peruvian nation and U.S. imperialism. It should be mentioned, however, that the revisionists, through their control of the major trade union council CGTP, played a significant part in the efforts of imperialism to weaken the stand of the Peruvian masses. The CGTP continued to support the government until early 1975 and supported strike action later in that year only "as a limited gesture not to end up overtaken by its own rank and file" (Desco, vol. IV, p. v.i.).

declared the Decreto Ley 17716, the constitution already provided for expropriation by payment in bonds since 1964 (Article 29, amended by article 1 of Ley no. 15242 of 28.11.1964).[23] Why the Velasco government succeeded in bringing about the social change earlier governments could not implement is the subject of the next section.

MASS SUPPORT FOR THE VELASCO GOVERNMENT

The question of mass support in a Third World country is the question of support by the peasantry. Peasants and *hacienda* laborers are the majority of Peru's population (Velasco Alvarado, in his Message to the Nation of 28 July 1971, gives the percentage of people working the land as roughly 50%[24]) and the links between the working class and the peasantry are particularly strong as many of the workers actually are ruined peasants. In a study[25] of the situation of 1092 workers in Lima, 92 arrived in Lima less than 4 years before the study, 182 between 5 and 10 years and 388 earlier than 11 years before the study, a total of 662. Of these 662, the occupations given for the time before arrival in Lima were as follows ("peasant" includes agricultural worker):

	Percent		
Activity	Men (531)	Women (131)	Total (662)
Peasant	36.0	2.3	29.3
Worker	6.0	0.7	4.9
Work not specified	8.8	7.6	8.6
At school	32.8	37.4	33.7
Did not work	11.3	45.1	18.0
No answer	5.1	6.9	5.5

Table 7, which only covers a minor fraction of the struggle for land that went on for two centuries, shows the peasants' determination with their fight. The wave of peasant uprisings finally brought about the constitutional changes of 1964 and another tentative Agrarian Reform. Like all earlier reforms, it was another failure. The 1969 Agrarian Reform was the first to succeed because it recognized the right of the peasants to uphold and develop their Indian nationalities, Quechua or Aymara. The importance of these two nationalities within Peru is quite clear from the population census figures given in Table 8. It should be noted that the table does not reflect the actual situation at any particular place.

TABLE 7. Peasant Massacres in Peru, 1956–1964

Type of Conflict	Number of Incidents	Number of Dead
Wage claim	7	12
Occupation of land	20	105
Trade union recognition, better working conditions	5	23
Liquidation of feudal debts	6	53
Others	8	18
	46	211

Source. Reference 26.

Nearly all Peruvian peasants are Indians, and in the agricultural areas the vast majority of the population does not speak or understand Spanish (an example of such a situation is given in Table 9). Consequently, no government could ever get any mass support from the peasantry as long as it continued with the colonial practice to try to destroy the Indian nations as such. The Velasco government was the first to accept the existence of Indian nationalities in practical terms, in particular the Indian system of communal land-ownership and organization of the work force. The officially recognized Indian communities in Peru in 1968 numbered 2290, but it is known that the actual number was much larger.[29] Earlier governments occasionally referred to the constitution which declares that "indigenous communities have legal existence and juridical status" and that "the state guarantees the integrity of the communities' property."[30] In practice, however, Indians were constantly threatened off their land. As an example, in a raid on the community of Cospán on 29 October 1968—19 days after the Velasco government came to power—250 families were dislocated by order of a judge, resulting in 7 dead, 34 wounded, and 18 arrested.[31]

The Velasco government, in an effort to win the peasants' support for its anti-imperialist policy, put an end to this and introduced the Decreto Ley 17716, followed in February 1970 by Decreto Supremo No. 37-70A, which introduced the Statute of the Peasant Communities and guaranteed their title to the land.

> Article 2: The Peasant Community is a group of families who possess and identify themselves with a determined territory and are united by common social and cultural links, by common labor and mutual help, and basically by the activities linked with agriculture. Article 4: The Community is the sole proprietor of its lands, and its members are their usufructuaries.[32]

TABLE 8. Population of Peru, 1876–1972[a]

	1876 Number	%	1940 Number	%	1961 Number	%	1972 Number	%
Population	2,669,106	100.00	6,207,967	100.00	8,235,220	100.00	11,790,150	100.00
			5,228,358	*100.00*				
Indians	1,554,678	57.60	2,843,196	45.86				
Quechua			*1,625,156*	*31.08*	*1,389,195*	*16.87*	*1,311,062*	*11.12*
Aymara			*184,743*	*3.53*	*162,175*	*1.97*	*149,664*	*1.27*
Whites and Mestizos	1,040,652	38.55	3,283,360	52.89				
Others	103,776	3.85	81,411	1.25				
Bilingual Spanish—Quechua			*816,967*	*15.63*	*1,293,322*	*15.70*	*1,715.004*	*14.55*
Bilingual Spanish—Aymara			*47,022*	*0.90*	*125,702*	*1.53*	*182,241*	*1.55*

[a]Numbers in *italics* include only the population of five years and over, percentages in *italics* refer to this population group. In the census of 1961 and thereafter, statistical data for ethnic groups were discontinued; figures for Quechua and Aymara are based on data on languages spoken.
Source. Reference 27.

TABLE 9.   Language Distribution in Some
Agricultural Districts

A. Departamento Apurímac

Group	Number
Population, 4 years and over	266,760
First language Quechua	243,324
First language Spanish	20,728
Illiterate Quechua	163,610
Illiterate Spanish	4,533
Monolingual Quechua	149,988

B. Population speaking both Quechua and
Spanish

Departments	Percentage
Apurímac	4.97
Ayacucho	6.39
Ancash	6.98
Huánuco	12.61
National Average:	20.0

Source. Reference 28.

In May 1975, finally, Quechua was declared an official language in addition to Spanish,[33] and schools and university courses in Quechua were established.

Although all these measures were the result of centuries of struggle by the peasants, they explain the mass support from the peasants that the Velasco government enjoyed, far more than support from the working class, as is indicated by the figures of Table 10. It was this mass support which actually enabled the government to carry out the Agrarian Reform.*

---

*For a full assessment of the relationship between Indians and the Alvarado Government it is important to note that the government was not a government of the Indian people but one which responded to their demands with important concessions. The government's attitude, after all, still was bourgeois-paternalistic: "In response to the cry for justice and to the rights of those who need it most, the Agrarian Reform Law has given its acceptance to this large mass of peasants who constitute the indigenous communities which from today will be called - abandoning a qualification of vicious racist habits and of unacceptable prejudice - Peasant Communities." (República Peruana. Oficina de Información, 1969, *op cit*, p. 7) The political aims of the Indians, on the other hand, go much further than the

**TABLE 10. Reported Outbreaks
of Class Struggle in Peru,
1974–1976**

	Number of Incidents[a]	
Year	Trade Union Movement	Peasant Movement
1974	75	23
1975	167	6
1976	122	12

[a]*Trade union movement:* disputes,
strikes, stoppages, factory sit-ins.
Peasant movement: disputes,
strikes, stoppages (includes agricultural laborers).
Source. Reference 34.

It remains to be seen whether it eventually resulted in a gain in national
independence.

THE SUCCESS OF THE IMPERIALISTS*

The declared aim of the Velasco government was the liberation of the
country from imperialist domination. Figure 1 shows the development of
Peruvian imports and exports and of the annual inflation rate. It is
obvious that during 1968–1974 a remarkable improvement of Peru's
economic situation occurred. Since 1975, however, the situation was

establishment of Peasant Communities; they have been formulated repeatedly on all-Indian
congresses, and a statement from a 1975 conference made quite clear that "the present-day
situation of the Indian population results from the conquest by invaders who enforced their
politics of exploitation on them. Our people, particularly the Indians, have as their final
aim the reconstitution of their power to control their own fate." (Die bolivianische
Delegation zur 1. Konferenz der Ureinwohner, Port Alberni 1975. pogrom 8 no. 50/51 p.
28, 1978)
*Numbers in square brackets in this and subsequent sections refer to Desco (Centro de
Estudios y Promoción del Desarrollo) *Cronología Política,* where events are numbered
consecutively. Numbers up to 2881 refer to 1974, up to 3972 to 1975, and up to 4829 to
1976.

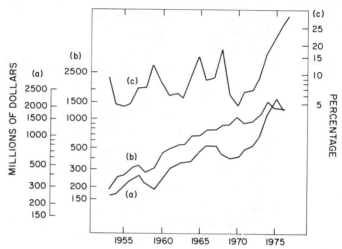

Figure 1 Imports (a), exports (b) and rate of annual inflation (c and right-hand scale), of Peru during 1953–1977. (Source: Nachrichten fur Aussenhandel, Bonn, several volumes.)

reversed again, and the imperialists regained control over all economic activities. The question to be asked is how this could have occurred.

The interest of the proletariat is to put an end to exploitation of man by man, by abolishing the system of private ownership of the means of production. Termination of foreign economic domination is one prerequisite for achieving this aim. The Peruvian government could therefore rely on the firm support of the working class for all measures of expropriation, both for the nationalization of foreign monopolies and for self-organization of any industrial plant as a workers' cooperative.[35]

The interest of the Velasco government, on the other hand, was to secure a larger share of the profits for the national bourgeoisie, and its actions prove that it was quite prepared to defend this share against any working class action. As a consequence, its relationship with the working class was a very ambiguous one. The expropriation of the Cerro de Pasco copper mining company was supported by all Trade Union Councils [2437], as was the government take-over of Gulf Oil and Gulf del Peru [3462]. When the Marcona Mining Company proposed to transfer the major part of its activities to the government in exchange for guaranteed free delivery of 14 million tons of iron pellets over a period of 10 years [3405], the working class supported the expropriation of Marcona with a number of actions, including the occupation of the trade union office which had been occupied by police [3405, 3485, 3496, 3702]. The involvement of the police, however, indicated that the government already used the instruments of state power to restrict working class activity. Its elaborate plan of workers' participation—a sophisticated product of social democrat reformism which cuts effective wages by making the workers shareholders of "their" company, forcing them to increase its capital by investing part of their wages—did not tie the working class to bourgeois interests because real wages were drastically reduced (Figure 2), and the government was increasingly confronted with factory occupations and strikes. Although many of these initially only urged the government to apply its own laws against capitalists found responsible for economic sabotage, the government increasingly responded by police and military force; public manifestations were forcibly broken up [2516], trade union leaders arrested [2565]. When at the beginning of 1975 workers in the copper mines and on large construction sites [3105, 3107] and parts of the police force [3168] joined the strikes for the restoration of sufficient wages, the government declared a state of emergency; 6 months later Velasco Alvarado was replaced by Francisco Morales Bermúdez, the former Finance and Prime Minister,

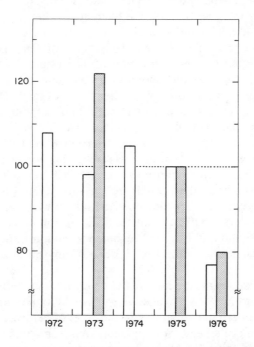

Figure 2 Index of effective wages and salaries in Peru at 1 January, 1972–1976, based on 1 January 1975 = 100. Left column: official minimum wage, right shaded column: national average. (Calculated from Desco, Cronología Política, volume IV, pp. iv–v.)

to continue the discredited politics. As will be seen, the Morales Bermúdez government continued the tendency of 1973–1975 of renewed sell-out of the country to the imperialists.

The crucial point in the failure of the anti-imperialist program was the strategy to boost industry at the expense of a self-reliant agriculture. Such a strategy was bound to lose the peasants' mass support which during 1969–1973 had made possible the Agrarian Reform. The bourgeois character of this reform became more and more obvious in its application: It favored the agricultural companies at the expense of the peasants' communities and cooperatives (Table 11), and it left the peasants with a large debt to be paid to the state for the distributed land (Table 12). The result was a radicalization of the peasant movement—which was not taken up by the government and turned into a weapon against imperialism because the agricultural unrest endangered its strategy of increasing cash crop exports for building up industry.

As a consequence, the peasants' demands for more land and for the cancellation of the agrarian debt turned against the government. In the elections for the Councils for Vigilance and Administration set up in the cooperatives by the government, the turnout was low and the number of abstentions or invalid ballots in some cooperatives exceeded ⅔ of all votes [2527]. Instead, the Peasant Federation of Peru (CCP) grew in strength, demanded the reorganization of agriculture for the service of the country's needs and the abolition of the agrarian debt [2662] and mobilized the peasants to occupy further areas of land. During 1974, in a large wave of land occupations, more than 20,000 peasants were involved in the take-over of 47 *haciendas,* which comprised more than 40,000 hectares [2830].[38] Such was the determination of the peasants that the government in some situations was obliged to agree to negotiations with the peasants' representatives of the CCP on the reduction or abolition of the agrarian debt [2745]. At the same time, however, it tried to ban and prevent marches and other activities wherever possible and detained the leaders [2810]. Finally, it declared the end of the Agrarian Reform for 1976 and had frequent recourse to military and police forces against the peasants. This marked the end of any mass support for the government.

During all these years, the imperialist countries had been watching the situation and had actively engaged in various activities to restore their control over the Peruvian economy to the level of before 1968. The main instrument was Peru's heavy foreign debt: Debt service payments, which in 1967 amounted to 11.1% of Peru's export-derived income had risen to 32.5% in 1973, the second highest debt service ratio in the world.[39] The International Monetary Fund (IMF) closely monitored this development

TABLE 11. Distribution of Land in Peru's Agrarian Reform until August, 1975

Type of Unit	Number	Distributed Land		Families Affected		Cattle	
		Hectares	%	Number	%	Number	%
Agricultural companies	51	2,293,377	40.56	55,688	23.81	869,746	66.65
Cooperatives	421	1,963,883	34.73	94,420	40.37	414,792	31.78
Peasant communities	151	514,822	9.10	43,348	18.53	17,107	1.31
Groups of peasants	331	751,490	13.29	22,529	9.63	2,809	0.21
Individuals		129,589	2.32	17,871	7.66	336	0.05
Total		5,651,161		233,856		1,304,790	

Source. Reference 36.

**TABLE 12. Balance of Account of Peru's Agrarian
Reform, December 1974 (in million soles)**

Amount to Be Paid According to Law by the Peasants		Amount Actually Paid by the Peasants		Amount Paid by the State to the Former Owners of the Land	
Capital	1100	Capital	180	Effects	700
Interest	1300	Interest	220	Amortization	70
				Interest	80
				Investments	150
Total	2400	Total	400	Total	1000

Source. Reference 37.

and in 1972 already pointed at future possibilities for profitable foreign investment, when commenting on the government's 5-year plan for 1971–1975:

> The lapse of time is not sufficient to permit a meaningful appraisal of the impact of the Government's stabilization policies. Moreover, the economy is now in a recessionary phase due to labor unrest. . . . In the longer run, government stabilization policies should result in improved allocation of resources, through more realistic pricing of foreign exchange.[40]

For a period of a few years, mass support for the Velasco government prevented the imperialists from achieving very much; they even lost control over important economic sectors such as mining, oil and fish meal production. However, by February 1974, the United States managed to obtain Peru's signature for a treaty in which Peru accepted to pay 76 million US dollars in compensation for the expropriation of a number of U.S. companies. In view of the fact that the original claims of the companies amounted to 300 million US dollars, the treaty seemed rather favorable (apart from the fact that the Peruvian government accepted the principle of indemnification for property which was used to rob the country of a multiple amount).[41] The crucial point, however, was that in order to enable Peru to pay the 76 million US dollars, Morgan Guarantee Trust of New York granted Peru an 80 million US dollar credit [2497].

The Velasco government tried to find relief through borrowing from Third World countries like Iran [3174], but the imperialists tightened their grip and organized a "Consultant Group" for the coordination of all outstanding debts.* Marcona, the U.S. company expropriated because of working class action in early 1975, announced only a few months later that "since expropriation, a new President took control of the Government of Peru, and the Marcona administration has resumed discussions with the new Peruvian leaders" [3854], and a year later the imperialist press could report "concessions made to foreign capital—like the important indemnification for Marcona . . . and the totally unforeseen license for new oilfields. . . ."[42] During the years of 1976 and 1977 also a "more realistic pricing of foreign exchange" came into effect, raising the value of the U.S. dollar from 45.00 soles in April 1976, to 80.88 soles in July, 1977.† In October 1977, "with a view to promoting the restoration of confidence"—as the IMF put it—the Central Reserve Bank of Peru freed the exchange rate for the *sol* which led to a further depreciation of 27% within a period of a month.[43] This was the appropriate moment for the imperialists to "approve a stand-by arrangement"—agreement on further credit in principle, provided the debtor meets certain conditions which are checked regularly on the spot whenever a new purchase of the loan is up. The Peruvian government is anxious to try and meet the requirements—wage restraints, transfer of capital and interests, increased taxation of the masses.‡ As a result, Peru has seen and is going to see general strikes and mass struggle on a large scale.

This development is not an inevitable one. It was inevitable given the particular circumstances: a strong anti-imperialist movement under a bourgeois leadership. Nowhere (and never) can liberation from imperialist domination be consolidated if it is not carried on toward social revolution, toward liberation from exploitation of man by man because, without this social revolution, mass support of the state leadership will dwindle away as the exploitation of the working masses intensifies—the case of Peru, after 1974, is just one example. The necessity to proceed from national to social liberation in order to consolidate the achievements of the struggle for national liberation calls for the leadership of the proletariat in this process because it is the only class which can free a country's productive forces from the strings imposed by imperialism. Such a situation did not exist in Peru. The struggle against imperialism was led by the bourgeoisie, and the proletariat did not have the strength yet to forge a united front of peasants, workers and national bourgeoisie

*The group met in February 1972, in June 1973, and in April 1974, and consisted of representatives from Belgium, Canada, Finland, France, Federal Republic of Germany, Italy, Japan, Holland, Spain, the United Kingdom, and the United States [3373].

†This development is carefully monitored in the IMF *Survey* of those years.

‡"Up to now, only 80,000 of the 5 million Peruvians of wage earner age are effective taxpayers," pointed out the *Neue Zürcher Zeitung* 2 December 1978.

under the leadership of the working class. Without this united front, imperialism ultimately gains control.

It would be wrong to assume that the situation of Peru in 1978 is the same as in 1968. Economically, it is worse. The debt service ratio reached 46% at the end of 1978, and the imperialists, in order not to strangle their victim, had to agree to a moratorium. In other respects, some permanent changes did occur. The large *haciendas* with thousands of hectares of idle land definitely disappeared. There are even some remainders of the attempts to replace imported foodstuff by national production; an example of interest in relation to the Peruvian fishery is the introduction of a new process to produce edible oil from by-products of the fish meal industry.* It is possible to build on these few remaining achievements by uniting forces with the other Third World countries. This will prepare favorable conditions for a new struggle for national and social liberation.

STRUCTURE AND CHANGES IN PERUVIAN FISHERY

The preceding sections discussed the economic and social development of Peru between 1968 and 1978. Before discussing the economic and social effects of the El Niño of 1972–1973, which falls precisely in this period, it is probably worth stressing that the above results of our analysis so far do not depend on factors such as El Niño. There were a number of factors that *influenced* the development in Peru to some degree but did not *cause* the regaining of the imperialists' control over the country. One such factor was the fall in copper prices during the period, which created difficulties because it considerably reduced Peru's foreign income. The El Niño was another such factor. Whereas the fall in the price of copper was to a large extent organized by the United States, this cannot be said of the El Niño.

It follows that what is being considered in the remainder of this chapter are the modifications imposed on the process described so far by a natural event, which is adverse to the Third World country and, as far as the relations between the imperialists and the particular Third World country are concerned, favorable for the imperialists. Before considering this problem, the structure of the Peruvian fishery and its work force will be briefly assessed, on the basis of data of the Northern Region of Peru where the relevant statistics are available for 1970. In 1968, 46.0% of

*In 1975, 30% of all edible oils and 80% of fats and lards came from the fish oil industry. (Ministerio de Industria y Turismo, p. 172).

Peru's total catch came from that region, which operated 35.7% of Peru's fishing fleet, and supplied 45.6% of Peru's fish meal, 48.3% of canned fish, 100% of frozen and salted and 63.6% of fresh fish.[45]

To understand the development during the years after 1972, it is appropriate to divide the work force into three groups: those engaged in the catch for human consumption (5793), those who work in the catch of anchoveta for fish meal (7872), and those employed in the fish meal factories (3570.)[46]* The catch for human consumption is taken nearly exclusively by traditional fishermen, "a large part of them living in a situation of pauperization."[49] In terms of class analysis they can be compared with family peasants who work their own land: many of them are owners of their instruments of production; they are rapidly overtaken by modern techniques, but they constitute one of the classes which create wealth and can therefore be close allies of the working class and the peasants. In Peru they operate from boats often not exceeding 3–6 m in length and in many cases rely on a sail only (Table 13). It is interesting to note that when their number increased as a consequence of the government's incentives to boost the catch for human consumption, the new work force was recruited almost entirely from other areas of the fishery (Table 14).

In the anchoveta fishery, on the other hand, the tremendous increase of the work force was largely supplied by the working class (Table 14).

It is said that one of the better paid activities is that of an anchoveta fisherman; and it is assumed that his living conditions should be better

TABLE 13. Work Force, Craft, and Production in Peruvian Fishery, Northern Region, 1970

	Human Consumption	Fish Meal
Work force	5,793	7,872
Craft		
Engine-powered	705	548
No engine	495	0
Catch		
Million metric tons	111,518.0	4,691,845.1
Thousands of soles	476,099.9	1,619,330.6

Source. Reference 48.

*The third group should include those employed in the plants for the production of frozen and salted fish. This number is small, and these workers are most likely in a very similar position.

TABLE 14. Occupation of Peruvians Working in Fisheries before Joining this Type of Activity, Northern Region, 1970

Type of Fishery	Fisherman	Peasant	Worker	At School	Other	No Information	Sample Size
Human consumption	373	44	29	11	21	11	489
Fish meal	106	160	301	71	144	6	788

Source. Reference 47.

than the average of the fishermen and should facilitate him to reach substantial improvements in the general conditions of living, housing, clothing, education, health, etc. However, it has to be taken into consideration that this activity served as an occupation of refuge for many who were displaced from other activities, which, together with the predominantly low levels of instruction, explains the depressing conditions which exist.[50]

Thus, the crews of the anchoveta vessels are in a position no better than the traditional fishermen, but they accumulated a large amount of working class experience and can be expected to have developed a fair degree of class consciousness.

The workers in the fish meal industry, which in the Northern Region constitutes 40% of all industries, are in a much better economic position; they are "of the better paid within the workers and employees of industry."[51] The fact that the fish meal industry workers were organized in the Union of Fisheries Workers of Peru (FETRAPEP), whereas the sea-going workforce was organized in the Union of Fishermen of Peru (FPP) enabled the bourgeoisie to create an important political split during the aftermaths of the El Niño.

Figure 3 shows the development of the work force as laid down in the government plan of 1971 for the following 5 years. The large increase in the human consumption sector reflects the plan's target to reduce foodstuff imports by increasing the consumption of fish; no increase in the catch of anchoveta was planned, and the small increase in the work force of this sector is restricted to the processing plants where the plan foresaw increased output only as a consequence of improved processing techniques.[52] It is likely that the number of vessels and consequently the

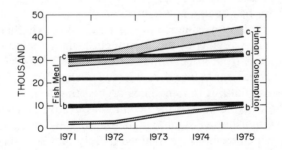

Figure 3 Projection of work force employed in Peru's anchoveta (dark) and other (shaded) fishery as planned in 1971. (a) extraction; (b) transformation; (c) total. (Source: Plan nacional de desarrollo 1971–1975, Vol. II, p. 199).

work force in the anchoveta sector was in excess of actual needs in 1971.[53] The major reason for the increase of the catch above scientifically recommended levels during 1971–1972, however, was the dominance of private capital in the fishing operations. It can be shown that nationalization of the processing of the anchoveta improved the fish/fish meal ratio. National control of the catch could have avoided the surpassing of recommended levels.

In the context of this study it is of utmost importance to note that the strategy of no increase in the anchoveta catch—which has its justification in itself because it helps to reduce Peru's dependence from the imperialists—also gradually reduces the economic impact of an El Niño: Because most of the species caught for human consumption are not affected by an El Niño, this sector of the fishery is based on much more stable natural conditions. (The development of the sector for human consumption continued without major interruption over the El Niño period; the plan for 1977–1978 lists 7 major projects under construction, including canneries, deep-freeze facilities, and handling of fresh fish.[54]) If the relative size of the work force employed in the anchoveta fishery is reduced, the percentage of the population directly affected by an El Niño becomes less and the possibilities to handle the problem increase. Since the implementation of the plan included improvement of the social conditions of the fishermen, such as the introduction of a social security scheme for the traditional fishermen and of a system of salvage and operational guidance,[55] it enjoyed their determined support. The General Law of Fisheries of 1971, which was supported unanimously by both the FPP and FETRAPEP, aimed at the implementation of the plan by bringing the marketing of fish meal under state control and providing the necessary infrastructure and further incentives for all fisheries for human consumption.[56]

In May 1973, after the El Niño, the whole anchoveta industry nearly collapsed. There has been some controversy as to whether the collapse occurred as a direct consequence of the El Niño or whether it was due to other reasons. A review of the anchoveta fishery[57] shows a dramatic fall in recruitment, from a mean index value of 438 for 1966–1971 to 56 in 1972, that is, one year before the observed changes in the hydrography.

Based on these data, it is sometimes argued that a recruitment failure rather than the El Niño caused the decline of the stocks. The view taken by the Peruvian fisheries scientists in the review is that the fall in recruitment—which cannot be explained merely on the basis of overfishing although this might have contributed—was an early indication of a major change in environmental conditions and that the process called El Niño commences earlier than indicated by temperature changes in coastal waters. There is no doubt that the heavy fishing before 1972 aggravated the consequences of El Niño, but it is also justified to say that the conditions which caused the collapse of the fishery were primarily natural ones, modified to some degree by human activity, and that they were attached to an El Niño.

The industry was nationalized in an attempt to guarantee employment in the sector.* It is most worth noting that during these difficult months, well into 1974, the government did not lose the fishermen's support. The Third National Congress of the FPP in August 1974, declared to "support and defend the achievements which resulted during the Revolutionary Process" [2727]. It was not before 1975 that a major split occurred, and this split was not so much the result of the difficulties created by the El Niño but of the general reversal of the relationship between the masses and the government during that time.

During 1975, the Ministry of Fisheries, encouraged by the government, began to lend its support to the Revolutionary Labor Movement (MLR), a bourgeois shock group within the working class which operated since 1973 but did not gain importance until 1975 when the reformist government decided to attack openly the working class. The MLR physically tried to take over trade union offices and on occasion did not abstain from violence. There is ample evidence that FETRAPEP which continued to declare its nearly unconditional support for the government [3370] collaborated with the Ministry of Fisheries in support of the MLR.[59] The FPP, on the other hand, while still defining the government as anti-imperialist and antioligarchist, declared at the Second Extraordinary National Congress in December 1975, that it will support all government action against imperialism and oligarchy "without losing our political class independence" and excluded several MLR leaders from its ranks [3972].

At the same congress, the Ministry for Fisheries announced an austerity programme for the state-owned fishing fleet PESCAPERU and a restructuring of the fishermen's social security system. It soon became clear that the government aimed at massive wage cuts and cuts in social security payments. A reduction of the work force by 5670 fishermen was announced in May 1976 [4340]. Next, PESCAPERU was partly returned to its old owners [4338] and a state of emergency was declared for the

---

*Decree published 5 August 1973, covering 105 plants and 1486 vessels. The owners received 10% in cash, 90% in nontransferable bonds at 6% interest redeemable in 10 years.[58] Considering the fact that they faced an appreciable loss over the El Niño season, this nationalization mainly served the capitalists who received good value on property that was virtually worthless under the given circumstances.

anchoveta industry. With the Decreto Ley 21558 of July the government hoped to get rid of the El Niño problem by proposing the preferential sale of PESCAPERU vessels to cooperatives set up by the fishermen themselves [4492]. The idea clearly was to disperse the sea-going proletariat into a large number of owner-operators who compete against each other and are ruined individually rather than dismissed in large numbers at a time.*

The FPP—which repeatedly declared its readiness to organize mass support for all measures to overcome the crisis if it is based on the national interest and not on the interests of the imperialists—denounced the attack on the working class:

> The Government declared a state of emergency for the anchoveta fishery, authorizing the most drastic measures of repression against anyone who tries to commit crimes against production. We understand that christian feeling and human treatment to give a solution to the minimal claims which we put forward have been forgotten. . . . The proletariat of the sea has been notified, the declared national emergency has put it on alert, we know from experience that no struggle of the exploited is easy. We are learning to rely on our own forces. [4253]

It denounced the Decreto Ley 21558 because it would inflict the loss of all achievements gained during thirty years of hard struggle [4500], a position which was publicly supported by several unions of other industries and by peasant organizations [4548]. Finally, at the beginning of the second anchoveta season it called for an all-out strike against the Decreto Ley 21558 (preferential sale of PESCAPERU vessels to cooperatives) which began on 18 October 1976, and lasted for nearly 5 weeks despite of government threats to sack all workers under emergency legislation [4694, 4701, 4710, 4829].

The Peruvian fishermen were defeated. On November 11th ten union bases declared the unconditional end of the strike:

> At the same time we call upon all comrades to understand this difficult situation for the proletariat of the sea and its families, and we urge them to remain solidly united around their union organizations. [4829]

Imperialism could win because it could exploit the existing split between

---

*As FPP pointed out in a leaflet during the strike, the law restricts the size of the new companies or cooperatives to three vessels. Since the new owners received a fixed price per ton, they would not be able to have enough income unless a large number of them is forced into bankruptcy [4710].

FPP and FETRAPEP which continued to support the government [4301] and considered the Decreto Ley 21588 to be "generally positive" [4633]. It was difficult for the FPP to organize massive support for its case from workers in other industries and peasants, under such conditions. It must also be kept in mind that the workers of the mining industry and the peasants were themselves involved in heavy struggles with only limited success. The situation in 1976 indeed called for a general strike, which did not materialize until 1978.

THE SIGNIFICANCE OF AN EL NIÑO FORECAST

We have seen that in the years that followed the El Niño the Peruvian masses suffered a severe setback in their living conditions; this was paralleled by and closely linked with increased imperialist control over the country. The heaviest struggles were waged by the peasants, the miners and the fishermen. The fishermen's struggle was initiated by particularly adverse natural conditions, the struggle of the miners by events which were the result of the capitalist system in its imperialist stage (the fall in the price of copper), the struggle of the peasants by the decision of the national bourgeoisie to prevent them from carrying national liberation further to social revolution.

The defeat of the peasants determined the outcome of all other struggles. A fall in copper prices can be counteracted by close unity of the Third World; its proposal to UNCTAD that the United Nations create a fund of basic resources to stabilize prices shows the way. The consequences of adverse natural conditions can be overcome in a similar way; certainly the Peruvian government would not have had major difficulties in obtaining financial or technical aid from Third World countries to readjust its fisheries. It is neither the El Niño nor the fall in copper prices that led to the reestablishment of imperialist control but the limitations that any national liberation struggle suffers under bourgeois leadership. The El Niño was a severe problem and a very untimely one for the process of national liberation but it did not decide its outcome. It did, however, under the given circumstances, considerably aggravate the situation of the masses.

What, then, can be said with respect to the "social value" of an El Niño forecast? Obviously, it is always desirable to be able to know changes in the natural conditions in advance; for the imperialists who could use knowledge of a coming El Niño for a better planning of their exploitation measures against Peru; for an anti-imperialist government who could take preparatory action, by negotiating early with Third

World countries on terms of loans, changes in its fishing fleet, temporary fishing licenses in other countries' waters, and so on; for the Peruvian masses who, through their trade union and political organizations, could urge the government to rely on the masses against imperialism under difficult conditions and, if the government rejects this line, could organize their ranks for a coming major battle. Overall, a forecast would not change the outcome. It would modify the details of the struggle, particularly if it is available to one of the opponents only, such as the imperialists. The factor which determines the outcome is the existence or nonexistence of a united front of workers, peasants and national bourgeoisie against imperialism which will never last very long unless it is lead by the working class. Masses make history; governments, as long as they rely on the masses and follow their main interests, can win against imperialism—otherwise, they lose. No El Niño forecast will change this.

This does not imply that one should not try to work for the establishment of a reliable forecast. All it says is that the forecast itself does not end or ease imperialist control over Peru.

Finally, let us just mention a few things on which other investigators spent hundreds of pages but which went without mention in this analysis. It is well known that the Velasco government and its successor were and are military governments. These and other aspects of Peruvian society have been treated *in extenso* in a fair number of studies. Mistakes of governments, personal styles of politicians, controversies between government and "industrialists" (which put the imperialist companies and the national bourgeois enterprises in the same category) enjoy comprehensive coverage, while actions of the imperialists generally remain unreported. The result, of course, is support for imperialism against a government which, despite all the oppression and crimes committed against the Peruvian masses, still plays an important role internationally in the efforts of the Third World to unite against imperialism and particularly against the superpowers. In a study of the main contradiction of Peru during 1968–1978, questions for example, relating to military or civilian government are of secondary importance and could therefore be safely disregarded.

REFERENCES

1 Octavio Diez Canseco B., ¿*Saldremos del Subdesarrollo?*, Librería Studium, Lima, 1969, p. 61.

2 Ibid., pp. 17–18.

3 Ministerio de Industria y Turismo, *La Industria en el Perú/Industry in Peru,* Oficina de Relaciones Públicas e Información, Lima, 1975, pp. 172–173.

4 Gustavo Espinoza R. and Carlos Malpica S.S., *El Problema de la Tierra,* Biblioteca Armauta, Lima, 1970, p. 283.

5 Ibid., p. 285.

6 Eduardo Anaya Franco, *Imperialismo, Industrialización y Transferencia de Technología en el Perú,* Editorial Horizonte, Lima, no year, p. 14.

7 Ibid., p. 13.

8 Ibid., pp. 45–54.

9 Octavio Diez Canseco, *op cit* p. 135.

10 Ministerio de Relaciones Exteriores del Perú, *Discurso Pronunciado por el Ministro de Relaciones Exteriores del Perú, General de División Edgardo Mercado Jarrín, el 18 de setiembre de 1970, en la XXV Asamblea General de las Naciones Unidas, en Nueva York,* Lima, 1970, p. 13.

11 Ministerio de Relaciones Exteriores del Perú, *El Perú en la Conferencia de Lusaka,* Lima, 1970.

12 Ministerio de Relaciones Exteriores del Perú, *El Perú en la ECLA, en 1970,* Lima, 1970.

13 Ministerio de Relaciones Exteriores del Perú, *El Perú en la XIV Reunión de Consulta de Ministros de Relaciones Exteriores en la OEA,* Lima, 1971.

14 República Peruana, Oficina Nacional de Información, *Juan Velasco Alvarado, Quinto Aniversario de la Revolución Nacional Peruana, Mensaje a la Nación 10 de marzo de 1973,* Lima, 1973.

15 República Peruana, Oficina Nacional de Información, *Mensaje a la Nación del Presidente de la República General de División Juan Velasco Alvarado, 24 de junio de 1969,* Lima, 1969, p. 14.

16 For the text of the Decreto Ley 17716 see F. Bonilla, *Reforma Agraria Peruana,* Editorial Mercurio, Lima, 1977.

17 Eduardo Anaya Franco, *op cit,* pp. 72–73.

18 Guillermo Lima, *Reforma Agraria y Lucha de Clases en el Perú.* Mimeo, Lima, pp. 26 and 34.

19 Ref. 15, p. 9.

20 República Peruana, Presidencia de la República, *Plan nacional de desarrollo para 1971–1975,* Volumen I: Plan global, Lima, 1971, pp. 75–76.

21 Ministerio de Relaciones Exteriores del Perú, *El Perú en la Conferencia de Lusaka,* Lima, 1970.

22 See, for example, Erik Thorbecke, *Some Notes on the Macroeconomic Implications of and the Cost of Financing Agrarian Reform in Peru,* International Studies in Economics, monograph 3, Iowa State University, Ames, IO, 1966.

23 F. Bonilla, *Constitución Política del Perú,* Editorial Mercurio, Lima, 1977, p. 14 and 83.

24 República Peruana, Oficina Nacional de Información, *Mensaje a la Nación del Señor General de División Presidente de la República, con motivo del sesquicentenario de la independencia nacional,* Lima, 1971, p. 7.

25 Guillermo Briones and José Mejía Valera, *El Obrero Industrial.* Instituto de Investigaciones Sociológicas, Universidad Nacional Mayor de San Marcos de Lima, Lima. 1964, p. 12 and 18.

26 Gustavo Espinoza and Carlos Malpica, *op cit,* pp. 236–237.

27 Gustavo Espinoza and Carlos Malpica, *op cit* p. 120 (upright numbers); Alberto Escobar, José Matos Mar and Giorgio Alberti, *Perú ¿país bilingüe?,* Tables 1 and 2, Instituto de Estudios Peruanos, Lima, 1975.

28 Alberto Escobar et al., *op cit,* pp. 118–119.

29 Gustavo Espinoza and Carlos Malpica, *op cit,* p. 296.

30 Gustavo Espinoza and Carlos Malpica, *op cit,* p. 113.

31 Gustavo Espinoza and Carlos Malpica, *op cit,* p. 113.

32 *Estatuto de Comunidades Campesinas del Perú,* Librería Distribuidora "Bendezu," Lima, 1970, pp. 14–15.

33 Alberto Escobar et al., *op cit,* p. 61.

34 Desco, Centro de Estudios y Promoción del Desarrollo, *Cronología Política,* several volumes 1975–1977, Lima.

35 For statistical evidence, based on a study of 1092 workers in Lima, see Guillermo Briones and José Mejía Valera, *op cit,* p. 102.

36 José Matos Mar, *Yanaconaje y Reforma Agraria en el Perú,* Instituto de Estudios Peruanos, Lima, 1976, p. 58.

37 Guillermo Lima, *op cit,* pp. 30–31.

38 For the history and impact of these events see Mariano Valderama, *Siete Años de Reforma Agraria Peruana,* Pontífica Universidad Católica del Perú, Lima, 1976, pp. 108–116.

39 *International Monetary Fund Survey Volume 4,* IMF, Washington, 1975, p. 324.

40 *International Monetary Fund Survey Volume 1,* IMF, Washington, 1972, p. 105.

41 Eduardo Anaya Franco, *op cit,* p. 30.

42 *Le Monde* (Paris), October 31, 1976. See also [4494].

43 *International Monetary Fund Survey Volume 6,* IMF, Washington, 1977, p. 367.

44 Ministerio de Industria y Turismo, *op cit,* p. 172.

45 Oficina Regional de Desarrollo del Norte (ORDEN), *Análisis general de situación, region norte, año 1970,* vol. 11: *Pesca,* ORDEN, Lima, 1970, p. 43.

46 Oficina Regional de Desarrollo del Norte, *op cit,* pp. 67, 75 and 134.

47 Oficina Regional de Desarrollo del Norte, *op cit,* p. 72 and 80.

48 Oficina Regional de Desarrollo del Norte, *op cit,* p. 113.

49 Oficina Regional de Desarrollo del Norte, *op cit,* p. 67.

50 Oficina Regional de Desarrollo del Norte, *op cit,* p. 75.

51 Oficina Regional de Desarrollo del Norte, *op cit,* p. 134.

52 República Peruana, Presidencia de la República, *op cit,* pp. 42, 65 and 106.

53 Garth I. Murphy, "Fisheries in upwelling regions—with special reference to Peruvian waters," *Geoforum,* **11,** 63–71 (1972). See also *La Prensa* (Lima), May 3, 1977.

54 República Peruana, Instituto Nacional de Planificación, *Plan global de desarrollo para 1977 y 1978,* Lima, 1977, pp. 180–181.

55 República Peruana. Presidencia de la República, *op cit*, p. 204.

56 Ministry of Economy and Finance, Office of Public Relations, General Fisheries Law (Decree Law 18810), Editorial VIRU, Lima, 1971. (Priorities and incentives are covered in Articles 52–55, state monopoly for fish meal marketing in Article 3).

57 Actas de la Reunión de Trabajo sobre el Fenómeno Conocido como "El Niño" Guayaquil, Ecuador, 4–12 de diciembre de 1974. FAO, Roma, 1976, pp. 80–93.

58 *Wall Street Journal,* May 9, 1973.

59 Desco, *op cit,* Volume 1975, p. vi.

PART IV
THE FUTURE

CHAPTER 15

On Predicting El Niño

COLIN S. RAMAGE

Abstract

Although weather cannot be forecast accurately more than a day or two in advance, meteorologists have responded to public need by attempting to make seasonal or even long period forecasts. They reason that large seasonal anomalies must surely be linked to or caused by recognizable antecedent conditions. Except possibly in cases of winter snow or ice accumulation, causes producing later prolonged effects have not been identified, and so most long-range forecasting depends on unstable statistical relations with no detectable physical bases. Performance is consequently disappointing.

In the case of El Niño, decreasing central equatorial Pacific trade winds as reflected in Easter Island-Darwin pressure difference, seem to "cause" El Niño-like conditions to develop in about two months. Quinn's[17] long-range forecast technique depended on this sequence and on the propensity for the trade winds, after becoming unusually strong, to then decline over many months. Quinn[18] and Wyrtki[19] forecast a weak to moderate El Niño in 1975 but the event was aborted. Since then, Quinn and his associates have developed a more complex statistical basis for forecasts, but are no closer to discovering causes of trade wind fluctuations which in turn seem to drive the El Niño cycles. To make matters worse, instead of sharply differing El Niño and anti-El Niño periods, we now find that the whole gamut of intensity, extent and duration occurs. Wright's[23] recent attempt to introduce ocean-atmosphere feedback is likely to fare no better since once more the essential physical link cannot be found.

El Niño and the closely related North and South Pacific trade winds are integral parts of the global atmospheric-ocean machine and might be successfully predicted only in the unlikely event of worldwide seasonal forecasts succeeding.

INTRODUCTION

Ideally, understanding a natural phenomenon usually comes after it has first been observed (or measured), described, and analyzed. At this stage a hypothesis is advanced which is then tested by using it to forecast a future state of the phenomenon. Forecast imperfections suggest modifications to the hypothesis, leading to a new forecast. Successive iterations, usually coupled with new observations, finally lead to a sustainable hypothesis. In most earth science disciplines, seismology, for example, hypothesis is rarely, if ever, tested by forecasting. However, for more than a hundred years public demand has forced meteorologists to predict long before observation, description, and analysis could provide the framework for adequate hypotheses. Little wonder the weather forecaster is sometimes wrong.

Besides wanting daily weather forecasts, the public, especially farmers, has a need for seasonal forecasts and this too the meteorologist has striven to satisfy. Since a season comprises the aggregate weather of that season and since weather itself cannot be forecast for more than a couple of days ahead, accurate seasonal forecasting would seem to be unattainable. Nevertheless, many meteorologists remain undaunted; in fact, seasonal and climate forecasting are being given more attention now than ever before (see, for example, Climate Research Board[1]). The reason for this can be found in the existence of *persistent* weather regimes, producing, for example, a devastating drought (the Sahel, 1968–1973) or a severe winter (United States, 1977–1978). Surely, the argument goes, such massive prolonged events must stem from equally significant antecedents; if these can be identified, prediction might be feasible. Reality is much less satisfactory. The aim of identifying generally plausible physical links and of testing them statistically has usually been abandoned in favor of developing purely statistical relationships between earlier and later events.

LONG-RANGE FORECASTING WITH A PHYSICAL BASIS

Blanford, after studying the Indian summer monsoon rains from 1864 to 1881, concluded that the "varying extent and thickness of the Himalayan snows exercise a great and prolonged influence on the climatic conditions and weather of the plains of NW India."[2] He hypothesized that more snow in spring, by delaying the advance of summer, resulted in less summer rain. Although the hypothesis worked in 1885, it failed in other years, was unconfirmed by statistics and so abandoned. Lack of success could have resulted from the inadequacy of a few spot measurements to delineate snow cover. The area covered by snow (but not the depth) can now be mapped by weather satellites.[3] Hahn and Shukla have reported that in 8 of the 9 years 1967 through 1975, winter snow cover over Eurasia was inversely related to Indian rainfall during the following summer.[4]

In a normal year, the early summer rains of the Yangtze Valley cease by mid-July as the polar front weakens and shifts north. In some years, however, the shift occurs by the end of June and the valley suffers a drought; in other years when the shift is delayed until the end of July, the month is noteworthy for disastrous floods.[5] In all of China, July rainfall in the Yangtze Valley is the most precarious with variability exceeding 60%. The flood of July 1931 stemmed from a prolonged severe Siberian winter.[6] This delayed the ice melt over the Bering and Okhotsk seas, thus maintaining the Okhotsk high for longer than usual. Over southern China warm moist air swept in from the Pacific high to meet the cold polar air along the active polar front. Lows developed, stagnated, and precipitated. The 1954 floods arose from similar causes.[7] In drought years the Sea of Okhotsk warms rapidly and the Okhotsk high becomes weak by the end of June. In flood years, ice melt is delayed over the Bering and Okhotsk seas and the Okhotsk high persists for most of July.

As spring advances the cold but ice-free coastal waters off southeast China warm about as fast as the air. By contrast, both the Himalayan snows and Okhotsk Sea ice "predictors" involve changes of state of water (melting, sublimation). These processes use heat (80 and 680 cal/g, respectively) and thereby slow the very atmospheric temperature increases which cause melting and sublimation.

It seems physically reasonable to postulate that the more ice and snow accumulates in winter, the longer summer weather could be delayed. However, the possible effects of Himalayan snows and Okhotsk Sea ice on subsequent summer weather awaits rigorous statistical testing.

LONG-RANGE FORECASTING WITHOUT A PHYSICAL BASIS

The concept of correlation to test whether preceding events almost anywhere in the world might have a significant relationship with subsequent seasonal rainfall over various Indian climatological subdivisions was introduced by Walker and made the basis of a new method of seasonal forecasting.[8] From thousands of computed predictor-predictand correlation pairs, those having the highest correlation coefficients were combined in linear regression equations designed to forecast subsequent seasonal (3-month) rainfall over subdivisions of India.

At the beginning, Walker warned that the probable error in the highest of a group of correlation coefficients is significantly greater than the probable error of a coefficient selected at random from the group.[9] For example, in a set of 50 correlation pairs ($n = 50$), the respective probable errors are 0.35 and 0.10. In the long and vigorous search for better predictors, Walker's disciples and perhaps even Walker himself appeared at times to ignore this important limitation of the method.

Walker was at the mercy of statistics, for he could find no plausible physical links between his predictors and the predictand. His regression equations performed poorly and as the dependent data sample grew, predictors were modified or even replaced by new predictors. Of those used by Walker,[11] four were rejected and only three retained through 1960 and the correlations of these varied widely through the years (Table 1).[10]

Bell reported similar secular changes in correlations.[12] Since 1959, the correlation coefficient between the January Irkutsk-Tokyo surface pressure difference and subsequent summer rainfall in Hong Kong amounted to −0.82. However, between 1925 and 1940 it was almost zero and reached +0.3 at the end of the last century.

The history of long-range forecasting is a sorry one. Physical links are

TABLE 1. Decade Correlation Coefficients of the Three Most Consistent Predictors of Summer Monsoon Rain over Peninsular India*

Period	Southern Rhodesian Rain	South American Pressure	Java Rain
1881–1890		0.48	−0.11
1891–1900		0.78	−0.68
1901–1910		0.42	−0.48
1911–1920	−0.72	0.27	−0.19
1921–1930	−0.24	0.02	−0.24
1931–1940	−0.19	0.50	−0.07
1941–1950	0.31	0.58	0.04
1951–1960		−0.73	−0.11
Entire period	−0.31	0.34	−0.21

*From Reference 10.

extremely hard to recognize across a span of months while even the best statistics have accounted for only a fraction of the variance and usually fail to forecast important extreme events on the wings of the frequency distribution. Even so, because many had anticipated the 1972–1973 El Niño from its earliest stages, hopes were kindled that the equatorial Pacific cycle of which El Niño is a part might be predictable some months in advance.

EL NIÑO FORECASTING

The surface air pressure difference between Easter Island (27°10'S; 109°25'W) and Darwin (12°26'S; 130°52'E), the two foci of Walker's[11] Southern Oscillation, was shown by Quinn and Burt[13] to be statistically related to rainfall over the central and western equatorial Pacific. From a study of the pressure difference curves, Quinn and Burt concluded that "if the pressure difference fell to 5 mb or less by the end of April or May, an extended period of abnormally heavy equatorial rainfall would be predicted to occur during the following 9–12 month period; otherwise a nonoccurrence would be predicted." In May 1972 the difference amounted to only 0.6 mb; for almost a year thereafter the central equatorial Pacific experienced extremely heavy rain. This coincided with the 1972–1973 El Niño, encouraging Quinn to apply his statistics to this phenomenon. At about this time Wyrtki's hypothesis of the origins of El Niño appeared.[14,15] He postulated a forcing atmosphere and a slowly responding ocean which in turn might influence the atmosphere. When the central equatorial Pacific trade winds diminished from previous high speeds, a train of events ensued embodying an El Niño cycle which might last one or even two years. Easter Island-Darwin pressure difference proved to be a good index of equatorial Pacific trade wind strength. Once the central Pacific trade winds had become strong enough (high index), they were sure to decrease. The consequent reduction in ocean surface stress then would lead (with about a 2-month lag) to surface waters warming off South America and along the equator. In the past, since data became available, the index fall and accompanying trade wind decrease were as prolonged over many months as was El Niño. The causal connection between trade wind strength and subsequent change in sea surface temperature appeared fairly well established and has received some statistical confirmation.[16] On the other hand, no physical reason could be adduced for a prolonged trade wind decrease.

To allow for oceanic lag and to highlight interannual changes, Quinn began using 12-month running means of the pressure difference.[17] The historical record indicated to him that "once the index peak ($\geq$ 13 mb) is reached and a downward trend sets in, one could expect an El Niño invasion about 9–13 months later, with January to March the following year as the inception time . . . [T]he outlook could be given 3 months or more in advance of the El Niño invasion." A forecasting opportunity soon arose.

THE 1975 EVENT

In October 1974, Quinn forecast "a weak El Niño occurrence in early 1975, unusually heavy central and western equatorial Pacific precipitation in 1975 and below-normal precipitation over Indonesia in mid-late 1975."[18] He qualified the forecast by saying that "this development may be underestimated" along the coasts of Peru and southern Ecuador. K. Wyrtki,[19] using essentially the same data, predicted El Niño to begin in early 1975 and to resemble the 1965 event which Quinn et al. classed as moderate.[20]

In February and March 1975 the research vessel *Moana Wave* made a detailed oceanographic and meteorological survey between 2°N and 14°S and between 95°W and the South American coast.[21] All the features of a developing El Niño were present—above-normal sea surface temperature, weakened trade winds and a weak or absent trade wind inversion.

However, a second survey during April and May 1975 showed the trend reversed with sea surface temperature as much as 5°C lower, freshened trade winds and much stronger trade wind inversion. The

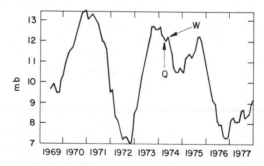

Figure 1 Twelve-month running means (points plotted at middle of 12 months) of the difference in sea level atmospheric pressure (mb) between Easter Island and Darwin, 1969–1977. The data prior to "Q" and "W" on the curve were available to Quinn and Wyrtki, respectively, when they made their forecasts of the 1975 event.

rainfall forecast fared no better. The author has tabulated 1975 monthly rainfalls for 10 near-equatorial stations stretching from Kapingamarangi (155°E) to Christmas Island (157°W). Of a total of 113 station months, 84 had rainfalls *below* normal. For the period July–December 1975, of 54 stations well distributed through Indonesia and Malaysia, 37 were *wetter* than normal and at five of these 200% of normal was exceeded.

Figure 1 shows why the forecast failed. In late 1973 the index reached 12.7 mb at the same time of year as the 1970–1971 maximum (which preceded the 1972–1973 El Niño) but 1 mb weaker. It began to fall in 1974. When the forecasts were made the fall had lasted 3 months. After plunging sharply the index leveled out at 10.5 mb and climbed back to 12.2 mb by August 1975, the first case of an aborted fall in the 35 years of the index. The ensuing fall to 7.3 mb was accompanied by a moderate El Niño in 1976.[20]

OUTLOOK

Subsequently, Quinn and his associates have replaced the 12-month running mean with a triple 6-month running mean filter on the monthly anomalies (from the long-term means).[20] They now use a much more complicated prediction scheme. However, it is still based on statistics relating the modified curve to El Niño events.

The problem can be simply stated—how to forecast Central Pacific trade wind strength months in advance? In 1973, the author, along with many others, held very simplistic views on the atmosphere-ocean system of the Central Pacific/coastal South America. We visualized two relatively stable, persistent modes—*El Niño* (large positive anomalies in sea surface temperature and in Line, Phoenix, and Gilbert Islands rainfall), *anti-El Niño* (correspondingly large negative anomalies), separated by relatively brief, unstable transitions. Recognizing onset of a transition, as was done in 1972, would be tantamount to making a successful long-range forecast.

Reality is much more complex. Within only the last 5 years we have observed aborted transitions (1975), prolonged transitions (1976–1977), a North Pacific El Niño (1976), and even a "half" El Niño. December 1977 was the wettest month on record at Canton Island and there were frequent tropical cyclones in the Central, North, and South Pacific. However, off the coast of South America corresponding anomalies were absent. Four years ago the author confidently listed six major El Niños between 1890 and 1972, choosing to ignore any other perturbations.[22] A recent careful study identified no fewer than 25 El Niño events during this period, classifying them as either strong, moderate, weak, or very weak.[20] If anti-El Niño events can be similarly classified, the historical record must include an unacceptably large random component and no amount of statistical manipulation could result in useful forecasts.

Quinn is apprently fated to repeat Walker's experience in trying to forecast Indian summer monsoon rainfall. As data on which to base statistics increase, so do the range of possibilities and the risk of failing to predict the important events.

Recently, Wright proposed a method for forecasting Central Pacific rainfall anomalies based on the statistics of their persistence.[23] He found that anomalies developing during May to September are most likely to persist until the following February–March. Realizing that statistical forecasts without physics do poorly, Wright tried to explain his findings in terms of positive feedback between ocean and atmosphere, the strength of the interaction being a function of season. Since feedback cannot yet be measured, the argument becomes circular with the statistics providing the only evidence for feedback and the only indication of its intensity. So once again the demonstrable physical link is missing.

CONCLUSIONS

The Central Pacific southeast trade winds wax and wane irregularly. Large amplitude changes from strong to weak lasting several months cause major El Niño events with sea surface temperature rise lagging wind decrease by about two months. Although this causal mechanism is fairly well understood, causes for the antecedent trade wind fluctuations are obscure. Presumably they originate in global-scale atmospheric interactions. Thus, successful El Niño forecasts might only be achieved in the framework of successful worldwide seasonal forecasting, a task of truly daunting magnitude. Even then, probabilistic forecasts might be the best we could hope for.

ACKNOWLEDGMENT

The work leading to this paper was partly supported by the National Science Foundation under Grant ATM 76-16737 A01. Contribution No. 79-4 of the Department of Meteorology, University of Hawaii.

REFERENCES

1 Climate Research Board, *International Perspectives on the Study of Climate and Society*, National Academy of Science, Washington, D.C., 1978.

2 H. F. Blanford, On the connexion of the Himalayan snowfall and seasons of drought in India. *Proceedings of the Royal Society, London*, **37**, 3–22 (1884).

3 D. R. Wisnet and M. Matson, Monthly winter snowline variation in the northern hemisphere from satellite records, 1966–1975. *NOAA Tech. Memo.* NESS 74, 1975.

4 D. G. Hahn and J. Shukla, An apparent relation between Eurasian snow cover and Indian monsoon rainfall. *Journal of Atmospheric Science*, **33**, 2461–2462 (1976).

5 C. Lee and M. W. Wan, The climatic types of floods and droughts in the valleys of the Yangtze and the Hwai Rivers (in Chinese). *Acta Geographica Sinica*, **21**, 245–258 (1955).

6 J. Lee, Die grosse Uberschwemmung des Jangtse-gebietes in Juli 1931. *Meteorologische Zeitung*, **49**, 234–237 (1932).

7 H.-Y. Chen, The circulation characteristics during the 1954 flooding periods in the Yangtze and Hwai Ho basins (in Chinese). *Acta Meteorologia Sinica*, **28**, 1–12 (1957).

8 G. T. Walker, Correlations in seasonal variations of weather, VIII. A preliminary study of world weather. *Memoirs of the India Meteorological Department*, **24**, 75–131 (1923).

9 G. T. Walker, Correlations in seasonal variations of weather, III. On the criterion for the reality of relationships of periodicities. *Memoirs of the India Meteorological Department*, **21**, Part 9, 12–15 (1910–1916).

10 K. N. Rao, Seasonal forecasting—India. WMO-IUGG symposium on research and development aspects of long-range forecasting, *WMO Tech. Note No. 66*, Geneva, 1965.

11 G. T. Walker, World weather II. *Memoirs of the India Meteorological Department*, **24**, 275–332 (1924).

12 G. J. Bell, Seasonal forecasts of Hong Kong summer rainfall. *Weather*, **31**, 208–212 (1976).

13 W. H. Quinn and W. V. Burt, Use of the Southern Oscillation in weather prediction. *Journal of Applied Meteorology*, **11**, 616–628 (1972).

14 K. Wyrtki, E. Stroup, W. Patzert, R. Williams, W.H. Quinn, Teleconnections in the equatorial Pacific Ocean. *Science*, **180**, 66–68 (1973).

15 K. Wyrtki, El Niño—the dynamic response of the equatorial Pacific Ocean to atmospheric forcing. *Journal of Physical Oceanography*, **5**, 572–584 (1975).

16 T. P. Barnett, An attempt to verify some theories of El Niño. *Journal of Physical Oceanography*, **7**, 633–647 (1977).

17 W. H. Quinn, Monitoring and predicting El Niño invasions. *Journal of Applied Meteorology*, **13**, 825–830 (1974).

18 W. H. Quinn, Outlook for El Niño-type conditions in 1975. *NORPAX Highlights*, **2**(6), 2–3 (1974).

19 U.N. Food and Agriculture Organization, Report of the Workshop on the Phenomenon Known as "El Niño," Guayaquil, Ecuador, 4–12 December 1974. *F.A.O. Fisheries Reports*, No. 163, 1975.

20   W. H. Quinn, D. O. Zopf, K. S. Short, and R. T. W. K. Yang, Historical trends and statistics of the Southern Oscillation, El Niño and Indonesian droughts. *Fishery Bulletin*, **76**, 663–678 (1978).

21   W. C. Patzert, T. J. Cowles, and C. S. Ramage, *El Niño Watch Atlas*. Scripps Institution of Oceanography, La Jolla, 1978.

22   C. S. Ramage, Preliminary discussion of the meteorology of the 1972–1973 El Niño. *Bulletin of the American Meteorological Society*, **56**, 234–242 (1975).

23   P. B. Wright, Persistence of rainfall anomalies in the central Pacific. *Nature*, **277**, 371–374 (1979).

CHAPTER 16

Considerations of the Societal Value of an El Niño Forecast and the 1972–1973 El Niño

MICHAEL H. GLANTZ

Abstract

There has been a great deal of research on the scientific aspects of the El Niño phenomenon in an attempt to reduce the uncertainties surrounding its environmental and biological impacts. However, little research effort to date (with a few notable exceptions) has been directed toward the societal value of an El Niño forecast. This chapter is an attempt to address that gap.

More than 60 experts in various fields were asked what they might have done in 1972–1973 if a hypothetically perfect El Niño forecast had been available 2–4 months in advance of its onset. The purpose of this approach (what might be called a case scenario) was to determine how much flexibility decisionmakers in Peru and elsewhere might have, given the hypothetical availability of an extremely reliable El Niño forecast. The study's findings suggest that while in theory many good uses can be attributed to the development of a reliable El Niño forecast, in practice it appears that its value may be quite limited.

INTRODUCTION

This chapter discusses some of the societal implications of a hypo-thetically reliable El Niño* forecast, primarily for Peru but for other

*For this chapter El Niño is defined as the occurrence from time to time of a unique meteorological-oceanographic event—an invasion of nutrient-poor warm water into the eastern equatorial Pacific from the western part of the Pacific Basin. It is important to note that there are scores of contending definitions of the phenomenon known as El Niño.

countries as well. Such a forecast would contain information on the general location, intensity (in terms of geophysical parameters), and lead time of this event. To date, much attention and research effort have been directed toward improved understanding of the physical and biological aspects associated with the Peruvian upwelling and El Niño. Some scientific researchers have suggested that such an improved understanding will eventually mean the ability to forecast reliably, some months in advance,[1] the economically devastating El Niño, thereby minimizing its adverse societal effects. On this, Klaus Wyrtki et al.,[2] among others,[3-5] have suggested that

> the economy of Peru is strongly influenced by its fisheries, as is the world market of protein for animal feed, and so a prediction of the occurrence of El Niño would be a valuable guide for long-range economic planning.

Aside from occasional, usually brief comments in the literature about the importance of the Peruvian fisheries to Peru and to the world, the main research interest with respect to that fishery has apparently been in forecasting what the state of the environment will be at some time in the future.

The regions where coastal upwelling generally occurs represent on the order of one-tenth of one percent of the ocean's surface area; yet in the recent past they have accounted for 44 to 50% of the commercial fish taken from the sea.[4] Upwelling, and therefore relatively high biological productivity, also occurs in midocean regions around the Equator and in the southern ocean off Antarctica.* (For an estimate of the potential yield from the upwelling areas of the world, see D. H. Cushing.[6])

The U.S. National Science Foundation (NSF), through its International Decade of Ocean Exploration (IDOE) division, has funded a 6-year multinational, multidisciplinary (with respect only to the physical and biological sciences) project called Coastal Upwelling Ecosystem Analysis (CUEA).† A stated goal of this project was "to understand the coastal upwelling ecosystem well enough to predict its response far enough in advance to be useful to mankind."[11] In reality the goal has been to

---

*There has been much recent interest in the exploitation and conservation of Antarctic living marine resources (i.e., krill).[7]

†For a general discussion of the IDOE goals see IDOE's second report.[8] Similarly, CINECA (Cooperative Investigations of the Northern Part of the Eastern Central Atlantic) was concerned with scientific aspects of the Northwest African coastal upwelling. (For comments on the scientific aspects of CINECA, see, for example, *Téthys*;[9] for comments on ideological and other problems associated with CINECA, see Tomczak.[10])

develop an ecological systems model by coupling a physical oceanography system with a biological systems model. The coupled model would then serve to improve the ability to forecast upwellings and, therefore, to forecast reliably the location of the most productive fisheries. Again, comments emerge on the proposed value to society of such a forecast capability.

> The Coastal Upwelling Ecosystem Analysis Program is predicated on the idea that the coupling exists, that it is understandable, and that the responses can be predicted far enough in advance and well enough to be useful in managing, preserving, and wisely exploiting the vast protein resources of the upwelling regimes.[11]

While the CUEA project has been multidisciplinary with respect to the natural sciences (i.e., biology, chemistry, physical oceanography, meteorology)—and despite statements in IDOE documents such as "amongst the oceanic phenomena affecting man, coastal upwelling has the most direct and largest impact on the social fabric"[11]—there has apparently been little effort to investigate the political, legal, economic, and environmental implications of the availability of a reliable coastal upwelling forecast in general or more specifically of an El Niño forecast. While brief statements alluding to the economic and social importance of an upwelling or an El Niño forecast appear in project reports, program descriptions, memos, and letters pertaining to the CUEA project, such statements have in the past reflected either an unassessed view that such forecasts have only positive benefits for humanity or a concern for some degree of justification of the project in societal terms. An example of this reasoning appeared in an article summarizing current scientific research efforts on upwelling and El Niño forecasts.[12] It was suggested that:

1   By gaining a deeper understanding of the physical and biological dynamics of upwelling, scientists, fishermen, and national legislators will have more potential control over the supply of fish stocks.
2   The major practical benefit of an upwelling prediction system for locations throughout the world is that fishermen will be able to obtain an above-average harvest with less time wasted searching for fishing locations.

Recently the minimal progress made on research related to the social application of some of these scientific findings has been acknowledged. For example, a report from a post-IDOE planning session acknowledged this by noting that

> when IDOE was established, its objective of achieving more comprehen-
> sive knowledge of ocean characteristics and more profound understanding
> of oceanic processes was linked to the more effective utilization of the
> ocean and its resources. . . . In practice, the IDOE investigations have
> leaned toward fundamental research.[13]

The report later called for continued emphasis on fundamental research
but asked scientific researchers to keep in mind the areas of eventual
application of research findings.

Often, the emphasis on the scientific aspects of such research activities
reflects a prevailing view that decisionmakers (economic, political, and
others) need more scientific information than they currently have available before they can make "rational" decisions about, for example, the
rates and levels of exploitation for a given living marine resource. This
author questions the view, often expressed within the scientific community as well as within policymaking circles, that "until more is
understood about the influence of temperature, currents, pollutants, and
weather on marine life, sensible decisions about the management of these
resources will not be possible." It can be argued that at some point
enough scientific information has been gained to make these necessary,
sensible decisions and that at such a time continued political and
administrative inaction coupled with a call for "more scientific information" is no longer justifiable.

Associated with the scientific research on eastern equatorial Pacific
upwelling and El Niño there should be a concern about the implications
of an El Niño forecast before it is developed, so that preparations, or at
least consideration of such preparations, could begin to deal with the
possible effects—both positive and negative—that might accompany the
development of such a forecast.

An assessment of the impact of an El Niño forecast is essential and
timely (though the forecast capability has not yet been developed)
because of the extremely sensitive nature of national and international
marine policies and, as can be seen from the Law of the Sea deliberations
concerning ocean resource management policy, as much lead time as
possible would be needed to prepare for the international and national
effects of the development of such a forecast. It is also especially timely
given the optimism that some researchers have placed on the ability of
the scientific community to develop in the not-so-distant future a "reliable" El Niño forecast.

It is also important to be aware of the possibility that a reliable El
Niño forecast might lead to adverse societal effects, if for no other
reason than by perhaps "producing false expectations with regard to the

value of fisheries wealth, thereby tempting entrepreneurs (including those not now involved in the fisheries sector) to go to greater lengths than desirable in order to acquire greater shares of the wealth for themselves."[14]

Of those countries in the eastern equatorial Pacific most directly affected by the 1972–1973 El Niño (Ecuador, Peru, and Chile), Peru has been the most noticeably affected, because its fishing industry had rapidly developed in less than two decades and had become a key factor in the Peruvian economy. Peru, therefore, can serve as a focal point for an assessment of the value to society of a hypothetically reliable El Niño forecast. Although all sectors of society, including agriculture, might be directly affected by a reliable El Niño forecast, primary attention is focused on its effect on the exploitation and conservation of the anchoveta fishery.

THE FISH MEAL BOOM

It was recently suggested that the Peruvian anchoveta fish meal boom, unlike earlier Peruvian economic activities such as guano, sugar, silver, copper, and cotton production, would not be another of those boom-to-bust economic ventures.[15] However, events (both physical and societal) since the late 1960s suggest that the boom-to-bust phenomenon may have claimed another victim.

The Peruvian fish meal industry based on the anchoveta underwent a meteoric expansion between the mid-1950s and the late 1960s. Its development began slowly in the early 1950s, with some impetus coming from the growing postwar demand for fish meal and from the collapse of the California sardine industry, and received a major boost in the mid-1950s with the development of nylon nets* to replace cotton ones. The sharp, almost exponential, increase in officially reported anchoveta landings is evident in Figure 1.

Opposing views concerning the reasons behind this sharp increase in the development of the Peruvian fish meal industry have emerged. Some authors have attributed it to social and economic forces within Peru (entrepreneurs, the availability of the markets, the seemingly unlimited availability of the anchoveta, its "unknown growth factor" as a feed supplement, the desire to use the fish meal export earnings to promote the development of the Peruvian economy, and so on). With respect to

---

*Nylon nets were relatively more durable, stronger, and lighter. In general they facilitated the fishing effort and lowered the operating costs associated with the fleet.

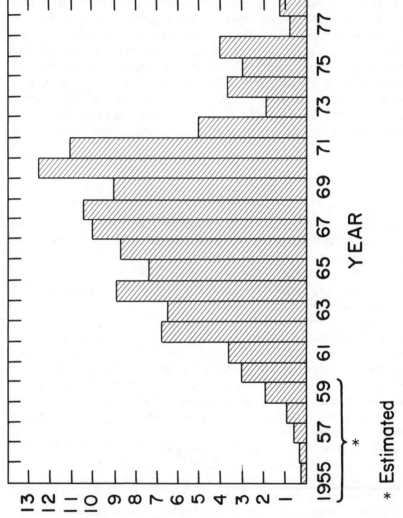

Figure 1 Peruvian anchoveta catches.

export-led development, it was believed that growth and development in one sector of the economy could finance growth and development in other sectors.[16]

Others have argued that the reasons for the sharp rise were external to Peru and centered around capitalistic or neo-imperialistic aspects of international economic relations.[17,18] It has been said that the industrialized countries (the United States, Western Europe, and Japan, primarily), have used the high-protein fish meal to support their increasing affluence and their inefficient use of protein by consuming meat rather than grains and fish—that is, converting the grains and fish to meal that is fed to livestock and poultry—or by using fish meal in the preparation of pet foods at a time when per capita food availability appeared to be declining worldwide. A respondent, a representative of the latter ideological perspective, has summarized this, noting that

> Peru is a country of the Third World, i.e., it depends on and is exploited by the imperialist countries. It follows that the question how to react to a coming El Niño can be answered from two points of view that are invariably opposed to each other: from the point of view of the imperialists, and from the point of view of the oppressed country.

Thus, it is important to keep in mind that one's ideological perspective will shape one's view of the societal value of an El Niño forecast.

The fishing effort increased into the early 1970s, despite some government measures to reduce that pressure.[19] Regulations proved to be ineffective. Reduction in fishing effort, for example, often coincided with the periods of naturally low availability of the anchoveta. Fleet and factory owners sought to circumvent regulations (i.e., limiting the number of trips per month encouraged fishermen to use larger vessels with larger hold capacities). Considerations of such nontechnological factors affecting potential yields as the learning factor bias of the fishermen, that is, experience gained in locating and landing fish, were left untouched.

The industry was considered to be greatly overcapitalized in most of its component sectors. The IMARPE report of the Economic Experts made note, in very clear and ominous terms, of this overcapitalization and its eventual devastating impact on the resource itself.[20] Even though the report made no mention of complicating factors such as El Niño, it suggested that there was a high probability the Peruvian achoveta fishery would collapse if the economic pressures were permitted to continue to dominate the exploitation of the fishery.

Commenting on an aspect of the economics of overexploitation of fisheries, C. W. Clark noted:

Renewable resources, by definition, possess self-regeneration capacities and can provide man with an essentially endless supply of goods and services. But man, in turn, possesses capacities both for the conservation and for the destruction of the renewable resource base.[21]

The anchoveta fishery serves as a useful example of the problems associated with the exploitation of a common-property renewable living marine resource (for this and other examples, see Bell[22]). It represents a challenge to economic modelers who attempt to balance commercial exploitation of this resource against its conservation, a balance sought in theory by rational economic man; it is also important, of course, to consider the possibility that "a corporate owner of property rights in a biological resource might actually prefer extermination to conservation, on the basis of maximum profits."[21]

The exploitation-conservation balancing act for the Peruvian anchoveta fishery is further complicated by El Niño. This invasion of nutrient-poor water is quasiperiodic, which means (to a social scientist) that it returns every so often but not necessarily at the same time or with the same magnitude, that is, intensity, duration, and geographic scope. The El Niño overshadows the coastal upwelling processes which ordinarily bring cold, nutrient-rich, deep water up into the euphotic zone, where those nutrients are photosynthesized and consumed at lower trophic levels of a food chain that culminates with the anchoveta, the guano birds, and with the Peruvian fishermen.

Ramage (1975) noted that in addition to the major one in 1972–1973, there have been two other major El Niño events since the early 1950s— in 1957–1958 and 1965–1966.[23] However, the societal impacts of these two earlier events were relatively minor compared to those of the El Niño in 1972–1973, perhaps because fishing pressures in the earlier years were apparently still below some maximum sustainable yield (MSY)* for the given environmental conditions and in part because the combination of fishing pressures and El Niño intensity did not adversely affect the standing stock of the anchoveta from which future catches would come. Also, in the years preceding the earlier El Niño events, anchoveta recruitment levels (the yearly number of small fish entering the fishery) were favorable. Such was not the case, however, in 1971.[27,28]

---

*For discussions of this extremely controversial concept see, for example, Edwards and Hennemuth,[24] Larkin,[25] Gulland,[26] and Bell.[22]

EFFECTS OF THE 1972–1973 EL NIÑO

At the least, one can argue that the 1972–1973 El Niño, in terms of its impact on society, was one of the worst El Niño events, primarily because the Peruvian economy had developed an economic dependence on the fishery, one which included such associated activities as ship building, gear manufacturing, net production, fish meal processing, factory construction, and market operations. By the late 1960s about one-third of Peru's foreign exchange was derived from fish meal exports. The 1972–1973 El Niño, combined with poor recruitment in the preceding year (1971), and excessive fishing pressures, changed all that.

The characteristics of the fishing industry have changed since 1973. For example, the government nationalized the fleet and the fish meal factories (it had already taken control of the marketing mechanisms). By 1977, after several years of relatively poor anchoveta fishing, it was forced to denationalize because it could no longer bear the costs of subsidizing the industry with little revenue coming from the fishing sector.[19]

There has also been a sharp increase in the availability and landing of sardines relative to the availability of anchoveta.[28] This prompted a shift of interest to the commercial exploitation of sardine. It has been suggested, however, that the MSY for the sardine may be considerably lower than for the anchoveta and that it may already be superceded.

Finally, another change in conditions was that, following the collapse of the anchoveta fishery, the United States for a few years did not import any fish meal from Peru, getting its supply from Canada or using substitutes. Only recently (1979) have imports resumed, and the Peruvian fish meal is now derived from a combination of anchoveta and sardine. There had also been a major shift in Peruvian exports to Eastern Europe and Japan.[30,31]

FORECAST VALUE

Most authorities who are aware of the Peruvian fisheries or of El Niño lament the fact that no reliable El Niño forecast exists: "if only we had an El Niño forecast," they insist, "we could prepare ourselves for the impact of El Niño."

As noted earlier, this study is an attempt to consider the value to society of a (currently hypothetical) reliable forecast of El Niño two to four months in advance of its onset.* To this end, experts knowledgeable

*Wrytki et al. suggested a possible 6-month lead time between forecast and El Niño.[2]

about El Niño, Peru, and the fish meal industry—in government, fisheries management, the physical sciences, economics, and public policy within and outside Peru—were asked in an open-ended mail survey what they might have done had they known two to four months before the onset of the 1972–1973 El Niño that it was coming. At that time they would only have known about the environmental conditions associated with the onset (its intensity, geographic scope, and magnitude of impact), but not its duration or its future characteristics. The available information was assumed to be extremely reliable and for this study hypothetically perfect.

"What Ought To Be" Versus "What Is"

This section is divided into two parts: one which discusses "what ought to be" the value of a reliable El Niño forecast based on the responses to the survey, and another which discusses "what is" more likely, the actual value of such a forecast, based on comments drawn from the survey and the literature about the limitations that societal and environmental realities place on the use of such a forecast.

The "what ought to be" section is further subdivided into strategic and tactical responses to an El Niño forecast. The responses discussed below are not all-inclusive but serve only to suggest a range of possibilities. Gross[32,33] distinguished between tactical and strategic responses as follows:

Strategic approach	Tactical approach
Broad scope	Narrow scope
Long time-horizon	Short time-horizon
Formulates problem	Seeks objective answer to problem formulated by others
Systems oriented	Operations research oriented
Considers relation of problems	Considers only solution of particular problem
Concentrates on desirability	Concentrates on feasibility and consistency

For this chapter, strategic responses to El Niño are considered to encompass long-term objectives related to the fishery and would provide the overall goals toward which tactical policy decisions should be geared.

In other words, those strategies determine the framework for decision-making regarding the commercial exploitation of the anchoveta.

Tactical responses to existing conditions at a given time should be (but in practice are not necessarily) guided by the general strategy. They may often be unplanned, *ad hoc* measures which may be no more than reactions to prevailing conditions.

There is a subjective element in determining which policies are strategic and which are tactical, depending on how one perceives the characteristics that define these concepts (for example, the narrowness of the scope requires subjective appraisal). Thus, one person's tactic may be another's strategy. The division between strategic and tactical suggestions of the experts was made by this author.

Strategic Responses to an El Niño Forecast

The primary strategic objective would be to minimize the effects of El Niño, as nothing could be done to thwart its occurrence. One must assume that in an ideal situation, the development and implementation of such strategies would have been based on the desire of Peruvian decisionmakers to exploit the anchoveta in such a way as to assure its availability to future generations. This, of course, is based on the assumption that decisionmakers are operating under conditions of "unbounded rationality," and that they will have considered all possible exploitation schemes and will have selected the optimal strategy for preserving the resource base. Such a situation may not have been the case for fishery policy in Peru. In that situation, decisionmakers may have operated under conditions of bounded rationality, "which takes into account limitations of the decisionmaker's perceptual and cognitive capabilities."[34]

In an attempt to balance the desire for unlimited exploitation against the need for conservation of the anchoveta, it was suggested by several survey respondents that a comprehensive, long-term scientific program be developed to know better the linkages between coastal upwelling, El Niño, and biological productivity of the fishery. Most scientific experts feel that these linkages have not been well established and that an El Niño forecast, no matter how reliable, might give little, if any, direct insight into the impact of El Niño on the fishery. This, of course, implies that the forecast would be extremely valuable as a fishery management tool if such linkages were known. That implication is speculative, however.

There is still much that is not well understood about the impact of El Niño on the Peruvian fisheries, and much of what is known is still controversial. While one respondent noted, for example, that "the year-classes associated with earlier El Niño events do not appear to be much reduced," another suggested that "it may be necessary to curtail fishing the year following El Niño so that the missing class can be substituted for a larger juvenile class." Another scientist raised some of these concerns that apparently were unresolved in his mind.

> When El Niño comes, do fish die or do they simply go away—to return and breed again when conditions are more favorable? *If* they're going to die, fish them out before El Niño strikes; *if* they will return and propagate, then halt fishing when El Niño appears.

Along the same lines a respondent specifically referred to the 1972 El Niño situation by noting that "there remains the nagging questions of whether the unusual failure of the fish stocks to bounce back after the 1972 event was due to overfishing or to some biological problem as yet undiscovered."

With respect to El Niño research activities, it was suggested that, when evaluating the overall benefits of such efforts, we should spread the cost and benefits across several generations, not just one. This would highlight the importance and need for a long-term research effort and might reduce the tendency toward the "stop and go" funding approach to, as well as data collection and management for, research activities in the eastern equatorial Pacific. A general belief was expressed that El Niño research could stand on its own merits as an important scientific problem, albeit one with definite societal implications, for Peru as well as for the rest of the world. Similar comments appear in the literature.[13] It is a research effort in need of constant long-range support so that a reliable data base can be developed to relate the El Niño phenomenon more directly to its environmental and societal impacts. Such a data base would remove at least some of the uncertainty that currently surrounds the impact of an El Niño "on the amount of fish, its natural survival and the success of its reproduction."

Some respondents did assume that there was a direct connection between El Niño, fishing pressures, and the decline in fishery productivity and specifically suggested that the stock could be preserved by reducing, if not eliminating, fishing effort at the first sign of El Niño. It was suggested that

> past failures to close or restrict the summer-fall fishery have been cited as the human factor most responsible for enhancing environmental effects in reducing catches to very low levels in the post-El Niño period. *This is*

probably the 'forecast response' of clearest benefit to the long-term success of the fishery.

Some respondents suggested tying up the fleet and financially supporting fishermen, industrial workers, and even industrialists, from a fund contributed to during favorable fishing periods.

> The world community should prepay into an impound account, according to accepted actuarial estimates, an amount sufficient to cover the cost of shutting down or restricting such an industry when the natural disaster occurs. This could be spread over the total fish meal sales as a tax or premium to be accumulated and held in trust for use later.

Similarly, it was recommended that "fishing cooperatives be established and fees collected from the fishing fleet during periods of good fishing." This is in essence a call for the implementation of a genesis strategy,[35] that of preparing in the good times for the bad which are, as in the case of El Niño, sure to come. This is an important strategy, especially in regions where the exploitation of a renewable resource is directly (often adversely) affected by human activities, as well as by quasiperiodic climatic variability.

It was suggested that a storage system be established to maintain fish meal stocks at a predetermined level so that in times of environmental stress on the fishery, fish meal stocks would be available, prices would be maintained (if not elevated), and fish meal customers could be retained.

A frequent suggestion was that there be one authority to oversee the exploitation of the fishery in order to remove it from the common property category. By making someone independent of both government and industry responsible for the exploitation and conservation of the anchoveta, observations and management decisions could be made objectively. The decisions of that sole authority would (in theory) be beyond reproach.* This, of course, is not a new attitude about how to manage a fishery (see, for example, Gordon).[36]

It would be necessary to have such a regulatory authority in charge of the fishery, for example, to be able to close the fishery to fishing when fish stocks or recruitment levels showed a decline, regardless of cause, whether that be El Niño, fishing pressures, or natural fish population fluctuations. The decisions of this authority would be crucial during El

---

*Similarly, scientific as well as economic data should be collected, analyzed, and stored outside the control of government or industry in a scientific as well as an objective way.

Niño events because fishing efforts would tend to concentrate in the few remaining pockets of upwelling which would have greater than usual fish population density. Such a condition would require regulation, because the "coefficient of capturability" (increased yields for the same or less fishing effort) would sharply increase. Succinctly stated, the government must "cut down on excess capacity so that regardless of what happened in the sea, the fishery would not suddenly make a very serious inroad on the fish stock."

Related to this would be the reduction in capacity within the various sectors involved in the fishery. This might include, for example, limiting fleet size, boat size, and gear type; closing inefficient factories; and maintaining a capability for fishing other species when the main one declines. Because the industry had been greatly overcapitalized before the 1972–1973 collapse of the anchoveta fishery, the report of the IMARPE Panel of Economic Experts, referred to earlier, called for a sharp reduction in capacity of the industry in order to avoid the collapse of the fishery resulting from economic pressure alone.[20]

Finally, as a strategic response to El Niño, some respondents suggested that the Peruvian government (or the authority in charge of the fishery) speculate in commodities by buying and storing soybean in the United States to be sold at elevated prices during times of El Niño crisis. This suggestion was based on the prevailing assumption that there exists a linkage between fish meal availability and soybean prices (see, for example, Kolhonen[37]; Vondruska[30]). It was also suggested that the Peruvian government speculate in the commodities market with soybean and with chicken (broiler) futures (since chickens are a major industrial consumer of fish meal). This of course, raises questions about who would develop and disseminate the forecast as well as about how its availability might affect its value for fisheries exploitation or commodities speculation. In a similar example, the experimental coho salmon forecast for Oregon coastal waters elicited mixed feelings from fishermen about its value. To some, the forecast had an economic value because "it gave an indication of where to begin fishing and thereby saved unproductive time spent in searching." Others, however, felt that "such a forecast system would benefit the less efficient fishermen at the expense of the most experienced fishermen."[12]

Tactical Responses to El Niño Forecasts

Each El Niño is different in its physical characteristics and in its environmental and societal impacts. In turn, its biological and, ultimately, societal impacts are apparently affected by increases in fishing pressures, even though all the linkages in the impact chain are not specifically known. Most respondents were aware of this and suggested that tactical responses to an El Niño forecast must relate in some way to the timing, location, and purpose of fishing efforts. The respondents could essentially be clustered as follows: those who primarily emphasized conservation efforts and those who did not.

Those interested in conserving the anchoveta stock suggested that the fleet be tied up until after the adverse effects of the El Niño had passed, with some even suggesting that fishing not be resumed for at least a year after El Niño, so the fishery could rebuild. This would reduce fishing efforts in anticipation of poor recruitment. As one expert noted,

> an important factor causing disastrous decline in catches was the lack of recruitment of young fish at the beginning of 1972. This fish was born in September 1971, before the first warm water appeared along the Peruvian Coast. Warm water started at the extreme north end of the Peruvian Coast in January 1972. This new incoming year-class was already very poor before warm water appeared. A recruitment failure similar to that in 1972 had never been observed before, also not for the year-classes born during the earlier El Niño events.

It was suggested that in addition to the El Niño forecast there be a forecast based on biological factors related to recruitment. "The recruitment forecast can be done with existing technology (acoustic surveys and fishing with small meshed nets) irrespective of our ability to predict the oceanographic event, or to understand the interaction between that event and the fish stocks." Such a forecast would have spotted the poor anchoveta recruitment that immediately preceded the 1972 El Niño and could have, especially if coupled with a reliable El Niño forecast, prompted proper fishery management, at least in theory. As another expert noted,

> what seems to be more important in relation to fishery problems is . . . to have had arrangements for predicting or observing the abundance of young fish before they were fully available to the fishery so that the failure of the year-class entering the fishery in 1972 could have been noticed earlier. . . . Fishing on the older fish still present at the time would have been cut down in time to maintain an adequate breeding stock in 1972 and 1973.

Other respondents felt that even with an El Niño forecast some months in advance, the stock should continue to be fully exploited; perhaps even more so, reasoning that "if the anchoveta were going to die would it not be better to fish them out anyway?", or that "the stocks should be thinned out to give the survivors a better chance for survival." This hypothesis was referred to by W. G. Clark, who raised the possibility that "the collapse of the stock [in 1972] resulted as much from insufficient fishing in 1971 as from excessive fishing in 1972."[28] Others suggested heavily fishing the anchoveta with little concern for their viability (survivability) as an economic resource, believing that the anchoveta would be replaced by other industrial species that could then be exploited commercially (as happened to a certain extent with the sardine).

Still other respondents seeking a middle ground suggested careful exploitation of anchoveta coupled with (if not guided by) constant monitoring of the standing stock, especially once an El Niño has been forecast. With the appearance of signs that the anchoveta was being overexploited in the face of El Niño, fishing activities would be halted until the danger to the standing stock passed. As part of this tactic, it was suggested that fishing pressures be redistributed in a "gradual but sustained manner" to reduce the impact of concentrated fishing effort, a tactic that might involve a ban on fishing for anchoveta in the northern fishing areas off Peru or in the southern fishing grounds (on the Peruvian-Chilean border) or both, with a simultaneous shift to the exploitation of other potential commercial (both industrial and food) species. The ban on fishing should also extend to parts of the ocean less directly exposed to El Niño to give the anchoveta a chance to regenerate as strongly (and, more importantly to the Peruvian fishing industry, as soon) as possible.

One expert noted the confusion for decisionmakers that might accompany a reliable El Niño forecast and how responses to such a forecast can be affected by one's philosophy toward fishery conservation and exploitation: "I can only suppose that if a prediction of El Niño had been made in 1970, the government and industry might have run to panic station, or taking the 1967 incident as precedent, have decided no action was necessary." This intimates that few guiding principles exist for the exploitation of highly fluctuating living marine resources. While these El Niño events occurred under different levels of fishing pressure, with the pressure being greater in 1972 than in 1967, it is suggested that there are generalizations that can be made (based on existing knowledge about the anchoveta fishery, for example) about levels of fishing effort in the face of large uncertainties surrounding anchoveta population fluctuations and environmental conditions.

Even if one believes that nothing should be done to minimize the effects on society of an El Niño, there should be a major scientific effort to document each reliably forecasted El Niño event and to assess its effects on biological productivity and on society.

El Niño forecasts might also be used to forecast other geophysical events that are seemingly teleconnected. As one scientist recently wrote,

> . . . meteorologists have coined the phrase "teleconnections" to describe the apparent correlation between El Niño, cold winters in the United States and Europe, weakening of the Indian monsoon, heavy rains and unusual hurricane activity in the Pacific, and droughts in the Sahel.[38]

For example, studies suggest such linkages with droughts in the Brazilian Northeast,[39] with geophysical anomalies in the southeast Atlantic,[40] and with droughts in the West African Sahel,[41] More specifically (and more locally), it is known that associated with El Niño, tuna stocks and other species move into Ecuadorian waters where they are relatively more accessible to Ecuador's fishing vessels.

With respect to agricultural activities, excessive rains along the western coast of Peru accompany El Niño, rains that could bring potential agricultural opportunities (regulation of dam water) as well as disasters (mud slides). Temporary pastures might be exploited by cattle or wetter areas farmed, given a reliable El Niño forecast. The use of an El Niño forecast in the agricultural sector was not fully explored, and further research would be required.

As noted earlier, the possible strategic and tactical responses to an El Niño forecast mentioned above are cited only to suggest the kinds of responses that might be considered, given a reliable forecast. However, there are some conditions that exist within society as well as with the physical nature of El Niño that may block the realization of the potential value of a forecast. Some of these are now presented.

"What Is"

Many comments on the value of an El Niño forecast were stated in such general terms as "an El Niño forecast would be useful in the Peruvian anchoveta fishery," or "if a reliable forecast of El Niño was available, then certain economically useful strategies could be devised." On closer scrutiny of societal and environmental constraints, however, it appears that the value of the forecast may indeed be limited by such factors as (1) specific characteristics of such a forecast; (2) the nature of the El Niño phenomenon; and (3) prevailing management practices, which in turn are strongly influenced by (4) the Peruvian policymaking process; and (5) by other factors.

FORECAST CHARACTERISTICS
Forecast Lead Time

The probable lead time for the forecast was suggested to be about two to four months (but no more than 6 months). It has been argued that this would allow little or no time for mitigation tactics to be undertaken. As one respondent noted, "I doubt whether 2–4 months' advance warning of the 1972–1973 El Niño would have helped the fisheries much; no matter what action had been taken—the damage had already been done."

Problems in dealing with lead time for the planning of agricultural activities also exist because of the lack of an infrastructure that might allow for prompt and effective action. The relatively short lead time for the forecast would encourage emphasis on defensive tactics that would attempt to reduce fishing effort and to establish catch quotas but such action would have to be taken shortly after the forecast was issued.

Forecast Reliability

Would the forecast be reliable? What might be the effect of an inaccurate El Niño forecast? Would it damage the credibility of later forecasts and the faith that the users might put into such a forecast? What would be the costs of an erroneous forecast? On forecast reliability one respondent noted that

> the source of the perfect forecast would have to establish a reputation on the basis of past forecasts and a high degree of confidence among relevant users in order for them to place a monetary or other value on the new forecast.

In fact, on the basis of a recent forecast of an El Niño (sometimes referred to as the Quinn-Wyrtki 1975 El Niño forecast), an NSF/IDOE field research effort was undertaken. This particular forecast remains somewhat controversial within scientific circles. One author suggested that it was a "bust" and that physical changes had taken place in less than a month, between two cruises, proving that while something on the order of an El Niño began (something that a forecast would include), a forecast would not necessarily be able to predict its duration, intensity, and geographic scope. Another respondent, however, suggested that "from a physical point of view, the Quinn-Wyrtki forecast was an excellent prediction. It remains to be seen how well an El Niño forecast can be shaped into a fisheries prediction." Finally, on the question of forecast reliability, an economist suggested that generally "no environmental forecast is reliable enough to overcome longer term economic imperatives of a particular situation."

El Niño Characteristics

Some characteristics of the El Niño event itself also might serve to constrain societal responses to a forecast. There are differences in the intensity, magnitude, and duration of El Niño events as well as in their impact on both the physical and the social environments. One of the most pessimistic responses about these constraints was as follows.

> The conclusive point (to me) is that no one can yet say what might be the best strategy to adopt in anticipation of an El Niño, or while El Niño conditions prevail. . . . This results from unsatisfactory classification of El Niño phenomena, which are of several kinds, varying in different ways, each in its physical and chemical facets, as well as in the area affected, duration, and sequence to which it belongs. . . .

That respondent then suggested that El Niño be divided into three components: the physical changes in the ocean (and atmosphere), the effects on the fish stocks, and the effect of the latter on the economy. While the scientific potential for developing an El Niño forecast of some useful degree of reliability may be relatively high (even this is controversial), little reliable scientific connection between that forecast and its potential impact on the fishery can be made at this point. "We don't know enough about the correlation of meteo-hydrographic [sic] data with fishery data to make prediction of fishery results from such data." Research progress is necessary in each of these three aspects of El Niño and its impacts.

Because each El Niño event is different from the others, from a tactical standpoint "you would need to know with 100% certainty whether the forecasted El Niño will be a major or a minor one." Once an El Niño can be forecast with some degree of reliability, it would then be necessary to translate that forecast into an impact statement on the fishery but, as noted above, there is a need for simultaneous research concerning the effect of El Niño on the biological productivity in the eastern equatorial Pacific to reduce the uncertainty surrounding that linkage. However, it appears that, as one respondent noted, ". . . the present knowledge of an accurate forecast of an oncoming El Niño would not help very much in making proper decisions on the fishery.''

Fisheries Management and El Niño Forecast

Perhaps the most important consideration of an El Niño forecast is how to translate it into an effective tool for rational management (and exploitation) of the Peruvian fisheries. Supporting the views of many of the respondents to the survey are comments in the biological section of the Guayaquil Workshop on the Phenomenon Known as El Niño[42] that seriously questioned whether El Niño could be simply related to the state of the anchoveta stock or to the amount caught. Again, "although the forecast would be of value, you should not expect much in the transfer of an El Niño forecast to (knowledge about) year-classes." Finally, "regardless of our ability to predict the physical events in the ocean, we are still a long way off from being able to predict the effect of a general type of physical event (El Niño) on the fish stocks."

Thus, at this time it appears that there is a general feeling that a forecast of El Niño would not necessarily translate directly into useful information about how to manage the fishery. It seems that only in a general way might the forecast be used, for example, to signal the need for a reduction in fishing effort throughout the fleet. An American fishery expert suggested that his pessimism about the use of a reliable El Niño forecast for fishery management arose "in part from the knowledge of the Peruvian situation and in part from corresponding factors in European fisheries management where a foreknowledge of the year-class strengths seems to have insufficient effect on general management." Similar findings exist for the Namibian/South West African fisheries as well.[43,44]

Policymaking Process

To the Peruvian decisionmaker the El Niño forecast will be only one piece of information on which he will base his decisions concerning the management of the fishery. How this information is viewed and weighted will also be affected by how scientists and their information are viewed by the decisionmakers. The record to date seems poor. One European noted that

the Peruvian authorities have always reacted too late. The main point to impress on the political authorities in Peru is that they should listen carefully to their own biologists and to encourage them to exert maximum vigilance in periods of El Niño.

This deserved confidence in the output of the Peruvian scientific community is something apparently lacking from time to time, at least with respect to advice related to the anchoveta fishery. For example, W. G. Clark noted that "by March 1972, when the fishery reopened after a 2-month summer closure, it was apparent to the IMARPE that there would be little recruitment. The Institute recommended that the fishery remain closed to conserve the spawning stock, but the government allowed the fishery to operate until June, by which time the stock was reduced to a very low level."[28] As another example, a meeting of foreign experts was convened by the Peruvian Minister of Fisheries in July 1977 apparently to evaluate the validity of the scientific information from the Peruvian marine science community (the results of the meeting were considered confidential). The panel concluded that the Peruvian scientists had performed their research tasks and assessments admirably and that political, economic, and fisheries management decision makers must pay careful attention to their research findings and must incorporate those findings into a rational management scheme for the anchoveta fisheries.[45] A European fishery expert reinforces this view at some length:

> There is, however, no doubt in my mind that if the authorities heeded the warnings from IMARPE, they could, in time, have avoided the most harmful fishing. . . . Political and financial considerations have, unfortunately, usually overruled the recommendations from IMARPE and its international advisory group. . . . The 1972–1973 El Niño illustrates the course of events admirably. There was no lack of early warning, but fishing carried on for months, without effective restrictions. . . . Furthermore, fishing was resumed much earlier than recommended by IMARPE.
>
> Political and economic considerations force the hand of the authorities, even though they are aware of the consequences (Peru is not an exception—it has happened all over the world—the California sardine, the North Sea herring, the Atlanto-Scandic herring in Norwegian waters).
>
> I have only pointed out what difficulties the authorities are up against, even if they do know El Niño is coming.

Again, it is important to stress that Peruvian scientists had made reliable biological assessments but that those assessments were apparently not used as a management guide for the fisheries. It was also suggested that "Peruvian fishery biologists felt that serious mismanagement of the fishery in the early years contributed more to disaster than did the El Niño event."

On the input of scientific information into the political decision making process, an international civil servant (scientist) questioned the value of reliable scientific information. "Quite explicitly it is not the case that given a prediction, Peruvian decisionmakers would have been in a position to make the 'right' decision, much less to be sure of avoiding what happened."

Other Factors

In addition to an El Niño forecast, there would be other inputs into the decision making process at a given time, such as the international price of fish meal and the availability and price of soybean on the international market, biological information on recruitment and the state of the standing stock, and political input on the internal economic and social situation (such as labor and employment conditions). One Latin American scientist suggested that "even if ecological disruptions trigger social and economic reactions, there are still so many other social, cultural, and ethnic variables that a predictable response to an entropy becomes very questionable." Still another expert noted, "by the time Peruvians (or other governments) could respond to a reliable El Niño forecast it would be too late to hedge against economic disaster by commodities investments."

There would also be major, indirect as well as direct, economic pressures to keep the fishery open, from factory workers, fishermen, entrepreneurs, bankers, the government, and even members of the international community. These pressures would also depend on such factors as the quantity of fish meal available on the international market (the forecast value would depend on whether fish meal is scarce or in oversupply). A respondent commented that "the insatiable worldwide demand (1972–1973) was bidding up the price of a closure (of the fishery) decision to the point where it was politically unthinkable, even if it were to have happened in the United States." It was suggested that even if Peruvian authorities did wish to stop the exploitation of the fishery, "such a decision might not have met with the approval of the U.S. and the European economic communities."

These pressures on Peruvian decisionmakers are in general heightened by the fact that no two El Niño events are expected to affect society in the same way. There is also the question of increased fishing pressures. If there were two identical El Niño events several years apart, their impact would in all probability be different in that fishing pressures would have increased. Also, the industry had been less overcapitalized during earlier El Niño events than when the 1972–1973 El Niño occurred.

Yet another very important consideration adding to the pressure on decisionmakers to keep the fishery open was that the forecast would come 2 to 4 months before the event at a time when little, if any, physical effect on the fishery would be evident. As one expert suggested, "while all those involved *might* be willing to curtail their fishing efforts to some degree in the midst of an El Niño, it might be difficult, if not impossible, to get them to show restraint on the basis of an El Niño forecast in advance of the impact of the El Niño itself."

Who has the forecast will also affect its societal value. If the forecast were in the hands of everyone—government officials, commodity speculators, bankers, entrepreneurs, fishermen—its value would depend on the ability of the different groups to react to such powerful information. If however, the forecast were in the hands of only a few—the government, or the entrepreneurs, or the bankers, or the fishing industry, or other countries—its value to those who have the information would be greatly enhanced.

OTHER "FORECASTS"

The assumption that a reliable El Niño forecast would be of great value reflects a widespread belief that more information about the future can only have positive benefits. Yet, a brief review of past literature raises doubt about the validity of that assumption.

An earlier article suggested that while a reliable El Niño forecast as such has not existed to date, there have been other reports and publications specifically concerned with the exploitation of the anchoveta that might be viewed as forecasts.[46] They presented statements based on general principles and existing knowledge (at the time they were written) concerning the potential of the anchoveta as a long-term commercial resource, given various rates and levels of exploitation.

Each study had its own message based on different expertise. Murphy,[47] for instance, reported that unless there was careful (as opposed to careless) and well-planned (as opposed to unplanned) exploitation of the anchoveta, both renewable resources—the guano birds and the anchoveta—would disappear. He proposed that any commerical exploitation of the fishery "should be within the limits imposed by nature (including climatic cycles), with safeguards for the future taking precedence over the demands of the moment." He suggested that man was an "insatiable predator" and could only be deterred by the depletion of the resource or the financial failure of the fish meal industry.

An IMARPE report[20] warned of the detrimental effects on the fishery of the economic overcapitalization of the fish meal industry and forewarned the government of the increasing possibility of the collapse of the anchoveta fishery resource from economic pressures.

Yet another publication added the effect of an El Niño to the economic pressures and constructed a scenario that approximated what actually occurred a few years later.[48]

These are only three such warnings that might have been viewed as forecasts. Little, if any, attention was apparently given to these reports. One can only question the assumption that more information about the future will necessarily be of benefit to society when evidence shows that existing information has often been ignored.

CONCLUDING COMMENTS

While there has been a great deal of research on the scientific aspects of the El Niño phenomenon in an attempt to reduce the uncertainties surrounding its environmental and biological impacts, little research to date has been directed toward the societal value of an El Niño forecast and the potential use of such powerful scientific information. This study is an attempt to address that gap, at least in a preliminary way. It suggests that while in theory many good uses can be attributed to the development of a reliable El Niño forecast, in practice it appears that its value may be quite limited.

One serious limitation is that there is little agreement within the scientific community about the usefulness of an El Niño forecast specifically for fisheries management. The survey responses suggest that even given a reliable forecast, there still exists controversy as to how that forecast might be related to changing environmental conditions and the productivity of the fishery, and in turn, made to serve as a useful fishery management decisionmaking tool. While some respondents felt that such a forecast might be useful at least in a general way for fishery management, others raised doubts (of a scientific nature) about that use. Still others suggested that, because of existing political and economic constraints, there was no assurance that good scientific input would automatically lead to proper decisions about managing a living marine resource.

Contradictory views on whether (let alone how) a forecast might be useful for fishery management can set the stage for policy makers to respond to the lack of unanimity within the scientific community by weighting all views equally and therefore allowing the scientific community, in effect, to cancel out its input into the policymaking process. The policy makers would then be free to make decisions they perceived to be politically and economically expedient, but not necessarily environmentally sound. This rejection (or at least neutralization) of the scientific advice is often coupled with verbal support for more scientific research in an effort to reduce some of the uncertainties related to the physical environment. However, there is no guarantee (and little reason to believe) that more research will lead to more unanimity of view.

It is also important to give consideration to the levels of fishing activity in the face of environmental and biological uncertainties. (For comments on this problem related to other fisheries see Cushing.[49]) Although some fishing experts question the assertion that there are adverse effects of El Niño on the productivity of the fishery, it is clear that El Niño events cause major changes, at least temporarily, in the environmental setting in which fishing activities occur. While the direct or indirect impact of fishing pressure on the anchoveta may remain unclear, evidence does suggest that heavy fishing pressures at the onset of El Niño apparently increase the risks to the standing stock. What is needed is an attitude toward the exploitation of the fishery that is based less on short-term economics and more on trying not to add to the natural risks that affect the anchoveta, given their fluctuating environmental and biological conditions. W. G. Clark summarized this dilemma when he wrote that

> . . . in the anchoveta fishery, and most others, the aim of management is to obtain something fairly close to the maximum surplus production by maintaining the stock at some optimum size, which can be achieved by regulating fishing. But surplus production is the difference between growth and recruitment on one hand and natural mortality on the other and in Peru the determinants of this difference may change greatly in both the short term and the long term in response to factors other than stock size. . . . At present, however, the factors other than stock size that influence natural mortality and reproductive success are largely unknown, and until these factors can be identified and predicted, it will not be possible to manage the fishery rationally.[28]

This case study should be of interest not only to Peruvians or to historians trying to sort out what happened to the Peruvian fishery and development prospects in the mid-1970s, but also to development planners and officials in those states such as Somalia and Mauritania that seek higher levels of development through the exploitation of their own coastal marine resources. The Peruvian case (especially when coupled with other examples such as the California sardine and the South West African [Namibian] pilchard fisheries) suggests that overcapitalization of a fishing industry must be avoided; that lower (as opposed to theoretical upper) levels of exploitation must be made acceptable until environmental uncertainties can be reduced; that export-led development has its inherent limitations and should be questioned as the primary path to development; and that more scientific information about future environmental conditions will not necessarily ensure that the "right" decisions will be made. With respect to Peru, it is apparent that the anchoveta needs a spokesperson to counter the challenges of those who favor all-out commercial exploitation based solely on market mechanisms (i.e., entrepreneurs, fishermen, and politicians, as well as foreign aid advisors). Given the large uncertainties associated with environmental and biological factors, it should be assumed, in the interest of protecting Peru's living marine resource base, that high levels of exploitation ought to be avoided until the uncertainties can be reduced. (For a discussion of management schemes, see Gulland[50]). Otherwise, such fishing pressures might only add to the environmental risks associated with anchoveta exploitation. From the Peruvian example, it can be reasoned that social, economic, and political dislocations caused by the collapse of a fishery (contributed to by overcapitalization and excessive exploitation) may prove more costly than whatever temporary benefits might be derived from exploitation of the fishery based primarily on short-term economic interests.

REFERENCES

1 W. H. Quinn, "Monitoring and predicting El Niño invasions," *Journal of Applied Meteorology,* **13**(10), 825–830 (1974).

2 K. Wyrtki, E. Stroup, W. Patzert, R. Williams, W. Quinn, "Predicting and observing El Niño," *Science,* **191**, 343–346 (1976).

3 C. N. Caviedes, "El Niño 1972: Its climatic ecological, human and economic implications," *Geographical Review,* **65**, 493–509 (1975).

4 "All that unplowed sea," *Mosaic,* **6**(3), 22–27 (1975).

5 C. P. Idyll, "The anchovy crisis," *Scientific American,* **228**(6), 22–29 (1973).

6 D. H. Cushing, "Upwelling and fish production," FAO Fisheries Technical Paper #84 (FRs/T84), U.N. Food and Agricultural Organization, Rome, 1969.

7 U.S. Department of State, *Final Environmental Impact Statement for a Possible Regime for Conservation of Antarctic Living Marine Resources,* U.S. Government Printing Office, Washington, 1978.

8 IDOE (International Decade of Ocean Exploration), Second Report, Office of the IDOE, National Science Foundation, Washington, 1973.

9 *Téthys,* "Analysis of Upwelling Systems: Second Conference," (Marseilles, France, 28-30 May 1973), **6**(1–2), (1974).

10 M. Tomczak, Jr., "The CINECA Experience," *Marine Policy,* **3**(1), 59–65 (1979).

11 IDOE (International Decade of Ocean Exploration), "Prologue," Office of the IDOE, National Science Foundation, Washington, no date.

12 "Where to fish," *Mosaic,* **5**(1), 32–33 (1974).

13 NRC (National Research Council), *The Continuing Quest: Large Scale Ocean Science for the Future,* National Academy of Science, Washington, D.C., 1979.

14 F. J. Christy, Jr., "Distribution System for World Fisheries: Problems and Prospects," in *Perspectives on Ocean Policy,* National Science Foundation, Washington, 1975.

15 B. B. Smetherman and R. M. Smetherman, *Territorial Seas and Interamerican Relations,* Chapter 4, Praeger, New York, 1972.

16 M. Roemer, *Fishing for Growth: Export-Led Development in Peru, 1950-1967,* Harvard University Press, Cambridge, 1970.

17 A. Quijano, *Nationalism and Capitalism in Peru: A Study of Neo-Imperialism,* Monthly Review Press, 1972.

18 M. Tomczak, Jr., "Prediction of Environment Changes and the Struggle of the Third World for National Independence: The Case of the Peruvian Fisheries," in this volume.

19 L. Hammergren, "Peruvian Political and Administrative Responses to El Niño: Organizational, Ideological and Political Constraints on Policy Change," in this volume.

20 IMARPE (Instituto del Mar del Peru), "Report of the Expert Panel on the Economic Effects of Alternative Regulatory Measures in the Peruvian Anchoveta Fishery," 1970, reprinted in this volume.

21 C. W. Clark, "The Economics of Overexploitation," in G. Hardin and J. Baden, Eds., *Managing the Commons,* San Francisco, W. H. Freeman, 1977.

22 F. W. Bell, *Food from the Sea: The Economics and Politics of Ocean Fisheries,* Westview Press, Boulder, Co, 1978.

23 C. Ramage, "Preliminary discussion of the meteorology of the 1972-73 El Niño," *Bulletin of the AMS,* **56**(2), 234–252 (1975).

24 R. Edwards and R. Hennemuth, "Maximum yield: Assessment and attainment," *Oceanus,* **18**(2), 3–9 (1975).

25 P. A. Larkin, "An epitaph for the concept of maximum sustained yield," *Transactions of the American Fisheries Society,* **106**(1), 1–11 (1977).

26 J. A. Gulland, "Ecological Aspects of Fisheries Research," in J. B. Cragg, Ed., *Advances in Ecological Research,* Vol. 7, Academic Press, New York, 1971.

27 D. H. Cushing, "Climatic Variation and Marine Fisheries," in *World Climate Conference Proceedings,* World Meteorological Organization, Geneva, 1979.

28 W. G. Clark, "The Lessons of the Peruvian Anchoveta Fishery," in *California Cooperative Oceanic Fisheries Investigations* (CalCOFI) Report #19, 1977.

29 "Too much too soon," *Fishing News International* (Heighway Publications, London), June 1979, p. 96.

30 J. Vondruska, "Postwar Production, Consumption and Prices of Fish Meal," in this volume.

31 NMFS (National Marine Fisheries Service), *Fish Meal and Oil Market Review,* NOAA, U.S. Department of Commerce, September 1979.

32 B. M. Gross, "Management strategy for economic and social development: Part I," *Policy Sciences,* **2**, 339–371 (1971).

33 B. M. Gross, "Management strategy for economic and social development: Part II," *Policy Sciences,* **3**, 1–25 (1972).

34 P. Slovic, H. Kunreuther, and G. White, "Decision Processes, Rationality, and Adjustment to Natural Hazards," in G. White, Ed., *Natural Hazards: Local, National and Global,* Oxford University Press, New York, 1974.

35 S. Schneider with L. Mesirow, *Genesis Strategy*, Plenum, New York, 1976.

36 H. S. Gordon, "The economic theory of a common property resource: The fishery," *Journal of Political Economy*, **62**, 24–42 (1954).

37 J. Kolhonen, "Fish meal: International market situation and the future," *Marine Fisheries Review*, **36**(3), 36–40 (1974).

38 J. J. O'Brien, "El Niño: An example of ocean/atmosphere interactions," *Oceanus*, **21**(4), 40–46 (1978).

39 C. N. Caviedes, "Secas and El Niño: Two Simultaneous Climatical Hazards in South America," *Proceedings of the Association of American Geographers*, **5**, 44–49 (1973).

40 R. Doberitz, *Cross Spectrum and Filter Analysis of Monthly Rainfall and Wind Data in the Tropical Atlantic Region*, Meteorologuisches Institut der Universitat, Bonn, 1969.

41 H. Flohn and H. Fleet, "Climatic Teleconnections with the Equatorial Pacific and the Role of Ocean/Atmosphere Coupling," *Atmosphere*, **13**(3), 96–109 (1975).

42 Guayaquil Workshop, *Report on the Phenomenon Known as El Niño*, U. N. Food and Agriculture Organization, Rome, 1975.

43 D. Cram, "Research and Management in Southeast Atlantic Pelagic Fisheries," in *California Cooperative Oceanic Fisheries Investigations* (CalCOFI) Report #19, 1977.

44 D. Cram, "Hidden Elements in the Development and Implementation of Marine Resource Conservation Policy," in this volume.

45 Consultative Group, "Report of the Consultative Group to the Minister of Fisheries on the State of the Stocks of Anchoveta and Other Pelagic Species and on the Courses of Action to Be Taken for the Management of the Fishery," Lima, Peru, July 1977.

46 M. H. Glantz, "Science, Politics and Economics of the Peruvian Anchoveta Fishery," *Marine Policy*, **3**(3), 201–210 (1979).

47 R. C. Murphy, "Guano and the Anchoveta Fishery," 1954, reprinted in this volume.

48 G. Paulik, "Anchovies, Birds and Fishermen in the Peru Current," 1971, reprinted in this volume.

49 D. H. Cushing, "A Link between Science and Management in Fisheries," *Fishery Bulletin*, **72**(4), 859–864 (1974).

50 J. A. Gulland, *The Management of Marine Fisheries*, Scientechnica, Bristol, 1974.

INDEX